W0037451

INSECTS AS HUMAN FOOD

An Arunta woman in the scrub of Central
Australia with her child, in search of food,
such as grass seeds, roots, insects, etc. She
carries a *pitchi* on her head, a digging stick
in her right hand and supports the child on
her hip with her left arm. From Sir B.
SPENCER, Wanderings in wild Australia.
London. 1928 I, fig. 33.

INSECTS
AS HUMAN FOOD

A CHAPTER OF THE ECOLOGY OF MAN

BY

F. S. BODENHEIMER

Professor of Zoology, Hebrew University, Jerusalem

Springer-Science+Business Media, B.V. 1951

ISBN 978-94-017-5767-6   ISBN 978-94-017-6159-8 (eBook)
DOI 10.1007/978-94-017-6159-8

Copyright 1951

PREFACE

Insects as human food have often provoked curiosity. But only recently has this topic been recognized to be of major importance in the nutrition of primitive man, especially in tropical countries. In the tropics the insects often fill gaps in the one-sided vegetarian diets of food gatherers, and they even do so in the regions of highly developed monsoon agriculture This in fact is the key to the true interpretation of the puzzling results obtained by a number of physiologists who have studied the diets of tropical races and found them to be deficient in animal proteins and fats, and yet the people were fit and obviously adequately fed. The constant eating of termites, caterpillars, locusts, etc. in substantial quantities was not taken into account. Either insects were not included in the questionnaires used or their consumption was ignored and they were considered as providing minor additions to the diet of negligible significance. Primitive man and his cousins, the monkeys, have never at any stage shown any aversion to entomophagy.

The factual material on this subject is so enormously scattered in journals and books on travel, ethnology, geography, medicine, zoology, etc., etc., that it is an almost impossible task to aim at gathering all available facts in this essay. Numerous short notes offer nothing but an almost endless reiteration of the facts reported here. Yet we hope that the material presented in the following pages will amply suffice to convince the reader that entomophagy is more than a mere curiosity. The large number of insect totems and their survivals emphasise the great importance of insects as food of primitive man throughout the tropical regions of the world.

The writer wishes to point out that apart from a few personal observations on mannas and locusts, all the reports gathered here are the original work of the authors quoted. The value of this material depends on their reliability and common sense. In fact only a very few facts have been quoted which have not found ample confirmation by cross references or by independent observations.

The writer is fully aware that very many interesting and important first-hand observations, both published and unpublished, have not been included in this short book and will be very grateful, indeed, for the communication of any additional facts, sources or photographs for inclusion – with proper recognition, of course – in a possible later edition.

It would be impossible to enumerate all the many colleagues and librarians who have given help to the writer in collecting this amazingly scattered material. Acknowledgement must, however, be made to the following who, without exception, have added important information: Prof. TH. MONOD (Paris), Dr. CHINA (London), Dr. THÉODORIDÈS (Banyul), Dr. LEEFMANS (Amsterdam), Dr. BRYGOO (Tonkin), Prof. HOEPLI and Dr. FEN (Peking), Dr. NICHOLSON (Canberra), Dr. McKEOWN (Sydney), Dr. TINDALE (Adelaide) and Prof. FAURE (Pretoria). To all these named and unnamed helpers our sincere thanks are due.

We acknowledge with gratitude the permission of Messrs. MAC MILLAN & Co Ltd. – London for the reproduction of 19 figures from the books by Sir B. SPENCER, and to I. F. A. N. – Paris for putting at our disposal a number of photographs from French East Africa. The other figures are all duly acknowledged in their proper places.

The author also wishes to express his gratitude to the publisher, his friend Prof. W. WEISBACH, and to Miss I. WEISBACH, for the keen interest they have taken in the publication and the production of this book, and to Mr. KENNETH SPENCER for his editorial corrections to the style.

I. INSECTS AS HUMAN FOOD

1. INSECT CONSUMPTION FROM PRE-HISTORY TO THE PRESENT DAY.

It has been known for many centuries that insects are eaten as delicacies in many parts of the world. Reports have come down from antiquity of insects, especially locusts, being eaten by primitive peoples. Honey has been known as a prized food from time immemorial. Yet entomophagy, apart from honey consumption, has always been regarded as a curiosity or as barbarism. Although a number of recent authors such as NETOLITZKY (1918/20), BEQUAERT (1921), HARDY et RICHET (1933), GOUROU (1947) and others have hinted at the actual and potential nutritive value of insects for primitive man, the present writer began his study a few years ago largely out of curiosity. It is actually astonishing that the real and basic importance of insects as food for early and primitive man has been so long ignored. One of the main reasons is that a fuller understanding of the requirements of a well-balanced diet and of its necessary vitamin content, over and above the mere calorific value of food, has only been gained in our days. The French Colonial Service has played a leading role in investigating the actual diet of tropical peoples, followed by those of the British and Dutch empires. These studies revealed that very many of the primitive peoples of Africa, Asia and America are underfed or live on unbalanced, entirely unsatisfactory diets, with a serious shortage of either animal fats, animal proteins or carbohydrates. This deficient or improperly composed diet is today regarded as the main reason for the low standard of life and vitality, for the lack of energy which so often reduces the vital standard of men in hot climates, and also for the lack of resistance to many diseases. The gathering of insects has often helped to supplement grave dietary deficiencies either at certain regular seasons of the year or in times of emergency such

7

as recurrent droughts. Their utilization as food was the proper instinctive response to the physiological need for animal proteins, fats or other substances which large quantities of insect food could provide. We have insisted upon the words 'in emergencies or in certain seasons.' The reason is simple. Most insects abound even in the tropics only in certain seasons. Very few are available throughout the year, and even if this is the case, as with ants or termites, the desired form appears only during a short season. This fact is important, as neither insects nor any other food is ever stored by the food-gatherers or by the primitive hunters. The fact that some insects may be available in the tropics at every season makes the group as a whole still more important as a source of food, since every individual insect species is usually of limited seasonal incidence.

It is rather doubtful whether primitive man ever felt an instinctive aversion against the eating of insects. Scores of writers have explained at great length how most of the vegetarian insects in themselves, by their environment and by their food habits, belong to the cleanest of animals, actually being much cleaner than most other animals which are served at our tables. There is no possible reason to explain why insects should be more repulsive to man than dead mice, snakes, snails or mussels, toads, shrimps or fish and many other titbits of human gourmets of every race. No evidence suggests that there is anything basically repellent about insects. If this were so, the almost worldwide incidence of insect consumption by men of all races would be difficult to explain. It is certainly a falsification of facts, as the later chapters of this book will clearly show, to assume that only extreme famine has reduced man here and there to such depravity that he was forced to overcome his natural aversion and to still the acute pangs of his hunger with insects. Even the stench of fermenting insects described so vividly by ROESEL and FRISCH in connection with decaying locusts would not suffice to explain their exclusion from our daily food. Man, and even highly civilized man, has overcome his aversion to unpleasant and repellent smells in many other types of animal or vegetable food. The proper preparation by roasting, cooking or drying will easily transform any insect into a dish which at its best is often acclaimed as a rare dainty, or recognized at its worst as an insipid, yet nutritive food.

Since the times of the Hellenistic writers many students have been convinced that insects eaten in large quantities or as the main source of food would bring about lowered vitality, diseases and a shortening of life. This attitude has continued down to modern times. Even men like RILEY and HOWARD felt the need to invoke special precautions, such as complete sterilization of the insects, before eating them. And even in 1939 MILLS and PEPPER felt the need to prove experimentally that the consumption of a few flour beetles (*Tribolium confusum* Duv.) did not cause the slightest physiological disturbance in the four experimenting heroes. We should, however, be careful not to ridicule lightly the conception that certain special articles of food may induce certain diseases. Nevertheless, at the Xth International Congress of Medicine at Berlin, J. HUTCHINSON could state, for instance, his opinion that a heavy fish diet may very probably be the main cause of the local spread of leprosy (vide C. H. ROBINSON 1900, p. 149.) This theory was, of course, soon shown by special research on the question to be entirely without foundation.

Certain insects are poisonous and are traditionally avoided. In some of the aposematic species avoidance is perhaps instinctive, as could be concluded from certain experiments of CARPENTER (1921) in which monkeys hesitated to pick up aposematic insect larvae offered to them or even never touched them at all, whilst all other insects were readily taken and crunched. Frequent reference is also made to poisonous honey. Another rather curious case of a sickness which may prove fatal after large-scale consumption of insects is mentioned by G. BOUVIER (1945). He reports that locusts are hunted at Lomani in the Belgian Congo at regular times by bow and arrow, the special arrows having four divergent points; and that they are caught by tons during locust invasions. On such occasions the quantity of these insects which are devoured whole is such that the legs with the sharp bristles on the tibias may cause a serious stoppage of the intestines which proves fatal if no surgical intervention is made. After some time these accidents grow more rare, as the Negroes then pull off the legs and wings before frying the insects in palm oil. Greediness may also prove fatal to monkeys. BOUVIER saw several dead monkeys, the autopsy of which revealed intestinal occlusion caused by masses of locust legs.

9

And yet there cannot be the slightest doubt that, under normal circumstances, the average white inhabitant of Europe or America will refuse with horror any offer or temptation to taste insects. Or, if he eventually agrees or decides to do so, he looks upon himself as a martyr or hero of science. This conspicuous aversion is a prejudice acquired incidentally to the progress of civilization. Also our aversion to eat raw meat, to eat the flesh of man, to eat raw fish, etc. is by no means a primary aversion. It falls into the same category as the refusal to eat insects. Equally, the eating of frog-legs, snails, oysters, shrimps, frutti di mare and similar animal food is just as repulsive to the great majority of people who are not accustomed to eat them, as would be the eating of insects. All over Europe, however, we find children sucking the sweet crops from the bodies of bumble-bees, honey-bees or certain ants, crunching the meaty legs or rumps of grasshoppers or beetles, or eating with delight the sweet honeydew excretions of aphids and other *Homoptera*, etc.

No, the aversion to insect food in Western civilization, though an established fact, is nevertheless not based on a hereditary instinct. It is established by custom and prejudice. But, of course, important contributory factors must have helped to develop and maintain this prejudice. In the conditions of modern European and North American agriculture, man has not only, as a rule, sufficient to eat without being driven to secure regular additions to his normal diet, but his food is also well-balanced and adapted to the physiological exigencies of his environment. Hand in hand with this establishment of a high quality and adequate diet based upon intensive cultivation, goes an enormous and progressive reduction in the number of primary articles of food. A primitive food gatherer cannot dispense with any gift which nature doles out to him; many hundreds of species of plants and animals make up his 'daily bread', which excludes nothing edible and easily available. But modern man is reduced to a very few species of cereals, tubers, fruits and vegetables, in addition to the various products derived from a limited number of domestic animals. More substantial food, such as game or fish, may be occasionally added, and in these, as in other fields, some rare dainties may be occasionally welcome. Yet, the smaller game and the less stable objects of food gathering have disappeared and with them insects, for two reasons: the well-balanced, normal diet eliminated any dietary deficiencies which led to the desirability of insect eating. And further, food-gathering becomes uneconomic when there is neither a market nor a favoured place on the table for the special crop to be gathered. Finally, in the moderate climates of Central Europe and North America the insect component of food has certainly become extremely reduced since historical times and perhaps never was developed to the same extent as in tropical and sub-tropical regions.

In order to become an established article of food the object in question has to be, at least seasonally or periodically, of sufficient volume to form an important part of the diet. If, as will be shown below, insects provide not only a sufficient bulk, but in addition make good precisely those basic substances in which the normal diet either at a certain season or in general is locally poor or deficient, their importance becomes relatively even greater than that of their bulk alone.

In all primitive human societies, from the early food-gatherers and hunters to the early agriculturists and animal-breeding nomads, special rituals are devoted to every important article of food to ascertain, for instance, its abundance and the fertility of seeds or to ensure a good supply of animals and game. Such rituals or magic ceremonies are of worldwide distribution. They have found their earliest documented expression in the famous pictures in caves or on rocks of hunters from the Palaeolithic Age onwards to those of the present day Bushmen of South Africa. But the ceremonial paintings of the Australian aborigines represent a still more primitive stage, that of the food-gatherers. In all primitive societies we find totem groups. These are groups within the tribe which regard a certain object, usually an animal, as their ancestor. It must be stressed that this belief of transformation is quite material. The traditions of the Australian totems leave no doubt that these ancestors were originally individuals of the totem animal, which at a certain date in the remote past transformed themselves by their own resolution into the first men-ancestors of that totem. This explains how the eating of the totem animal or plant is forbidden to the members of each totem, or reduced to a very moderate consumption in connection with certain totem ceremonies. Yet the totem rituals for bringing about fertility and abundance of the totem anima or plant are regarded as of vital importance to the whole tribe, whose other totem groups will thus be certain to enjoy ample food. Every totem, while being extremely restricted in the consumption of its own totem animal, contributes thus to the total food production of the tribe as a whole. In all food-gatherers the drawings of the totem animals are rather rough, primitive and unnaturalistic, in contrast to the wonderful drawings of game animals of the Palaeolithic and even of more recent hunters. In the latter, the naturalistic achievements of the pictures were apparently in direct correlation with the magic effect of these drawings on the yield of the hunt.

It is of the utmost importance for the understanding of our problem that the knowledge not only of a number of totem names referring to insects have been preserved from many Australian tribes, but that mainly through the devotion of SIR BALDWIN SPENCER, who has written an imposing array of books on the topic, many of the rituals, the ceremonies and the traditions of a number of these insect totems have been recorded and preserved. These totem ceremonies are by far the best criterion for judging the role of any food in the regular diet of primitive man. Only basically important articles of food have been accorded the rank of totem animals or plants. To illustrate the great role which insects played in Australian totemism and hence in Australian nutrition, we have had to devote some space to the ceremonies of the insect totems of Australia, as well as to those of other continents, about which, unfortunately, far too little is known. The work of SPENCER has definitely established the basic role of insects as food, mainly in the arid interior of Central Australia, while in the fertile tropical North of that continent honey-bags still form the main important insect totems.

NOYES (1937, pp. 226 ff.) states that it is safe to assume that throughout the ages primitive people have eaten termites in tropical countries. This agrees with much current experience. He extends this experience into the undocumented past: 'Though we have no means of proving it, *Pithecanthropus*, shambling through the jungle, must surely have drawn largely on the termitary for his food supplies. It is only reasonable to suppose that since Pleistocene times there has been no diminution of the practice on the part of those who live, rawly, in the immediate neighbourhood of termites; apart from what we can observe, all the evidence points to this conclusion . . . . Whoever contrives to break open a termite hillock and abandon it to the tender mercies of attentive monkeys or baboons lurking in the neighborhood, may witness their interest. Chattering at the prospect of a feast, snatching, when the time arrives at the unfortunate workers and alates, careless of the soldiers' grip upon their marauding paws, they cram the insects indiscriminately into their greedy mouths. But, scrabble as they may, they never reach the sanctuary of the queen, the chiefest delicacy of all. That privilege is reserved for man and the ant-bear'. NOYES even suggests that the idea of tools first occurred to man through the difficulty his ancestors found in opening termite hills: when finding their talons unequal to the task the utilized sticks to force an entrance, from which the evolution of the lever would have been a simple matter.

This latter theory is, however, too far fetched and probably incorrect. Termite eating is rare in Australia, where the most primitive man was studied and where termites abound in the drier areas. Termite hills are only very rarely broken into and this in times of famine. In tropical Africa, Asia and America, termites are eaten in large quantities when the winged sexuals swarm out of the nests. This restriction is physiologically well founded, as the swollen abdomen, especially of the females, is rich in fats and proteins, in contrast to the much poorer body composition of the neutral castes. These swarming sexuals are easily caught without breaking the hard crust of the termite hills. And the thorough breaking deep into these hills in quest of the large and much prized queens is doubtless a very late development in search of delicacies or of medicine.

Yet, just as monkeys, together with scores of other mammals, birds, lizards, and other animals are very fond of locusts, swarming termites or ants, caterpillars and grubs, early man was doubtless fond of all available insects with no repellent taste, especially if he could gather them in quantities. We have no reason to assume that *Pithecanthropus* differed in this respect from the Australian aborigines or the inhabitants of primeval forests. There is no reason to doubt that locusts were readily eaten by prehistoric man, whenever they

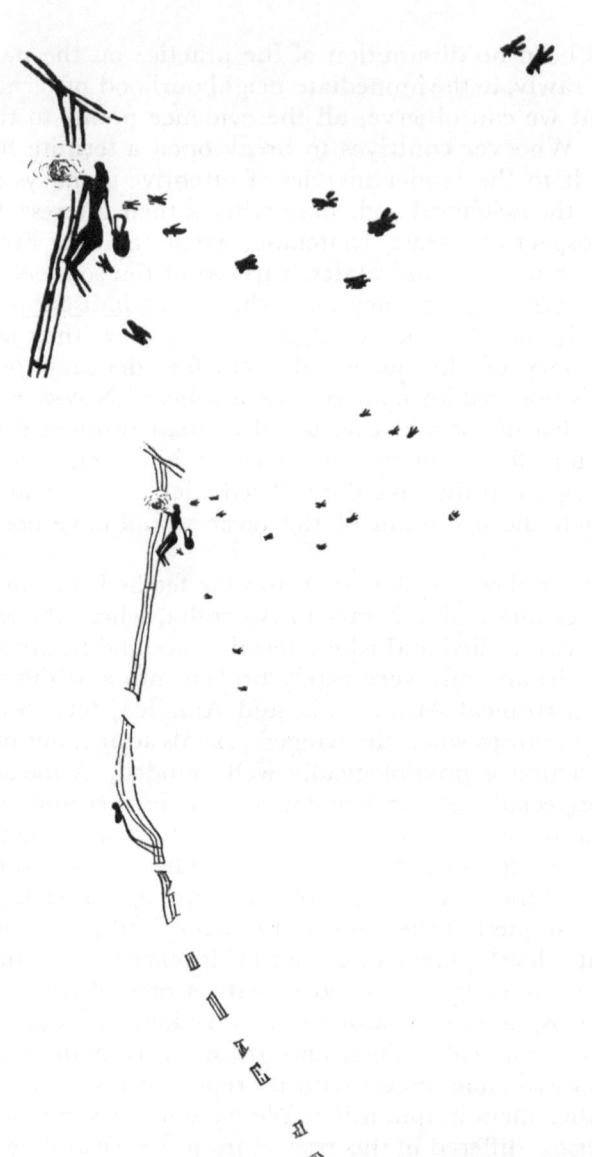

Fig. 2. Total view of the Palaeolithic cave drawing at Araña of honey-hunting. Above: The honey-hunter and the bees enlarged. From F. HERNANDEZ-PACHECO (1921).

14

were available in numbers. We may even suppose that he ate them roasted, as a rule, as all his animal food was roasted, before he took at a relatively late stage of his development to fishing. Fish were apparently the first animals eaten raw apart from occasional insects and molluscs. For honey-hunting we are by no means forced to base our conclusions upon parallels from recent ethnology. A Palaeolithic cave drawing from Araña in Spain gives a vivid picture of a honey-hunting Palaeolithic man who climbs along a ladder down a rock to collect the honey from a nest of wild bees (H. PACHECO, 1921.) This picture is the more convincing as it is almost identical with a recent sketch which SPITTEL (1924) has made of a Vedda taking honey from a rock in the forests of Ceylon (see fig. 33).

In view of the general importance of the discovery of honey-hunting as portrayed in the Palaeolithic art first described by F. HERNANDEZ-PACHECO (1921) in the caves of Araña (Spain), we give here a full description of these drawings. Many figures of men and animals, such as deer, wild goats and bulls, horses and a few extinct carnivores are drawn on the walls. Superimposed upon them are other, clumsier figures from a later period. Some of these latter designs illustrate scenes from the life of Palaeolithic man. They are brick-red on the yellow-reddish rock.

From two heavy horizontal lines above, three fine parallel lines descend over the length of the entire wall, here and there connected by short horizontal lines. Close to the upper lines there is a small natural cavity in the rock, over which the vertical lines pass. At the level of this cavity there is the figure of a man, surrounded by various small figures on the wall. Much farther below another man is portrayed in contact with the vertical lines.

In the neighbouring ravines close to this site, bees are still common, making their nests in crevices in the rocks. And even today, at the approach of the cold weather of winter, the local peasants collect the honey from any nests which are easy of access. This makes it a simple matter to interpret the picture: two men climb up one of the ropes which hang down from above, fixed to a pole, to collect the honey from a bee's nest. The three ropes make possible the ascent, forming a primitive ladder by means of the occasional horizontal connections. The man above, close to the entrance of the bee's nest, holds himself to the ladder with his legs and with one of his arms. The other, outstretched arm holds a vessel, perhaps a skin bag, with a distinct handle. The 16 small surrounding figures represent flying bees, and on some of them the head, the abdomen, the legs and the extended wings are recognizable. The man below carries a similar vessel on his shoulder, thus keeping his arms and legs free for the ascent. Both men are naked and apparently without protection against the stings of the bees; but it is well known, writes HERNANDEZ–PACHECO, that in winter the bees are lethargic from the cold and not so inclined to sting as on warm summer days.

In the neighbouring cave of Alpera there is a contemporaneous, similar design of a climbing man, whom BREUIL (1921) describes as climbing in the manner of the Australians, i.e. without making use of his knees. Espartograss still abounds in that region, offering an ideal material for the construction of such ropes. This first Palaeolithic insect illustration in a region where honey-hunting is still practised shows the general similarity which must exist between the local climate at that time and today.

In the Magdalenian grotte of Trois-Frères (Ariège) the graving of a cave-grasshopper on animal bones was found, wich L. CHOPARD could determine generically (BERLAND 1942 p. 112). The highly exact and artistic execution indicates the interest of the artist, and quite possibly it was also connected with food magics.

To return to our cousins the monkeys, D. and R. KATZ (1930), when studying the feeding habits of *Cercopithecus sp.* and of *Cebus sp.*, agreed that they behaved similarly to a young human child. When presented with seven different acceptable foods in small quantities simultaneously, the same monkey would eat them nearly always in the same order. The differences were small between the individuals or even between the two species. But after eating a very large quantity of grapes, their favourite food, for instance, this item might be changed from the first to the last place in the next meal. In general, whatever was eaten in quantity, was subsequently rated lower or altogether rejected. 'Their feeding behaviour reminds us of our own table manners . . . . All the rules controlling the feeding behaviour of monkeys have their parallels in human beings'.

Most experiments on the insect-feeding habits of monkeys were

16

performed to discover whether aposematic insects with warning colours or warning behaviour were rejected as food (COTT, 1942, pp. 256, 271, 290). Extensive experiments in this direction were made by G. H. D. CARPENTER (1921) in East Africa with *Cercopithecus sp.* This monkey definitely recognized a difference in palatability between various insects which it encountered when out hunting with its master or which were offered to it. Insects in general were readily and spontaneously eaten. Yet some species were eagerly pounced upon, others neglected, according to its experience of their suitability as food. Various experiments carried out by HEIKERTINGER and others have suggested fairly conclusively, in particular with aposematic behaviour in insects which display their warning colours conspicuously, that they are instinctively refused by many monkeys. The possibility cannot, however, be excluded, though this is by no means certain or even probable, that experience of unpleasant tastes in the past have influenced this attitude.

A large and varied assortment of insects offered or spontaneously found by the monkeys were classified by CARPENTER according to conspicuousness and edibility, as indicated by the monkeys' reactions. In 615 individual observations the *first* individual of 244 different insect species gave the following result: of 143 aposematic species 120 were distasteful, 23 edible; of 101 cryptic species 83 were edible and 18 distasteful. CARPENTER concludes that warning colours are, in general, a sign of distasteful qualities.

Various cryptic grasshoppers (*Cyrtacanthacris spp.*, *Catantops spp.*) formed the staple diet of *Cercopithecus*. Yet nothing would induce the monkey to eat aposematic grasshoppers, such as *Dictyophorus productus*, a heavy, bloated, sluggish insect which freely exposes its grey and bright red abdomen to view. On one occasion, when one of the monkeys saw from a distance that a grasshopper was being brought to him, he became very excited. However, he lost his excitement when getting a closer view of *Dictyophorus*. He picked it up from the ground, smelt it, and put it down again. To encourage him, CARPENTER pretended to taste it. The monkey then licked it, only to get a taste of the yellow froth which it exuded. He then shook his head, as if trying to get rid of a disagreeable taste and would have nothing more to do with this grasshopper. Shortly after this he seized with great eagerness and ate a huge, ten centimeter long cryptic *Cyrtacanthacris*. Four days later, before he had eaten anything, he was offered another *Dictyophorus;* this was examined and licked, then dropped uninjured. When out hunting ten days later, the monkey caught and ate another large *Cyrtacanthacris*, absolutely ignoring a *Dictyophorus*, which was on the ground just in front of him. *Zonocerus elegans*, another large, bright yellowish-green grasshopper with short reddish elytra and with the antennae altenately ringed black and orange, was similarly refused. When offered one, the monkey just looked at it and took no more notice of it. He was then shown another large cryptic grasshopper, upon which he leapt at once, seizing it and eating it with extreme haste. The large aposematic, green and red *Phymeteus viridipes*, when attacked, erects its wings vertically to display their red and black colour. On two occasions, when the monkey had begun to examine the grasshopper, this display prevented any further interference; he never would eat this species. An unidentified aposematic grasshopper was avoided by a Slow Lori in Burma (MacKenzie, 1930).

Guy Marshall (1902) confirms this experience for baboons with the big aposematic caterpillar of the hawk moth *Chaerocampa osiris:* 'The female baboon ran forward expecting a tit-bit, but when she saw what I had brought she flicked it out of my hand on to the ground, at the same time jumping back suspiciously; she then approched it very cautiously, and after peering carefully at it from the distance of about a foot, she withdraw in alarm, being clearly much impressed by the large blue eye-like markings. The male baboon, which had a much more nervous temperament, had meanwhile remained at a distance surveying the proceedings, so I picked up the caterpillar and brought it towards him, but he would not let me approach, and kept running away, until I threw the insect at him. His fright was ludicrous to see; with loud cries he jumped aside and clambered up a pole as fast as he could, into his box . . . . On concealing the larva I managed to coax him down again, drew him slowly towards me holding up the larva in the other hand; he simply screamed in abject terror'.

The observations of W. W. A. Philips, 1931, (also Poulton, 1932), carried out in Ceylon on a lemur (*Loris tardigradus*) in captivity, which was offered a variety of butterflies and moths, are

18

likewise in agreement with the rule that cryptic insects were accepted, aposematic ones rejected. The importance of these and other experiments for us lies in the unavoidable conclusion that not only the lemurs, but also the larger monkeys and apes readily eat insects in greater or lesser quantities, either spontaneously or if offered them. This removes the last possible doubt that primitive man, from *Pithecanthropus* to the recent food-gatherers, was in no way prevented by instinct from consuming insects. On the contrary, at least the occasional eating of insects belongs to his phylogenetical tradition.

The fact that many primates readily eat the most varied insects spontaneously can no longer be doubted. Some of the tiny lemurs, such as the Malayan Spectral Lemur, even exist primarily on insects. Furthermore, during a study of the influence of aposematic insects on feeding habits of various monkeys, it has been incidentally proved that the higher monkeys can make them their staple food. Therefore, we cannot agree with BRUES (1946, p. 418 f.), who is sceptical about the food value of 'these tiny creatures,' which make a major contribution towards satisfying the Gargantuan requirements of such complex social organisms as men only on rare occasions. He agrees, however, that certain insects on occasions form a real source of food, of which he quotes various instances from the life of the American Indians. Yet, considering the use of insects as tit-bits, hors d'oeuvres or medicines, etc., the wealth of information is so overwhelming, ranging from termite queens to minute stink-bugs, that it could only be catalogued by a librarian. We will, nevertheless, try in the following pages to demonstrate that at certain seasons, under certain sociological and ecological conditions, insects have formed and remain an important part of the all-the-year-round food of primitive man; they do not represent a mere dainty, but a physiological necessity by virtue of their qualitative as well as quantitative importance.

A. MAURIZIO (1932, p. 1) in his 'History of Vegetable Nutrition' defines the gathering of various animal and vegetable foods and primitive hunting as the lowest stage of human nutrition. Characteristic of this primitive stage is the great and ceaseless effort which the search for food requires. No part of a utilizable plant is wasted by the collectors and the same is true for all available animals, from the highest to the lowest: nothing is refused.

MAURIZIO mentions the excellent humour into which an African native is transported when gathering a handful of lice from the head of one of his neighbours. He also quotes the dramatic description of v. WISSMANN (1890, p. 168) who saw dense swarms of milliards of tiny gnats over Lake Tanganyika forming living clouds known as *kungu* by the natives. The natives follow these swarms as soon as they appear above the ground and collect them while the gnats are resting after having crossed the lake. A flour is made from them, which is used for roasted cakes, a highly appreciated food. The poorest and most primitive peoples, like the Australians of the deserts or the pygmies of the primeval forests, gather everything edible in their environment. Unfortunately, we are less well informed on the animal food of prehistoric man, apart from big game, than on his vegetable food. The meat of big game has been apparently roasted or grilled from the time of the first available records. Fish was, therefore, the first substantial animal food to be eaten raw. We have little information about insects and shell-fish, yet it may be safely assumed that in many cases early man ate them raw. G. RENARD (1931) also holds that the Australian aborigines well illustrate primitive conditions. In order to illustrate the primitive methods and at the same time the ingenuity of their search for food, he mentions that the Australians sponge honey from flowers for eating purposes, when no wild honey from bees is available. He likewise stresses the need to work hard throughout the year for their daily food supply. Nothing is known about primitive man making stores of food.

2. SURVEY OF ENTOMOPHAGOUS HABITS ON THE VARIOUS CONTINENTS.

The most primitive people who have been studied from the point of view of entomophagy are the Australian aborigines and they will be discussed in detail later. Here a few general quotations may be in place, in order to illustrate the role of insects in the diet of these primitive food-gatherers.

Keith C. MCKEOWN (1944, p. 170) says: 'Nothing in the way of additions to the food supply ever came amiss to the native. Throughout the continent the insects were exploited and played a prominent part in the aboriginal commissariat, varying often only with the presence or absence of any of the insects in particular districts'. Or J. MATTHEW (1910, p. 88): 'In the territory of the Kabi and Wakka of Queensland, food was plentiful and in great variety. The animal food embraced almost every living thing from a fly to a man'.

B. SPENCER (1922, pp. 66 ff.) who is, perhaps, the most competent judge, states in his 'Guide to the Australian Ethnological Collection at Melbourne' that among the higher vertebrates practically every mammal, bird, reptile, frog and fish that has enough flesh to make it worth eating, serves as an article of food in some part of the continent or another. Among invertebrate animals, shellfish of various forms, mussels, cockles, etc. are eaten in numbers, their empty shells lying in heaps beside the cooking places, forming on many parts of the sea coast shell mounds of great size. Various forms of insects, such as *Bugong*-moths and the larvae of other moths and beetles, as well as ants, are much relished and wherever obtainable, the honey-comb of wild bess is a favourite item of diet. *Mupingalu*, pounded up termite hill, is eaten as a cure for colds by the natives of the Kakadu tribe.

We conclude these few general quotations with two more: EYLMANN (1908, p. 278 f.) is the only observer to state that the natives of South Australia eat only few insects. He explains this by the poverty of the local insect fauna in edible species. We will see, however, that his own lists of insects are by no means small. The absence of observations on locusts is nevertheless remarkable. Recent research seems to indicate that in South Australia locust plagues are largely a more recent consequence of the changes in vegetation induced by human colonization. The South Australian diet is mainly of animal content. Vegetable food is unimportant, as it is either slight in quantity and rare, or unhealthy. During the common drought periods vegetable food is especially scarce and fat game is wanting. The aborigines of the interior then live on a diet rich in proteins but poor in fats and carbohydrates. During such periods, they lose in body weight, in consequence of the small quantity and the unbalanced composition of their food. Generally their diet lacks non-proteins, as is demonstrated by their perpetual eagerness to search for fats and carbohydrates. With regard to his

21

animal food, the native is by no means selective. He eats all verte-brates and the larger molluscs. Yet many insects are apparently not to his taste. He even refuses locusts and grasshoppers.

R. Semon (1903, p. 233), who has made no special notes on insect food of the Australian natives, makes the following observations on their diet: 'The great leanness of the native Australians is mainly based upon their prevalently animal food: Marsupials and *Echidna*, birds, snakes and lizards, turtles fish, beetle-grubs, eggs of birds and reptiles, shrimps and shellfish. Some of the Queensland tribes are not above eating human flesh. Hunting for game is the task of the men, whilst the women dig in the scrub for edible roots, collect mush-rooms and nuts of palms, pods of leguminous plants, seeds of grasses, sweet gum and manna of eucalyptus. Yet the flora of Australia is poor in edible fruits and mealy roots, which makes their diet poor in carbohydrates'.

Africa is perhaps today the continent where insects still play the most important role in native diet. M. Briault (1943, pp. 82 ff.) has given us a remarkable review of the diet of the African Negroes: 'Before penetration by the Arabs and white men, primitive Africa was a continent almost bare of basic foods, especially those of vegetable origin. The only native vegetable foods were bananas, citronella, gourd, beans and peas, sorghum, millet, durrha and perhaps taro. The natives of the primeval forest, including the Negrillos, have no plantations at all and live exclusively on game and fishing . . . . Meat, fish, pastes of caterpillars or big palmworms are wrapped into a large leaf, which is made supple by passing it over a fire. Some salt, some spice, some drops of wild lemon, and a pleasant meal is taken from the hot ashes, yet it never lasts long.... When game is insufficient or lacking, the worries of supplying the camp with provisions pass to the women. They go out to explore the rivulets, the caves, the ponds, the shrub, the bog, and return with rats, lizards, snakes, occasionally even with a small monkey or porcupine, or with a collection of caterpillars, palmworms, tad-poles or young fish. These dishes of lean days are stewed with salt, spice, and lemon, which are added to every native dish. Neverthe-less, famines are not rare'.

Bequaert (1921, p. 193) ascribes the wide use of insects as food in Africa more to necessity than to choice. He maintains that the climate and the small scale of animal husbandry reduce the amount of meat eaten, even that of chicken and dogs; hence the perpetual craving for animal food, which is one of the causes of cannibalism. Crops often fail. Thus, repeated famines are a contributive factor towards the inclusion of insects in the regular diet. The reader will have observed that, while BRIAULT's description refers to primary food-gatherers and primitive hunters, the remarks of BEQUAERT refer to agriculturists. But even in this case the conclusion that entomophagy is primarily induced by famines is inaccurate. The wide use made in the agricultural societies of tropical Africa of termites, locusts, and honey contradicts this conclusion. BRYGOO (1946, p. 27), for example, has pointed out, quite properly, that, while it would be exaggerated to talk in Africa about termite civilizations, the primary importance of these insects as part of the diet is very great. The origin of many ceremonies can be traced back to them and they may even eventually determine a certain rhythm of life, even among primitive agriculturists.

In tropical Asia all sociological and economic stages of development, from primitive food-gatherers to highly specialized agriculturists, are still co-existent. Corresponding to this variety of social conditions one finds a variety of insect feeding habits. Among the Veddas of the Ceylonese primeval forests strong reminders and remnants of an almost pure 'honey-civilization' from a not very remote past are still found. Another group of Asiatic pygmies inhabiting the Andaman Islands, who are also primitive food-gatherers, depend so much upon insect food that two of their months are named: 'the month when caterpillars abound' and 'the month of honey abundance'. The writer greatly regrets having lost the reference to this most important statement. The level of other food-gatherers, such as the Jakuns and similar groups in the forests of Malaya, is not much higher (FAVRE, 1865, MARTIN, 1905, etc.).

R. KIPLING tells us in the first chapter of his 'Second Jungle Book': 'Mowgli, who had never known what hunger meant, fell back on stale honey, three years old, scraped out of deserted rock-hives. He hunted too for deep-boring grubs under the bark of the trees, and robbed the wasps of their new brood'. These lines are certainly taken from tales heard and from observations made by the writer in India.

23

In the highly organized agriculture of the monsoon regions, such as Java, Malaya, Siam, etc., the ill-balanced, monotonous vegetable diet induces a craving for animal proteins and fats. This is easily satisfied, both by small game and fishing, and also over wide areas by gathering insects which are roasted or included in sauces, thus forming an important supplement to the daily rice. BURR (1939, p. 209) quite correctly states that the teeming millions of India, China, Japan and Malaya could hardly imagine life without their daily meal of rice. The agricultural Negro tribes of Africa feed almost entirely upon mealies, manioc, bananas or Kaffir corn. Such people feel an overwhelming hunger for meat and do not even scorn rats and grubs. The report of VAN DER BURG (1904) for Dutch Indonesia, of BRISTOWE (1932) for Siam, of NGUYEN-CONG-TIEU (1928) for Indo-China and many others, which will be discussed in detail later, also support this thesis.

The extremely frugal peasants and urban proletariat of China and Japan also improve their daily rice meals by the addition of small quantities of *any* kind of animal, from toads and mice to insects. Silkworm pupae which remain ready cooked after the reeling of the silk from the cocoons are highly appreciated as food. Grasshoppers are eaten in Japan. In China they form an important food during the famines, for which they are partly responsible. They are always a part of the food of the coolies and in Peking they are offered boiled in salty water in the restaurants. In Burma fried locusts are a delicacy. Many other insects are used to flavour the rice.

Locusts and wild honey are still appreciated in the Middle East as items of food for nomads and mountain peasants, who also delight in partaking of the various forms of manna.

FROST (1942, p. 63 f.) remarks that in North America the primitive idea of economy suggested the use of insects as food. When insects emerge in great numbers, such as cicadas from the ground or mayflies from water, they may be gathered with little effort. He even suggests that insects may supply salt when this is deficient in the diet. The scarcity of food might at times lead certain populations to select insects as food. In the Eastern United States insects were not frequently eaten, chiefly because the rainfall was generally sufficient to produce good crops and insect food was not essential.

In the West, where famines were more common, the Indians resorted to any kind of available food. Again, here also famine is not the only and not the most potent of the factors inducing entomophagy. The arid plains of the Western areas did not favour agricultural development and big game was fairly rare. Thus, the entire level of economic development on the two sides of the Alleghany Mountains was very different, resulting from the general ecological conditions. It will be shown below that the Indians of the Eastern states were by no means averse to eating insects.

Reports from the early days of occupation testify to widespread eating of insects in the West Indies. PETER MARTYR (1612, p. 121 f., vide F. OVIEDUS) says that in the houses of the inhabitants they found great chests and baskets made of twigs and leaves, which were full of grasshoppers, crickets, crabs, crayfish and snails, together with locusts which destroy the fields of corn, all dried and salted. The Indians explained that they kept these insects to sell them to their inland neighbours. And in the account of J. H. LINSCHOTEN (vide WANLEY, 1806, II, p. 373) we read that the inhabitants of Cumana eat 'horse-leeches, bats, grasshoppers, spiders, bees and lice, roasted or raw. They spare no living creature whatsoever, but they eat it'.

DE WAVRIN (1937, pp. 58, 122 ff) concludes that the South American Indians are, in general, far from strict as vegetarians. Many tribes are exclusively flesh- or fish-eaters, others are omnivorous. All like bees'-honey very much and honey-bees abound in certain parts of the forests. It is a general rule that all tribes eat almost every mammal, from the largest to the smallest. The Piaroa eat many fruits and vegetables, every fish and every 'reptile' from toads to poisonous spiders. Ants, caterpillars, palmworms, etc., are accepted as food and from the primeval forests of Paraguay VELLARD (1939) has described a real 'honey-civilization'.

3. THE NUTRITIVE VALUE OF INSECTS AS HUMAN FOOD.

The most recent work on the food value of insects has been carried out by French scientists, especially in West Africa. G.

HARDY and CH. RICHET (1933), in a book devoted to the native diet in the French colonies, come to the conclusion that most natives there are undernourished: in wide areas chronically, in others seasonally. The best measure of the soundness of a people's diet is its mortality (p. 41). Termites and 'earthworms' are quoted from the Lower Ivory Coast as sources of proteins (p. 163). All forest peoples are in need especially of proteins, as game in this habitat is very difficult to come by. Even if some people are apparently overfed, such as some tribes of Togo according to a survey of A. CHEYSSIAL, where the average daily diet contains 4000 to 8000 calories, it may be deficient in proteins or in fat (p. 289).

M. SARRE, Professor of Geography at the Sorbonne, in his 'Biological Basis of Human Geography' (1943), devotes a special chapter to the geography of dietetic regimes (pp. 247–290). 'The regime of a group is the combination of foods, either produced locally or imported, which assure its daily existence, satisfies its taste and assures its preservation in a given complex of life conditions. This is a norm. The medical view of the dietetic regime has an individual character'. The diet is composed of basic foods plus complementary and luxury foods. Primitive dietetic regimes, such as those of the aborigines of Australia, show an extraordinary diversity. They pass their life in a perpetual search for food. Everything edible is acceptable to them: emu-eggs, crocodiles, tortoises, ant-maggots, caterpillars and other insect-larvae, grasshoppers, moths and snails. The flora yields them water-lily bulbs, a wide variety of grains to make soups, roots and buds of *Xantorrhoea*, flowers of *Banksia*, in all not less than 300 species of edible plants. Among animals, kangaroo, opossum, tortoises, frogs, rodents, snakes, many birds including the emu, freshwater crabs, etc. are eaten when available. Close to the sea-shores whales, seals, sea fish, molluscs, etc. are added. The preparation of the food is governed by complicated rites (cf. WARNER, 1937). VELLARD's description (1939) of the Guayakis in Paraguay is quoted, as well as the similar food-habits of the Veddas in Ceylon. The absence of a permanent domicile and of stores make necessary a daily search for food, and often result in hunger combined with occasional overfeeding. During times of famine human flesh is a last resource. Many regimes are one sided, as often in hunting or fishing peoples, in the animal-breeding nomads or where one vegetable is the dominant food. In Africa and in the Oriental monsoon regions mixed diets with a predominant vegetarian tendency are often also one-sided. SARRE goes into the causes and effects of undernourishment and of famines (pp. 277 ff.) and into the effects of ill-balanced diets, which may produce diseases such as beri-beri, pellagra, scurvy, etc.

The best survey on the problems of native nutrition in tropical regions has been given by P. GOUROU (1947, pp. 76 ff.). Nutrition is mainly vegetarian in an environment little favourable for animal breeding. And even many pastoral people have an essentially vegetarian diet. Thus, the Betsiléo of Madagascar, despite their herds of cattle, live on rice, manioc, sweet potatoes, maize, spinach and fruits. Meat is an exceptional dish and milk is not consumed; *cockchafer grubs, caterpillars, locusts and small fish are the main animal food.* Even among the Foulah, the Hindus of India and other people who eat animal products, we find that the diet is deficient in animal proteins, fat, calcium and vitamins. In general, animal products form only 4 to 5 % of the total calories. The ingenuity of the natives of the warm and humid zones in utilizing the resources of the vegetable kingdom is great. The large-scale reduction in the number of nutritive plants so characteristic of modern Western civilization has not yet begun. GOUROU mentions from the plateau of the Gold Coast 114 kinds of fruit, 46 kinds of leguminous seeds, 47 kinds of vegetables. The essential food is always a porridge of carbohydrates (cereals or roots), seasoned by essentially vegetarian sauces, oil, fat or spices. Some of the vegetables, such as bamboo (*Elaeis*-palm), are prepared in the most manifold manner; in Northern Ceylon *Borassus flabellifer* is prepared in eight hundred and one different ways. And on the coasts of Guinea, when the preparation of *Elaeis* has been completed, the palmworms – the grubs of *Rhynchophorus ferrugineus* – are also extracted. The consumption of a daily rate of 1700 calories in tropical countries is nothing exceptional, a rather insufficient quantity! Inadequate nutrition is especially marked at the end of the agricultural year, when reserves are exhausted. Then food gathering in the savanna and in the forest commences: wild gramineous seeds, wild tubers and fruits, mushrooms and caterpillars are collected. Women searching for these foods are a common sight in the African scrub. And by a coincidence this season of want usually occurs at the time of the heaviest agricultural work. Thus, it is not surprising that the undernourished natives of the tropics concentrate their thoughts on food. The discovery of a nest of caterpillars is long remembered.

This situation is caused by the poverty of the soils, the irregularity of the rains and locust invasions. Regions with rains throughout the year are better off than those with seasonal rains. Malaria and similar diseases sap the natives' energy and reduce their capacity for work. In spite of the greater stability of root diets over those of cereals, the latter are everywhere preferred by habit, tradition and taste. GOUROU warns against accusing these people too hastily of improvidence, since the poor soil would be exhausted by more intensive cultivation and give reduced crops in the future.

Yet, for us the most important thing is that these diets are not only inadequate in quantity, but deficient as a rule in animal proteins, fat, vitamins and often in minerals, and also in the so-called protective foods which help to preserve health. Thus, lack of calcium and of vitamin E is held to cause the low fertility of the Haussa women (WORTHINGTON, 1938, p. 572). Taboos often prevent the use of milk and meat. Tropical diets are usually poor in minerals, such as phosphorus, calcium, iron, iodine and sodium. Lack of salt is often seriously felt. All these deficiencies are apparently at the base of the many cases of geophagy (earth-eating). And again this shortage of protective foods is accentuated towards the end of-the dry season and the beginning of the rainy season, the period of the heavy agricultural work.

This brief survey is sufficient to illustrate the great deficiency of tropical diets in animal components. Here is the main cause and the main importance of the widespread, large-scale and habitual use of insects of all kinds by all tropical native populations. Hence, this diversion clearly shows that insects as food in these regions are by no means dainties, but offer to primitive man just those elements in which his basic food is deficient. This will now be easily understood from the few adequate analyses of nutritive insects which we have at our disposal.

The first report on Nutrition in the British Colonial Empire (1939: I, p. 67) comments on the general character of colonial diets: 'The more normal sources of food supply are supplemented in many cases by wild animals and insects. We have already mentioned the consumption of such things as grubs, caterpillars, locusts and flying ants (read: termites). Many kinds of wild animals are consumed, though not usually with great frequency. In the more closely populated countries they are of some significance'. The second part of the same report (1939, pp. 19, 25, 29) mentions locusts, grasshoppers and white ants as universally eaten as delicacies. In Northern Rhodesia grubs, woodlice, caterpillars, flying (white) ants, honey and dried fish are widely used as local foods. In Nyassaland, although meat is relatively cheap, its price places it out of reach of the average villager, who depends on game, small rodents, caterpillars, flying (white) ants, locusts, etc., augmented by fish, for his intake of first-class proteins.

Perhaps the most elaborate chemical analysis of termites has been made by L. TIHON (1946). TIHON was stimulated by his often repeated observations of the arrangements made by natives of the Belgian Congo for catching termites for immediate consumption and for trade, dried in the sun or slightly fried. The propriety claims of families or villages on specified termite hills underline the great importance of this insect crop. In May one usually finds at the Léopoldsville market baskets full of fried termites, which are sold for 54 centimes for a small handful. Most natives eagerly devour the donge-termites, from which they also gain a colorless oil of good quality, excellent for frying. It was with such slightly fried, brownish-looking, oily and aromatic-smelling termites from the market at Léopoldsville that the analyses were made.

Moisture	6.0%	Ashes	6.4%
Dry matter	94.0%	Fat (petroleum, ether 50%)	44.4%
		Nitrogen	5.8%
		Protein (N × 6.25)	36.0%
		Chitin	5.1%

100 gr. of fried termites have a value of 561 calories, which ranges them among the richest foods; they are superior to other animal foods and approach the value of groundnuts. These termites are most important as a source of proteins which are too often deficient in the usual native diet. Does not this need of protein provide the main and instinctive reason for the eating of termites, caterpillars, adult beetles and their larvae? The analysis of the ashes yielded:

29

Insoluble	60.6%	NaO	2.2%
Iron and aluminium oxides	11.8%	P_2O_5	7.0%
Manganese	traces		
CaO	1.4%	Chlorides	8.2%
MgO	0.3%	Sulphates.	traces
K_2O	7.8%		

The insoluble component is mostly sand. The termites are rich in phosphates and potash, poor in sulphates.

Termite oil extracted by petroleum ether below 50° C. from lightly fried insects from the same source was brown and of a typical aroma. It contained:

Density at 28° C	0.906	Contents of brown, solid fat:	
Index of saponification . .	191		
Index of acidity	18.6	Titer	38° C
Acidity in oleic acid . . .	9.3	Melting point: beginning . .	40° C
Index of Hehner	94.1	,, ,, : end	41° C
Index of iodine Hubl . . .	55.4	Point of solidification .	38.5–39° C
Index of refraction at 40° C	1.461	Index of iodine	57.9
Soluble acids. Planchon in		Index of saponification . .	198
butyric acid	0.64	Average molecular weight . .	283.5
Insaponifiable	3.76	of the corresponding acids	

This non-siccative oil has a somewhat high acidity. It yields a clear, fairly hard soap. Even though this oil may never be of commercial interest, its study has some theoretical value. The termites themselves form an important part of the diet of the natives in most districts of the Belgian Congo and their study is of definite practical interest.

CH. AUFFRET and P. TANGUY (1947/48), of the French Colonial Medical Service, have fully confirmed these analyses of TIHON. They observed in French West Africa that the Soussou soak the termites in cold water, dry them in the sun and put them into an iron pot or a jar. The Fullah boil the termites first, before drying them in the sun, and later repeat the process. Then the insects are fried. They are eaten as they are or they may be incorporated into a sauce. Dr. PALES brought back from his journeys samples of fried termites from Pita in the Fullah country and living termites captured while swarming near Dakar. Their analysis permitted a comparison with other common foods of the same region:

	Moisture	Fat	Proteins	Ashes	Calories of 100 gr.
Living termites	44.5%	28.3%	23.2%		347
Fried termites	13.0%	36.2%	45.6%	5.0%	508
Meat (beef)	75.2%	6.6%	16.9%	1.3%	127
Dried salt-fish	32.4%	3.1%	43.7%	20.8%	203
Groundnut oil (Senegal)	7.0%	49.4%	27.6%	2.7%	598

No vitamin A was found. The samples were too small to permit an analysis of the ashes. The calorific value of the termites as compared with that of the other foods is surprisingly high. In addition the termites offer for certain peoples a most important source of protein, of fat, and of calories. The writers conclude that their nutritional value can explain an apparently paradoxical observation, namely, the healthy condition of some natives compared to their poor diet, if unknowingly certain dishes, such as the termites, are omitted from the analysis.

ANTONIE has analysed smoked caterpillars from the Upper Sangha, which contain 15.7% moisture, 40.% ashes, 13.7% fat, total N 6.1%, carbohydrates 13.9%, phosphates 0.7%, chlorides 0.4%, chitinous remainders 13.5%. A second sample yielded a slightly lower nutritive value. ANTONINI calculates the calorific value as 268 cal., of which 258 cal. are digestible.

An analysis of silkworm pupae by CH. AUFFRET yielded: 60.7% moisture, 23.1% protein, 14.2% fat, minerals 1.5%, calorific value 207 cal. Vitamin A was fairly ample. Both these analyses have been made available through the habitual kindness of Prof. TH. MONOD, Paris.

Y. YONEZAWA and K. YAMAFUJI (1935) give the following analysis of the eight days old pupa of the silkworm (*Bombyx mori L.*) (female), in gr. per 1000 individuals:

Water: 1191 gr; dry matter: 362 gr.

Protein 207.5; fats 90.5; carbohydrates 23.2; ashes 18.8; total phosphoric acid 7.8; organic phosphoric acid 4.3; total nitrogen 45.0; non-proteinous nitrogen 12.2 gr.

The amino acids of the pupa (E. KATAYAMA 1917) are alanin, valin, leucin, prolin, phenylalanin, glutamin acid, histidin and tyrosin in 0.05 to 0.67% of the live weight.

The properties of the fats of the pupae are (YONEZAWA and YAMAFUJI 1935): acid indicator 14, ester indicator 182, Hehner

31

indicator 95, saponification indicator 196, iodine indicator 137. The glyzerides of the oil of the pupae (YAMAFUJI 1933) are in % of the oil: dioleo-linolenin 0.6, insolin-oleo-dilinolenin 4.3, triolein 44.7, palmito-diolein 8.5, palmito-oleo-linolenin 16.7, tri-linolenin 8.3.

No data are available for the minerals of the pupae but those of the larvae are in % of the total ashes (LENZ 1867): P_2O_5 28.7; K_2O 48.1; SiO_2 0.6: CaO 5.9; Fe_2O_3 0.7; MgO 8.4; SO_3 6.2; Na_2 1.3.

All these data are quoted after K. YAMAFUJI (1937).

Some analyses of a few locust species are contained in the following table. In addition Dr. J. K. GUGGENHEIM informs me that in the specimens of *Schistocerca*, which he analysed, 100 grams con-

BODY COMPOSITION OF VARIOUS LOCUSTS

Species	Nomadacris septem-fasciata			Schistocerca gregaria		Locusta migratoroides			
Author	Hemsted 1947			Das 1945	Gug-gen-heim	Imp. Inst. Anim. Nutr. 1936			
Condition	DM.	DD.	W.	SD.	DM.	W.	SD.	SD.	SD	F.
	%	%	%	%	%	%	%	%	%	%
Water	—	8.0	70.6	5.0	—	10.5	—	—	—	7.1
Protein	63.5	58.4	18.7	61.4	75.0	46.1	61.1	55.0	49.7	47.5
Fat	14.1	13.0	4.1	17.0	3.4	9.6	10.1	15.0	18.4	22.9
Carbohydrates					7.5		6.4		9.6	6.8
Fibre	13.5	12.4	4.0	10.0		12.5	4.9			4.9
Total ash	8.7	2.6				5.0	4.6			
CaO	0.1	0.1	0.04	0.6			0.4			
P_2O	0.2	0.2	0.1	1.2			1.6			
Cl							1.2			
K_2O				0.8						

DM.: Dry matter; SD.: Sun dried; W.: wet or fresh weight; F.: locust flour.

tained 1.75 mg. riboflavin and 7.5 mg. nicotinic acid. This shows that even the vitamin content of insects may not be negligible and the same is true of certain minerals such as phosphates.

B. P. UVAROV (1948) has carefully compiled the available data on the chemical composition of grasshoppers. The complete mineral ash content of sexually mature adults of *Schistocerca gregaria* (E. B. UVAROV 1931) and of the two sexes of *Oxya velox* (ICHIKAWA 1936/37) is:

	Schistocerca	Oxya males	Oxya females
SiO_2	11.9%	6.64%	9.61%
CuO	0.16	—	—
Fe_2O_3	2.06	1.59	0.69
MnO	0.21	0.26	0.31
Na_2O	6.2	17.20	20.21
K_2O	18.2	21.78	19.23
CaO	6.2	1.05	0.74
MgO	4.9	5.44	3.85
TiO_2	0.16	—	—
NiO	0.009	—	—
P_2O_5	32.4	40.10	46.26
SO_2	2.56	0.67	0.27
Cl	0.40	—	—
C	2.4	—	—

The desert locust is richer in calcium, sulphur and iron content but poorer in sodium and phosphorus than *Oxya*. Other students found also lithium, barium and strontium in locusts which had been boiled in salt water and which were being sold on a Moroccan market. The organic constituents of *Oxya spp.* (KORIGAWA 1934) were moisture 11.78, total N 10.85, crude protein 67.79, protein N 7.45, protein 46.61, amino N 0.78, crude fats 4.52, crude ash 3.78, 'chitin' 9.25, glycogen 0.01. The high protein content of 46% is confirmed by other analyses: in *Nomadacris septemfasciata:* 46.1 protein and 9.6 fats; in *Schistocerca gregaria:* protein 56.7 in the males, 42.3 in the females (LAPP and ROHMER 1937). The protein analysis of *Oxya velox* (ICHIKAWA 1938) yielded for the females: water soluble N 36.49, albumino N 6.68, globulin N 8.51, protamine N 1.51, gluten N 21.84, chitin N 25.00.

TIMON-DAVID (1930) summarized the data on fats and oils of insects. He found 13% of oil in the larva of *Ergates faber L.*, 22.3% in that of *Rhynchophorus palmarum L.* For grasshoppers he gives the following data:

Species	% of wet body weight	iodine index	saponifica- tion index	reference
Dociostaurus maroccanus	3.3	109.9	181.0	Timon David 1930
Anacridium aegyptium	2.6	97.6	198.9	ditto.
Schistocerca { male	—	106.7	224.0	Lapp Rohmer 1937
gregaria { female	—	92.7	224.0	ditto.
Schistocerca paranenis	—	65.2	194.5	Tewithick Lewis 1939

Oxya japonica	3.0	122.6	171.5	Tsujimoto 1929
Taeniopoda ⎧ male	2.5	111.5	196.5	F. Giral 1941
auricornis ⎩ female	2.8	101.3	185.0	ditto.
Melanoplus mexicanus	—	75.3	216.3	J. F. and M. Giral 1943

The role of fats and their bio-chemical changes within the individual's development has not yet been studied. The ranges of the indices given above may equally well refer to changes due to maturation as to differences between species. MATTHÉ (1945) obtained a considerably higher fat content in the gregarious than in the solitary phase of *Locusta migratoria* and *Locustana pardalina*. For further information it may be added that chitin, or rather the indigestible exoskeleton, froms 4% of the wet, and 10% of the dry weight of adult locusts. This means that the indigestible part is not large enough to diminish to any considerable extent the protein value of the insects as food. B_1 and B_2 are the vitamins most commonly found in grasshoppers.

All these chemical analyses may mean very little to the layman, yet they confirm that precisely those insects which are eaten in large quantities are rich in animal proteins, in animal fats and in calories. They include a fair number of mineral salts, which may under certain circumstances be valuable for human nutrition. The importance of insects as a source of vitamins has apparently been overestimated by some authors. The richness of insects in animal proteins and fats and the richness of certain animal products (honey, honey-dew) in carbohydrates suffices to explain their international importance in the poor and ill-balanced tropical diets. And although vegetables in the tropics are especially poor in vitamins of the B-group, it may well be that this relatively small vitamin content has an even greater importance than its small size might indicate.

Honey is not only one of the few concentrated sweets to which primitive man had access but, with its almost pure solution of monosaccharid sugars, represents one of the most stimulating foods. Physiology has well established that carbohydrates are the most perfect food for the production of working energy in man and domestic animals. The nectar in the crop of the honey-bee worker, as well as in that of most other strongly-flying insects, is the main source and reserve of the energy expended during flight. The exhaustion of this store will speedly immobilize the insects or in the bee-worker bring it to the border of starvation. There is perhaps no other food, excluding stimulating drugs, comparable to honey for the prevention of fatigue or for the restoration of strength after thorough physical exhaustion. The warriors of the Masai and of the Bushmen relied almost exclusively on honey as food during their longer expeditions and wars. Many explorers in Africa and Australia have praised the invigorating qualities of a mouthful of wild honey This was also well known in Biblical times. In I. Samuel 14 : 27 we read that Jonathan dipped the rod in a honey-comb and put his hand to his mouth; *and his eyes were enlightened*. It is almost impossible to describe more vividly the immediate restorative effect of honey, when one is fatigued. Possibly the war with the Philistines would have ended differently if Saul and his army had all participated in the meal of honey instead of fasting. Athletes from the age of Greek glory down to the mountaineers of our days, pilots and deep-sea divers, they all by tradition and by belief eat large quantities of honey during their periods of training, during their actual efforts and immediately afterwards. Physicians, dieticians and bee-keepers have always strongly supported this claim of honey as a restorative of energy and as a means of preventing fatigue (for example, see BECK and SMEDLEY 1947). The following table gives a condensed compilation of the main facts known about the composition of bees'-honey.

Bees'-honey has been thoroughly analysed, especially by the Agricultural Department of the United States of America. Samples from all over the world were included in these analyses, in order to establish the standard requirements of commercial honey. The results are (BECK and SMEDLEY 1947; ECKERT 1936):

I. *Main components:*	Water	about 17.7%	
	Dextrose	34.0 (24–37)	Dextrose and levulose are variable,
	Levulose	40.5 (39–49)	but their combined total is fairly stable.
	Sucrose	1.9	The total of the 3 basic sugars: 76.4%.
	Dextrins and gums	1.5	
	Total:	95.6%	

35

II. *Ash contents:*	The minerals of the ashes amount to about 0.18%: calcium about 5 mg., phosphorus 15–17 mg., iron 0.4–1.0 mg. Traces of the following are also present: silica, copper, manganese, chlorine, potassium, sodium, sulphur, aluminium and magnesium.

III. *Vitamin contents:*

Vitamin A	0
Vitamin C	1–6 mg.
Thiamine (B_1)	0.0021–0.0091 mg.
Riboflavine (B_2)	0.035–0.145 mg.
Pyridazine (B_6)	0.210–0.440 mg.
Pantothenic acid	0.047–0.192 mg.
Nicotinic acid	0.04–0.94 mg.

IV. *Miscellaneous:*

The following substances form a total of 4.1%.
Proteins (mainly from pollen) 0.3–0.5%
Fats   0 to traces.
Acids (formic, acetic, malic, citric, succinic, amino)
Pigments (carotene, xanthophyll)
Chlorophyll (decomposition products)
Bees'-wax
The following occur in even smaller quantities:
Enzymes (invertase, diastase, catalase, inulinase)
Aromatic bodies (terpenes, aldehydes, esters)
Higher alcohols (mannitol, dulcitol, etc.).
Other sugars (maltose, sometimes melezitose, etc.).

The calorific value of 100 grams of honey is 280 to 330 calories, the normal pH value about 4.5.

An examination of this list shows that there is no rapidly energizing food-component in honey to explain its quick physical effect apart from the dextrose and levulose. The great and easy absorptivity of these sugars may be considerably enhanced by the effect of other components such as vitamins, calcium, phosphorus and others. None of these latter has the rapid effect of the sugars. It is the habit of primitive man to eat the honey in the combs, together with the bee-maggots and bee-pupae and with a large quantity of pollen. The bees'-wax itself has no digestive value. But the proteins of the bee-maggots and of the pollen, as well as the high concentration of vitamins in the latter, add considerably not only to the calorific value of honey but also result in a greater balance in its dietetic value. This raised value gains in importance as the significance of honey increases within the general diet. Thus with the Guayakis, Veddas and African Pygmies, among whom honey is the principle food, this habit of consuming maggots, pupae and a certain amount of the stored pollen assumes great significance and importance.

Among the many medicinal virtues of honey its anti-haemorrhage and its bactericidal properties are undisputed. Some over-enthusiastic statements and claims, however, such as found in BECK and SMEDLEY (1947, pp. 120 ff.), should be accepted with great reserve. Honey mead, i.e. a fermented watery solution of honey was the nectar of the gods, not only in ancient Greece. Another refreshing drink is hydromel, a fresh mixture of honey and water.

Honey, whenever and wherever available, has always been a most prized food since Palaeolithic times and was rightly considered a highly effective restorative and guarantee of physical strength and a welcome guardian of health. Many primitive people, such as Pygmies, would not be able to survive in their last refuges in the primeval tropical forests, if they were not able to live mainly on honey. Half a kilo of honey provides their daily calorific requirement, while the addition of bee-maggots as a rich source of proteins and also of fats, as well as vitamins and minerals, make honey a wellbalanced food, especially as considerable quantities of pollen are taken simultaneously. Thus honey became one of the most important items of food throughout the tropics, before man reached a high degree of efficiency in hunting or the level of primitive husbandry.

A number of similar sweets play a more restricted role in native nutrition: certain honeys of social wasps in South America, those of the 'honeypots', various desert ants (*Myrmecocystus* in America, *Melophorus* in Australia), and the honeydew excretions or mannas of various cidadas, plant-lice and scale insects in various parts of the world. Even if the quantity of these sources of honey is limited, they are easily digestible and of great value as a restorative from fatigue especially if bees'-honey is not available.

These analyses and their comparison with the calorific and nutritive value of other basic items of food lead us to the unavoidable conclusion that insects are most nutritive and only to be compared with the most valuable types of food. Only where big game or a thoroughly balanced, rich diet in highly agricultural nations are available in adequate quantities, can the dietetic importance of insects be ignored and neglected. Among the primitive food-gatherers, who cannot disregard anything which nature may provide in the way of edible food, insects form of necessity an important part of the diet. In this case the only limiting factor is the quantity available. All insects which are neither poisonous nor aposematic (these are a small minority) are eagerly gathered for food, whenever they are present in numbers and of a sufficiently substantial size. Selection by taste becomes a quite secondary motive. In all higher stages of sociological development, man is often rather selective. The sporadic consumption of ants and the slight use made in many parts of Australia of grasshoppers and termites are indicative of such selective taste. Yet even for higher primitive men BEQUAERT (1921, p. 197) still reports that caterpillars are often appreciated as food in direct proportion to their size and abundance. Some of the descriptions given in the following pages from Australia and Africa, together with a number of photographs from African markets, not to mention the many reports of locust gathering for food, will convince the reader that in the tropics there is no lack of edible insects. At least in certain seasons, and to some degree the whole year around, insect life is sufficiently abundant to offer adequate quantities for human nutrition. In times of drought and famine these insects may become the decisive food which permits survival. The especially important role of honey for all primitive peoples cannot be over-emphasized. Particularly with locusts quantity has never been a limiting factor. Today a large amount of locust flour is sold as manure. 'World Trade' (1936, p. 42) announced that almost 3000 tons of locust flour containing 9.7% nitrogen and 12.4% fat was available in the Argentine for export. Various authors have discussed the value of locusts for manure in South Africa (ADLER 1934) and in India (DAS 1945).

At the outset of the following discussion we can thus be fully convinced that insects when eaten in quantity are an important source of food. In cases where the normal diet is poor in animal proteins or in animal fats, they offer a necessary supplement to the main food. Honey provides a most important source of immediate working energy. Vitamins, salts and minerals represent other possible additions to the value of insect food. We shall now examine the question of the extent to which primitive man has utilized this potential food in his normal diet.

II. HISTORY OF ENTOMOPHAGY

1. FROM ANTIQUITY TO AD-DAMIRI

The earliest contact of man with insects was utilitarian and there is little doubt that he was initially attracted by their potentialities as food. This early development from *Pithecanthropus* to the primitive tribes inhabiting the tropics today has already been discussed. Our scanty knowledge of these beginnings of entomophagy is largely derived from analogies and circumstantial evidence. Yet in the oldest civilizations we find mention of the consumption of various insects and their products, mainly locusts and honey. The beginnings of honey-hunting and of primitive bee-keeping will be discussed later. Most of the early documents concern civilized peoples with a high agricultural standard. In such cases they are merely consumed as dainties or perhaps occasionally during famines. Where large-scale eating of insects is noted, it is described as a curiosity of barbaric tribes.

Greek literature contains a number of references to the eating of cicadas, in which these were regarded as delicacies. The first quotation, by ARISTOTLE of Athens (Historia animalium 556b 1.5 ff.), hints that they were by no means an uncommon food in Attica: 'The larva of the cicada on attaining full size in the ground becomes a nymph (*tettigometra*); then it tastes best, before the husk is broken (i.e. before the last moult) .... (Among the adults) at first the males are better to eat, but after copulation the females, which are then full of white eggs'. PLINY of Rome (Naturalis Historia XI : 26), in addition to the last sentence quoted from ARISTOTLE, mentions that they are eaten in the East. ATHENAEUS of Alexandria (Deipnosophistoi IV p. 133b) quotes them as dainties in banquets, being served to stimulate the appetite. AELIAN (Historiae Animalium XII : 6) reports with discontent that he saw people selling small parcels of cicadas for food. That many Greeks of the

Hellenistic period rejected cicadas for consumption becomes clear from the following lines of PLUTARCH (Symposiaques B. 8): 'Consider and see whether the swallow be not odious and impious . . ., because it feeds upon flesh and kills and devours especially cicadas, which are sacred and musical'.

Eating of lice is almost cosmopolitan. HERODOTUS (History 4: 109) states that the tribe of the Burdini nomads of the middle Volga are louse-eaters (*phtheirotrageousi*). BURR (1939, p. 211) compares this report with the following observation of the well-known Russian traveller and naturalist NAZAROFF about the Kirghiz and the Kazaks, the descendants of the Budini: 'I was witness of a touching, if barbarous scene of wifely devotion. Our host's son was deep in sleep of the just in the usual Kirghiz way, quite naked under his blanket. Meanwhile his affectionate and devoted wife profited by the opportunity to clean his shirt of the vermin (lice) swarming in it. She performed this operation in a manner that was as effective as it seemed to me original. She systematically took every fold and seam in the shirt and passed it between her glistening white teeth, nibbling rapidly. The sound of the continuous cracking could be heard clearly. This strange scene reminded me of the words of HERODOTUS, which accurately describe the habits of the same tribes 2200 years ago'.

We have a great deal of information about locusts as food in the Middle East in antiquity. In the palace of Asurbanipal near Nineveh (8th Cent. B.C.) we find among the servants bringing the various dishes to a royal banquet one who carried locusts arranged on sticks. In Leviticus (XI : 22) we read: 'These you may eat; the *arbeh* after his kind, the *sal'am* after his kind, the *chargol* after his kind, and the *chagav* after his kind'. The traditional translations are erroneous. The *arbeh* is the adult of the desert locust (*Schistocerca gregaria* Forsk.), the same locust which appeared in the banquet at Niniveh, while the other names doubtless refer to various stages of development of the same locust. Other grasshoppers may also have been eaten, as we read (Leviticus XI : 21): 'You may eat of all winged creeping things that go upon all fours, which have legs above their feet, wherewith to leap upon the earth'. Yet none of them was sufficiently abundant to provoke differentiation of names for their various species. This permission to eat locusts is nothing more than a codification of a habit existing since oldest times among the nomads of the Middle East, which, as we will see, has lasted down to our day. In the New Testament, St. John is mentioned as nourishing himself upon locusts and wild honey (Matthew III: 4). And in the Mishnah (translated by DANBY 1933, p. 518) we find in the treatise Hullin (3 : 7) the following interpretation: 'Among locusts (these are clean): all that have four legs, four wings, and jointed legs, and whose wings cover the greater part of their bodies. Rabbi Jose says: Or (all) that are called by the name 'locust'.' And in the treatise Kelim(p. 640 : 24 : 15) there is mention of the leather gloves worn by locust-catchers. The Jerusalem and the Babylonian Gemarah also talk about locusts and their gathering for food.

Among the Greek historians and geographers we find references to a people called *Acridophagi* or locust-eaters. DIODORUS of Sicily (2nd Cent. B.C.; Historia III : 2) talks about the *Acridophagi* of Ethiopia as small, lean and spare, and extremely black men. When in their country in spring the south winds rise high, they drive out of the desert an infinite number of locusts, of an extraordinary size, with dirty wings of an unpleasant colour. These provide plentiful food and provisions for them all their days. They are caught as follows: in their country there is a large and deep vale, extending for many miles; all over this they lay heaps of wood and other combustible material. When the swarms of locusts are driven thither by the winds, some of the inhabitants go to one part of the valley, some to another and set the grass and other combustible matter on fire. Thereupon arises a great and suffocating smoke, which so stifles the locusts as they fly over the vale, that they soon fall down dead to the ground. This destruction of the locusts is continued for many days together, so that they lie in great heaps. As their country is full of salt, they gather these heaps together and season them sufficiently with salt, which gives them an excellent flavour and preserves them a long time sweet, so that they have food from these insects all the year round. This people dies early as a result of this food. They are very short-lived, never exceeding an age of forty. And when they grow old, winged lice breed in their flesh, divers kinds of horrid and ugly shapes. This *phthiriasis* begins first in the abdomen and breast, consuming in a short time the whole body.

41

STRABO's report on the *Acridophagi* is obviously a resume of DIODORUS' description (XVI : 4 : 12). A special treatise on locusts by AGATHARCHIDES (2nd Cent. A.D.) has not been preserved, but quotations about the *phthiriasis* indicate that he also relied very much upon DIODORUS, but he apparently places the peoples concerned in the extreme South of Libya. SOLINUS also obviously goes back to DIODORUS. The great physician DIOSCORIDES (1st Cent; 2 : 47) recounts how the Afri inhabiting Leptis only eat the larvae (*asiracas* et *oni*) in any quantity. PLINY (1st Cent. A.D.; XI : 26) states that the people of the East use grasshoppers as food including even the wealthy Parthians.

O. KELLER (1913, p. 455) has compiled a number of references to Attica. ARISTOPHANES (Acharnai 889) quotes poulterers who sell 'four-winged' fowl on the market. These four-winged fowl are grasshoppers, which apparently were cheap and consumed by the poorer classes (see schol. Aristoph. Acharnai 1130.) On the fields the young shepherds enjoyed eating them (Theophyl. epist. 14).

PLINY in his Historia Naturalis has repeatedly referred to the *cossus* as a dish highly appreciated by the Romans. He writes: 'Insect larvae do not infest all trees, but most of them. The birds recognize their presence by the hollow sound of the beaten bark. The large grubs in oaks are included among the most delicate dishes as *cossus*. They are fattened by feeding them with flour' (XVII, 57). 'Not only the *cossus* takes birth from it, but also the *tabanus* comes out of the wood' (XI, 38). 'The *cossus* which originates in the wood heals ulcers' (XXX, 39).

Many naturalists have tried to identify this *cossus*. The main determinations are included in the following list:

Cossus cossus L. by Linnaeus (Fauna Suecica 1761, p. 265).
Rhynchophorus palmarum L. by Geoffroy (Histoire abrégée des Insectes, 1800 II, p. 104).
Cerambyx heros L. by Olivier (Entomologie, 1789, II, p. 6).
Oryctes nasicornis Ill. by Swammerdam (Biblia Naturae, 1737, I p. 318) and by Frisch (vol. V, p. 7).
Lucanus cervus L. by Roesel von Rosenhof (Insekten Belustigungen, 1742, 2.1, p. 31) and by Latreille (Hist. Nat. des Crust. et Insectes, XII, p. 245).
Melolontha vulgaris L. by Latreille (Cours d'Entomol., p. 60).

ST. JÉROME in his Treatise against Jovinian, Festus and others occasionally mentions that certain people eat wood-boring larvae as delicacies, while others would refuse with disgust to eat them. MULSANT (1841) has in our opinion said the last word on the question of the identity of the *cossus* of PLINY. The caterpillar of *Cossus cossus* can be excluded, as it does not or only very rarely live in oaks, while being common in willow-trees. The tropical *Rhynchophorus* can be excluded, as not occurring in Mediterranean oaks. Nor do the grubs of *Oryctes, Lucanus* and *Melolontha* feed in oak. The only grub which satisfactorily fulfils all characteristics is that of *Cerambyx heros*. Such common grubs as those of *Melolontha* would certainly be found mentioned in some of the extensive cookery books of later antiquity, while precisely the rarity of *Cerambyx* together with their pleasing taste would make them a desired tit-bit on the tables of the gourmets. Thus we conclude that almost certainly the larvae of *Cerambyx heros* from oaks are the *cossus veterum*.

AELIAN (An. Hist. XIV : 13) tells of a dessert of a similar kind on the table of the King of India from the tawny palms, which the Greek do not enjoy. He delights in a worm which originates in that plant, which is very sweet. Another delicacy is mentioned by ST. HIERONYMOS (Adv. Jovin. col. 102). On the Pontus and in Phrygia many families earned a good profit by collecting certain fat, black-headed, white larvae of insects, which occur in decaying wood, selling them as highly esteemed tit-bits.

KAMAL-AD-DIN AD-DAMIRI (1341–1408) in his great zoological Lexicon (Hayat al-hayawan, Ed. London-Bombay, 1906, I, pp. 413 ff.) while not insisting upon locusts as food in Arab countries, fully discusses the 'Lawfulness and unlawfulness' of eating them from the point of view of Islamic tradition: 'All the Muslims are agreed that its eating is permitted. Abdallah ben Abi-Awfa said: 'We went with the Prophet on seven military expeditions and we used to eat locusts'. Abu-Dawud, al Bukhari and the Hafid Abu-Nu'aim states: 'And the Prophet used to eat them with us'. Ibn-Majah recalls in the name of Anas how the wives of the Prophet used to send them locusts on trays as presents. Omar was asked about locusts and replied: 'I wish I had a basketful of them to eat.' .... Yahya ben Zakariya used to eat locusts .... The four imams state that the eating of locusts is equally permissible when they have died a natural death or have been killed lawfully or have died after being hunted by a Majian or a Muslim, and whether or not any part of them has been cut off. It is said in the name of Ahmed that if they have died from cold they ought not to be eaten, and the school of Malik holds that if their heads are cut, they are lawful, but otherwise unlawful. The proof, however, of their being lawful under all circumstances is the statement of the Prophet: 'Lawful for us are two dead (animals) and two bloods,—liver and spleen, fish and locusts.' . . . . . Our theologians and other sages differ as to whether locusts are land or sea game . . . . It is permissible to make an advance of money or property and to receive payment for it in locusts and fish, both alive and dead, when they are to be had in abundance, but every article has to be named for what is worthy of it (in return) . . . . . Among the proverbs we read: 'A date is better than a locust'.'

ABU OSURAN writes in his treatise on zoology: 'I do not know a more delicate dish than their meat. Roasted locusts have the same taste and smell as have roasted scorpions, both of which resemble that of chicken meat'. (BRYGOO, 1946, p. 38).

2. FROM ALDROVANDI TO THE PRESENT DAY

The New Age of entomology begins with the appearance of the compendious folio volume 'De Animalibus Insectis Libri Septem' (1602) by ULYSSE ALDROVANDI. ALDROVANDI mentions the use for food (usus in cibo) of various insects. The native inhabitants of Cumana eat bees. The ashes of bees are one of the ingredients for inducing the speedy growth of a beard (p. 107.) A long chapter (pp. 118–155) is devoted to the use of honey in rituals, as food and drink, in medicine, for preservation, etc. On another occasion we read that a certain learned man informed him that the German soldiers in Italy repeatedly and with obvious delight ate fried silkworms (p. 295). Ancient authorities are quoted for the consumption of cicadas (p. 340) and of locusts (pp.' 348–441), but not for that of beetles (pp. 468, 478, 488). Contemporary travel reports were also utilized by ALDROVANDI, when he states that ants are eaten in certain parts of India and in the Genusucian Islands (p. 531), or that lice are consumed, boiled or raw, in the Cumana Islands, as well as by the *Phtheirophagi* of HERODOTUS (p. 559).

The 'Insectorum sive Minimorum Animalium Theatrum', pre-pared from the manuscripts of Conrad Gesner with some additions by E. Wotton and T. Penny and edited by T. Moufet (1634) is the important competitive volume to Aldrovandi's great work. As the latter's volume, however, appeared in Latin only, while an English translation is available of Moufet's 'Theatrum', Moufet is far better known than Aldrovandi. Moufet's sources for locusts as food, all taken from antiquity, are imposing. The following peoples of hot regions are mentioned specifically as eating them and their eggs: Aethiopi, Tagetenses, Parthi, Arabes, Lybii, Mellenses, Zemenses, Darienses, Afri (Leptis), Azanughi, Senegenses et Mauri-tani (p. 124). Eating of cicadas by the Greeks, as well as by the *phtheirophagi* of Herodotus, is mentioned (p. 132), and the Spanish reports of lice being eaten by the Indians of Cuene in the West Indies are also given (p. 204). The use of honey is, of course, ex-tensively described (pp. 24–33), and he also refers to the paucity of honey to be gained from the nests of humble-bees (p. 54).

One of the earliest modern reports on locust eating in North Africa and Arabia is contained in the Description of Africa by the converted Mohammedan native of Morocco Leo Africanus (1805: I, p. 587 f.), who wrote in the first half of the 16th century. He describes the enormous locust invasions which often bring about famine, especially in Mauretania. But in contrast to these peasants, the nomads of Arabia and of Libya greet the appearance of locust swarms with joy. They boil and eat them, dry others in the sun and pound them into flour for future consumption.

De Réaumur in one of his superb 'Mémoires pour servir à l'His-toire des Insectes' (1737: vol. II, 2, pp. 113–120) discusses the edibility of insects, while dealing with the enormous ravages pro-duced in France by *Plusia gamma*. During that period some people who had eaten these caterpillars with salad or in soup claimed that they were poisonous. Yet, as with all other smooth caterpillars, they are actually harmless. However, the prejudice against this insect has been so great that when one of its caterpillars has been swallowed, it has been immediately held responsible for any symp-toms of poisoning. 'One may eat as many of our vegetable cater-pillars as one wishes without fearing the slightest damage, swelling or inflammation'. Réaumur continues to discuss the general problem of entomophagy: 'If large, smooth caterpillars were here as common as are locusts in certain regions, and especially if they were abundant in a year of famine, perhaps the peasants of France would eat them as locusts are eaten in Africa. And perhaps they would subsequently be regarded as an agreable and wholesome dish! We know a number of wood-boring beetle grubs which appear much less palatable than smooth caterpillars, yet the ancient Romans regarded these *cossi* as a first-class delicacy. We need not even go back as far as that. Similar beetle grubs, which also live in the interior of trees in our West Indian possessions, are considered when fried as a succulent and splendid meal. And the grubs of the common *Oryctes*-beetles, which are white, plump and fat, like those of the *Cerambyx*-grubs or *cossi*, would perhaps make an excellent entremet, if our prejudices would permit us to introduce them into our menus. One would look for these grubs in the soil, as one looks for truffles, and the number of the beetles of this injurious species in this way could be much diminished.

We could perhaps in due time overcome our repugnance at eating insects and accept them as part of our diet, and then realize that there is nothing terrible about them and that they may perhaps even offer us agreable sensations. We have grown accustomed to eating frogs, snakes, lizards, shell-fiish, oysters, etc. in the various provinces of France. Perhaps the first urge to eat them was hunger. In conclusion, while leaving the caterpillars for the time as a food for the birds, we need not accuse them of poisoning. In 1735 thousands and thousands of these caterpillars have been eaten by cattle, horses, sheep, asses, etc., which suffered no harm as a result'.

ROESEL VON ROSENHOF in his 'Insecten-Belustigungen' (Dutch ed. 1779, vol. 2, para. 37, p. 297 f.) has found two recipes among the older reports about locust-eating. The one is to tear off the legs and the wings, and to dry them in the sun (i.e. in the weak sun and humid air of the North!), until they ferment. They are eaten and it is claimed that they make an agreable dish. The other way is to boil the locusts in salty water and then to eat them seasoned with vinegar, salt and pepper. I have tried to prepare them in both ways. Yet when they began to ferment after preparation by the first method, the smell of the fermenting insects is so bad that any desire to eat them disappears; this has also been reported by FRISCH of locusts which died in the fields. The second method, however, was equally unpleasant to my palate. They then smell like shrimps, but their taste is repellent and unpalatable. When ROESEL was once busy boiling locusts for one of these feeding experiments, one of his friends entered his house. This friend had always encouraged ROESEL to overcome his prejudices and to taste them. Thus he asked him if he would accept an invitation to a dish of boiled locusts. He agreed but when the famous dish was set upon the table, he lost his appetite, and as did other friends, he lost all desire to taste them. Some affected to feel no repugnance, yet immediately the locusts had entered their mouths, everyone of them spit them out or

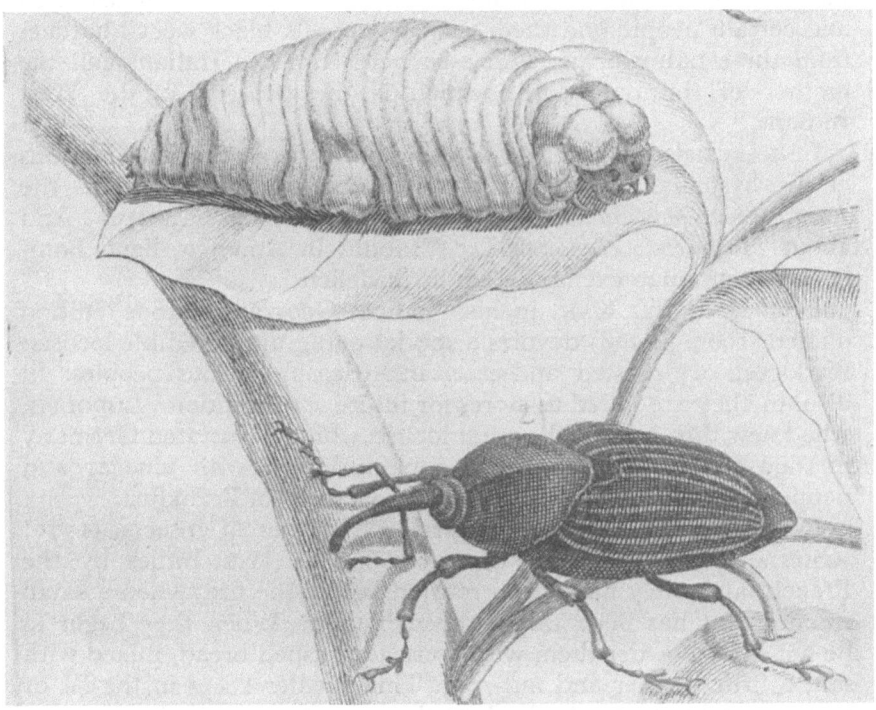

Fig. 3. The palmworm (*Rhynchophorus palmarum* L.) as drawn by Maria Sibylla Merian in her 'Insects of Surinam' (1771, pl. 48).

vomited them, while their faces showed their fear of swallowing them. It was just as if they had taken a drug for vomiting.

MARIA SIBYLLA MERIAN, the famous entomologist, talks about the palmworm in her General History of the Insects of Surinam (1771, pl. 48): A whitish worm creeps in the centre over a green leaf, the palmworm, which feeds on this tree .... The natives say that it grows for fifty years, before becoming fully-grown. Then they cut it from the base of the leaves; they also cut open the stem of the tree at the height of a man, where it begins to grow soft. They cook it as we cook cauliflour. The taste is more agreeable than artichokes. Certain worms burrow in the trunk of this tree, feeding on its marrow. To begin with they are no larger than cheese-mites, but grow to the size of that illustrated on the plate. They are fried and certain people find them very delicate. A black weevil hatches from these palmworms, as drawn here; this the Indians call the mother of the palmworm, which is the *grou-grou* of the West Indies.

LINNAEUS also mentions the eating of insects in the later editions of his Systema Naturae. He writes about the palmworm, the *Rhynchophorus palmarum* L.: 'Larvae assatae in deliciis habentur'. And about *Macrodontia cervicornis* L.: 'Habitat in America, ligno Bombacis larvae quae exemptae edulis in deliciis'.

Even IMMANUEL KANT in his Physical Geography (1802, edition quoted, 1905, p. 238) devotes a special paragraph to edible locusts: Big locusts are roasted and eaten in Africa by various peoples. In Tonkin they are salted as stores for future consumption. LUDOLPH, who knew this, cooked the great locusts which devastated Germany in 1693 like crayfish, ate them, perserved them with vinegar and pepper and with this dish treated the Council of Frankfurt.

BROOKES in his 'On the properties and uses of Insects' (1772) writes: 'The palmworms are eaten in the West-Indies by the French, after they have been roasted before the fire, when a small wooden spit has been thrust through them. When they begin to be hot, they powder them with a crust of rasped bread, mixed with salt, a little pepper and nut-meg. This powder keeps in the fat or at least sucks it up. And when they are done enough, they are served up with orange juice. They are highly esteemed by the French as excellent eating'.

Foucher d'Obsonville in his Philosophical Essays on the Habits of Various Strange Animals (1783, pp. 43 ff.) recounts that locusts are eaten with relish by most Africans, some Asiatics and especially the Arabs. On their markets they appear roasted or grilled in great quantities. When salted, they keep for some time in storage. They are used for supplying ships, where they may be served as dessert or with coffee. This food is in no way repugnant to look at or by association. It tastes like prawn, and is perhaps more delicately flavoured, especially the females when filled with eggs. Certain people assume this food to be the cause of the eye-diseases which are so common in some of these regions. D'Obsonville says he could easily imagine that excessive use would impoverish the blood and have dangerous consequences; but blindness and eye-diseases are probably caused by the salty and fiery particles transported by the winds. The Turks, Persians and Christians, who in the same regions do not eat locusts or only rarely, are subjected to the same eye-troubles, while some African peoples who eat locusts in great quantities have excellent eye-sight.

K. Illiger (1804) devoted a paper to edible insects, compiling a collection of curiosities. He begins with recipes for preserving may-bugs (*Melolontha*), partly taken from old cookery books. The butter gained from the fat of palmworms in the West Indies is highly appreciated. *Cossus* and locusts have been known since antiquity. The Barbary pirates include locusts among their regular provisions. The Bushmen consume termites, as well as locust hoppers. On Amboina a large 'praying mantis' is eaten.

W. Kirby and W. Spence (1823: I, pp. 331–371) in a special chapter on the direct benefit from insects, begin quite properly with the role of insects as food. They tell of the delight in palmgrubs and locusts, in caterpillars and silkworm pupae, in termites, bees, ants, lice and gall-apples, and last but not least in honey. This early collection of curiosities has served many later compilers as a source and model.

In a didactic poem in six songs 'L'Art Entomologique' (1814 p. 61 f.) Le Roux talks about entomophagy, as follows (2nd song):
"Que de privations subiroient nos gourmands,
S'ils se trouvoient sevrés d'Insectes quelque temps!
C'est au charmant auteur de la Gastronomie

A leur signaler ceux que le plus on envie,
A nous rendre ces mets des anciens si vantés,
Dont nos palais nerveux semblent épouvantés.
Les Romains, en ce point comme en d'autres, nos maitres
Des peuples sensuels orgueilleux thermomètres.
Gourmets par excellence, en tout voluptueux,
Nous font peu regretter leurs repas somptueux.
D'Insectes rebutants pour leurs propres familles,
Dans leur difformé état de larves, de chenilles,
Ils savoient composer des mets délicieux,
Dignes de faire envie aux plus friands des Dieux,
Et de flots de nectar venu des bouts du monde,
Teignoient en leur honneur les coupes à la ronde.
Le savoureux cossus, réservé pour les grands,
Vit leur or conspirer contre ses descendants,
Quand la Fourmi sucrée, à l'acide agréable,
Leur faisoit convoiter sa nymphe délectable.
La hideuse araignée, inspirant de l'horreur,
Trouva de son fumet maint appréciateur.
L'heureux rival de Hershed, qui regrette Uranie,
Avoit de les manger le bizarre manie;
De son gout dans un cercle il tiroit sanité,
Sans consulter l'effroi qu'en avoit la beauté;
Et l'on soit de l'histoire en feuilletant les pages:
Des peuples décorés du nom acrydophages;
Mais quoique notre siècle abonde en Lucullus,
Nos sens de ces ragouts ne s'accommodent plus.
L'art de nos cuisiniers, sans frein et sans méthode
Parviendra-t-il jamais à les rendre à la mode,
Quand l'effet merveilleux de l'eau de coladon,
Qui fait moins digérer que manger un glouton
Pour vanter leurs attraits, jusqu'alors sans puissance
N'a pu de nos gourmands vaincre la répugnance?'

J. J. VIREY, physiologist and pharmacologist at the beginning of the 19th century and a member of the French Academy, has often discussed entomophagy in his anthropological treatises. Thus he writes in his 'Histoire Naturelle du Genre Humain' (IX, I, p. 256): 'Temperature regulates our diets, determining whether much meat or more vegetables are eaten. I shall not discuss locusts and other insects which are consumed by the Arabs and some

Africans as good meals. At Tonkin also man and monkeys eagerly seek insects as food. PLINY and other ancient writers claim that the acridophagous peoples are weak, thin, precocious and do not live longer than 40 years (PLINY, Hist. Nat. 6 : 29 and 4; STRABO 16). It is certain that the insects which are usually eaten are unwholesome and an irritant. In general, foods which have the same type of organization (i.e. mammals for man) prove the most satisfactory'. And in the second edition of the same book (1824, II, pp. 319, 329) VIREY refers to entomophagy, mentioning the cicadas of the Athenians, the locusts of the Arabs and Africans (which may cause phthiriasis cf. Journ. Complém. vol. 15, p. 1), the *cossus* of the Greek, the palmworms of the Indians and natives of America. He writes: 'The Moor, starving in his deserts, devours locusts or feeds on the gum of its acacias, or on some pinches of cous-cous'.

Unfortunately, the writer was unable to discover in the great libraries of Paris VIREY's often quoted paper: 'Whether man may eat insects and whether he should eat them*'. We quote its concluding passage taken from J. BEQUAERT (1921, p. 200): 'Man may eat insects: nothing in his anatomical organization or his physiological functions is opposed to it. He should eat insects: in the first place, because his cousins the monkeys and his ancestors the bats, or in short the primates, eat them. In the second place, because insectivorous animals are superior to the other species of their order, both in their more perfect organization as well as in the superiority of their intelligence'.

An extensive paper 'Observations respecting various insects which at different times have offered food to man' by the learned Reverend F. W. HOPE (1842, pp. 129–150) begins with quotations from the Bible, HERODOTUS and DIODORUS, and continues: 'Insects live on cleanly diet, and consequently afford us more wholesome food than some of the animals that are usually served at our tables. It is not my intention here to recommend insectal food to nations living in northern climates, the supply of which should be scanty in winter.

---

* For the guidance of future students the following quotation from the Index Litteraturae Entomologicae (HORN und SCHENKLING 1929, p. 1265, no. 22773) may be given; however, its incompleteness indicates that HAGEN, as well as HORN and SCHENKLING have taken it from a secondary source, without themselves seeing the paper: „De l'Entomophagie. Journ. compl. Dict. Sc. Méd. 15'.

I see no reason, however, why in the warm and well-wooded regions of the world they should not be eaten, as the supply there is generally abundant. The New Hollander or even the European settler in those parts, may derive much benefit by adopting the larvae of insects as food, for the very worms regaled on, if left to themselves, in time might multiply so as to endanger the crops of future years, entailing ruin on the grower, and perhaps famine on the settlement. In our regions in famine insectal food may be adopted. Probably want and hunger were the original cause of introducing to notice several of the insects which have been taken as food, although I am unable at present to substantiate this'.

After the discussion of beetles which are eaten, locusts are dealt with, such as *Locusta migratoria* L., commonly eaten in the Crimea together with *L. tatarica* L.. In Egypt and Arabia the 'red locust', the 'light locust', the 'Dubbe locust' and *L. gregaria* Forsk. are consumed, together with *L. cristata* L.. *L. cernensis nov. sp.* is reported from Madagascar, *L. devastator* Licht. from S. Africa, and in India *L. mahrattarum nov. sp.*, and a few others are occasionally eaten. *L. persarum* Mor. is the yellow̄ locust from Bushire. Most of these species are identical, of course, with the common locusts (*Locusta, Schistocerca, Nomadacris*). It is amusing how freely HOPE creates new species, based only on reports of great swarms occurring in Madagascar, India or Persia. He concludes (p. 140): 'To mitigate the injuries of locusts, 1) the legislative powers should recommend locusts generally as an article of food; to employ people and children to hunt and destroy their eggs; to collect the hoppers with the aid of police and military. Where large swarms menace famine a levée en masse of the population is indicated. The locusts required for food may be prepared for future use, the others buried. A poll tax of a bushel of locusts may be required from every inhabitant and thus by considerably reducing their numbers future famine and pestilence might in some instances undoubtedly be prevented. Much good may thus be caused in Egypt or in Crimea'.

With cicadas the creation of new species is continued, as the *Tettigonia antiquorum nov. sp.* of the ancient Greeks, the *T. parthorum nov. sp.* for that reported by PLINY as food of the Parthians, *T. bennettii nov. sp.* for the galang of N. S. Wales (vide BENNETT). The consumption of caterpillars (*Lepidoptera*) refers to observations in

52

Africa and Australia, in addition to that of silkworms in China and the *cossus* of the Ancients. Finally, honey-bees, ants, termites and flies are discussed. HOPE concludes (p. 146) with another argument: 'I have little doubt, that insects may eventually afford us medicines more powerful than those of trees and herbs, and certainly less deleterious than those derived from minerals'.

P. L. SIMMONDS has devoted two books to the animal resources of nature. In one, 'The Riches of Nature' (1877), he discusses the useful insects (pp. 313–344) under the following headings:

  I. Insects producing silk.
 II. Insects producing wax, honey, etc., including the white wax of China and Mexico.
III. Insects as sources of dyes (*Coccoidea* and Cynipid galls).
 IV. Insects producing manna (*Cicada orni*, Sinai manna, *Trehala*).
 ᴠ. Edible insects.
 VI. Insects in medicine (Cantharids only are mentioned).
VII. Insects as ornaments (An interesting compilation: pp. 341 ff.).

Among the edible insects (p. 338)-SIMMONDS mentions the *hautlé* of Mexico and the insect larvae eaten in China, India and Madagascar; locusts from Algeria, where they are also used as bait for sardine fishery; termite oil from Gaboen, etc.

The second and more elaborate book 'The Animal Food Resources of Different Nations' (1885) deals in his Chapter X (pp. 347–375) with 'Various Insects eaten as food'. We give here a few selections from this extensive and useful compilation, while many other quotations will be arranged in geographical order. SIMMONDS opens with the amusing story, how a few years ago at the Café Custoza in Paris a great banquet was given for the special purpose of tasting the white grubs or cockchafer worms. This insect, it appears, was first steeped in vinegar, which had the effect of making it disgorge the earth, etc., it had swallowed while still free; then it was carefully rolled up in a paste composed of flour, milk and eggs, placed in a pan, and fried to a bright golden colour. The guests were able to take this crisp and dry worm in their fingers. It cracked between their teeth. There were some fifty persons present, and the majority had a second helping. The larvae or grubs generally, not only of the cockchafer, but also of other beetles may, according to some naturalists, be safely eaten.

In the 'Gazette des Champagnes' (cf. MIOT, 1870 p. 89) the following prescription is described as being in use in certain parts of France: Roll the white grubs, which are short and fat, in flour and bread crumbs with a little salt and pepper, and wrap them in a stout piece of paper, well buttered inside. Place it in the hot embers and leave it to cook for twenty minutes, more or less according to the degree of heat. On opening the envelope a very appetizing odour exhales, which disposes one favourably to taste the delicacy, which will be more appreciated than snails and will be declared one of the finest delicacies ever tasted. The bulk of the following parts of the treatise concerns locusts, while other insect food, such as silkworm, termites, ants, etc. are also mentioned. SIMMONDS concludes: 'All nations do not participate with us in the neglect of insects as food. Not only savage nations indulge in these gastronomic dainties, but amongst the more civilised races those which were the cradle of refined civilisations have not scrupled to indulge in various insects as food'.

Among these curiosities of eating the white grubs of may-bugs (*Melolontha melolontha* L.) we may add a few more which are found in BRYGOO (1946, p. 61). When fried in butter or oil with parsley or chopped garlic the Reverend SHEPPARD found them more delicate than snails. Doctor GASTIER, once a French representative of the people, delighted in eating beetles. He cleaned them like shrimps, and no gift could be more pleasant to him in spring than a box with living may-bugs. On the 13th February, 1878, the French Senator TESSELIN published the following presription in the Journal Officiel, in contesting a proposed law for the destruction of agricultural pests and the preservation of birds: 'Catch the may-bugs, pound them, put them through a sieve. For making a thin soup, pour water over them. For making a fat soup, pour bouillon over them. This gives a delightful dish, esteemed by the gourmets'.

M. W. DE FONVIELLE, Vice-President of the Société d'Insectologie de Paris, proposed in his opening speech at the Exposition d'Insectologie in 1887 the destruction of the may-bugs by 'absorption', and illustrated this by swallowing some before the audience, giving signs of high satisfaction, as if he had taken some excellent chocolate lozenges.

The writer CATULLE MENDÉS was partial to eating may-bugs.

He tore off the head and wings, and crunched the remainder, the taste of which he compared to a dish of chopped chicken in butter taken with salt.

In 1885 a small booklet of 99 pages by V. M. HOLT was published in London entitled: 'Why not eat insects',* with the motto: 'The insects eat up every blessed green thing that do grow and us farmers starve. Well, eat them and grow fat!'

HOLT explains in the preface that he is fully conscious of the difficulty of battling against a long-existing and deep-rooted public prejudice. Yet he asks his readers for a fair hearing, for an impartial consideration of his arguments, and for unbiassed judgement. 'If these be granted, I feel sure that many will be persuaded to make practical proof of using insects as food. *My* insects are all vegetable feeders, clean, palatable, wholesome and decidedly more particular in their feeding then ourselves. We shall some day right gladly cook and eat them!

I. Why not? Indeed, why not, as it appears from chemical analysis that the flesh of insects (among which HOLT includes molluscs as well as all arthropods) is composed of the same substances as are found in that of the higher animals'. All the insects proposed by HOLT for food are strictly vegetarian. Philosophy bids man not to neglect any wholesome source of food. And a good dish of fried cockchafers or grasshoppers would be a pleasant change in the poor man's unvariable diet. Wholesome food cannot be neglected. At the same time the collection for food would be an excellent control measure for many pests. 'The collecting children would be doubly rewarded by nourishing insect dishes at home'. The use of insects in folk-medicine shows that in the poorer classes no strong prejudice exists against swallowing insects. Hence we must discard unjustified prejudices. Cheese-mites (read: maggots) are freely eaten by many persons as 'part of the cheese'. In the same way cabbage worms are only part of the cabbage. HOLT points out the many and varied animal products in a Chinese menu (p. 25) which are eaten in Europe as a curiosity and regarded as fashionable. 'In these days

* This curious booklet has now almost disappeared. In London it was apparently only available at the British Museum, where it was destroyed by bombing. The only copy to be found by the author was in the University Library at Oxford. Therefore we give a condensed report on the lines of thought of this forgotten book.

Fig. 4. The dust-cover of HOLT's Book,
Why not eat Insects? London. 1886.

of agricultural depression we should do all we can to alleviate the
suffering of our starving labourers. Ought we not to exert our influence towards pointing out to them a neglected food supply?'

II. *Insect eaters.* This chapter is devoted to many records from
antiquity and from 'uncivilized' people, with which our reader is
by now fully acquainted.

III. *Insects that are good to eat, and something about their cooking.* The
Rev. R. SHEPPARD found locusts in Morocco excellent, which HOLT
confirms by his own experience. Raw they are pleasant to the taste,
cooked they are delicious with no limits to improvements by their
preparation. ERASMUS DARWIN guesses that cock-chafers are a delicious food, if properly cooked. In HOLT's experience they are deli-

cious. He recommends curried cock-chafers and also wasp-grubs. Those of ants were eaten by the 'Swiss Family Robinson'. According to DARWIN the caterpilars of hawkmoths as eaten by the Chinese are very palatable. Following a recommendation of Miss ORMEROD to handpick the caterpillars of *Pieris brassicae*, HOLT asks: 'Why should we not eat them?' as well as those of *P. rapae*, *Mamestra brassicae*, and others.

'Let not the labourer say: 'We starve. Meat is too dear; bread is almost as dear because the wireworm, the leather-jacket and the may-bug worm have thinned the crop'. Yes, but the wheat crop would would have been twice as thick if the wireworms, the leather-jackets and the luscious white chafergrubs had been diligently collected by you for food!' HOLT concludes that his advice is a measure for restoring the balance of nature and for preventing the undue increase of pests. Finally, he gives two menus of insects, which of course are unnaturally crowded with insect items. Yet their aim is to show how such dishes may be usefully introduced into the chief courses of an ordinary dinner:

I.	II.
Boiled Cod with Snail Sauce	Snail Soup
Wasp Grubs fried in the Comb	Fried Soles with Woodlouse Sauce
Moths sautés in Butter	
Braised Beef with Caterpillars	Curried Cockchafers
New Carrots with Wireworm-Sauce	Fricassée of Chicken with Chrysalids
Gooseberry Cream with Sawflies	Boiled Neck of Mutton with Wireworm-Sauce
Devilled Chafer Grubs	
Stag Beetle Larvae on Toast.	Ducklings with Green Peas
	Cauliflowers garnished with Caterpillars
	Moths on Toast.

These menus were obviously not planned to provide dishes for the poor!

Mr. RILEY (1877), when studying the biology of the Rocky Mountain locust (*Melanoplus spretus*), confesses that it had long been his desire to test its value as food. Thus he did not lose the opportunity of an invasion into the Mississippi Valley states. 'I know well enough that the attempt would provoke to ridicule and mirth, and even disgust, the vast majority of our people, unaccustomed to anything of the sort, and associating with the word insect or 'bug' everything horrid and repulsive. Yet I was governed by weightier reasons than mere curiosity, for many a family in Kansas and Nebraska was brought to the brink of the grave by sheer lack of food, while the St. Louis papers reported cases of actual death from starvation in some sections of Missouri, where the insects abounded and ate up every green thing the past spring. Whenever the occasion presented I partook of locusts prepared in different ways and one day ate of no other kind of food, and must have consumed, in one form or another, the substance of several thousand half-grown locusts. Commencing these experiments with some misgivings and fully expecting to have to overcome their disagreable flavour, I was soon most agreably surprised to find that the insects were quite palatable in whatever way prepared'.

'The flavour of the raw locust is most strong and disagreable, but that of the cooked insects is pleasant and sufficiently mild to be easily neutralised by anything with which they may be mixed, and to admit of easy disguise according to taste and fancy. But the great point I would make in their favour is that they need no elaborate preparation or seasoning. They require no disguise; and herein lies their value in exceptional emergencies, for when people are driven to the point of starvation by these ravenous pests, it follows that all other food is either very scarce or unattainable. A broth, made by boiling the larvae for two hours in the proper quantity of water, and seasoned with nothing in the world but pepper and salt, is quite palatable, and can scarcely be distinguished from beef broth, though it has a slight flavour peculiar to it, which is not easily described. The addition of a little butter improves it, and the flavour can of course be modified with mint, sage and spices, as desired. Fried or roasted in nothing but their own oil with the addition of a little salt, they make by no means unpleasant eating, and have quite a nutty flavour. In fact, like most peculiar and not unpleasant flavours, it is something one can soon learn to get fond of. Prepared in this way, ground and compressed, they would doubtless keep for a long time. Yet their consumption in large quantities in this form would not, I think, prove as wholesome as when made into soup or broth; for I found the chitinous covering and the corneous parts, especially of the larger species where the heads, legs and wings are carefully separated before cooking; and in fact some of the mature insects prepared' in this way, then boiled and afterwards stewed with a few vegetables and a little butter, pepper, salt and vinegar, make an excellent fricassée.

Lest it be presumed that these opinions result from an unnatural palate, or from mere individual taste, let me add that I took pains to get the opinions of many other persons. Indeed, I shall not soon forget the experience of my first culinary effort in this line – so fraught with fear and so forcibly illustrating the power of example in overcoming prejudice. This attempt was made at an hotel. At first it was impossible to get any assistance from the followers of the ars coquinaria. They could not more flatly have refused to touch, taste or handle, had it been a question of cooking vipers. Neither love nor money, could induce them to do either, and in this respect the kitchen hands were all alike, without distinction of colour. There was no other resource but to turn cook myself, and operations once commenced, the interest and aid of a brother naturalist and two intelligent ladies were soon enlisted. It was most amusing to note how, as the rather savory and pleasant odour went up from the cooking dishes, the expression of horror and disgust gradually vanished from the faces of the curious onlookers, and how at last the head-cook, a stout and jolly negress, took part in the operations; how, when the different dishes were neatly served upon the table and were freely partaken of with evident relish and many expressions of surprise and satisfaction by the ladies and gentlemen interested, this same cook was actually induced to try them, and soon grew eloquent in their favour; how, finally, a prominent banker, as also one of the editors of the town joined in the meal. The soup soon vanished and banished silly prejudice; the cakes with butter enough to hold the locusts together, disappeared, and were pronounced good; then baked locusts without condiments; and when the meal was completed with dessert of baked locusts and honey à la John the Baptist, the opinion was unanimous that that distinguished prophet no longer deserved our sympathy, and that he had not fared badly on his diet in the wilderness'.

Professor H. H. STRAIGHT of the Warrensburg (Mo.) Normal School, who made some experiments for me in this line, wrote:

'We boiled them rather slowly for three or four hours, seasoned the fluid with a little butter, salt and pepper, and it made an excellent soup, actually; we would like to have it even in prosperous times. Mrs. and Professor JOHANNOT pronounced it excellent. I sent a bushel of the scalded insects to John BENNET, one of the oldest and best known caterers of St. Louis. Master of the mysteries of the cuisine, he made a soup which was really delicious, and was so pronounced by dozens of prominent St. Louisians who tried it'.

P. L. SIMMONDS (1885, pp. 365 ff.), basing himself upon this and other reports, says that in the experience of many persons, including members of the London and Paris Entomological Societies, locusts have been pronounced far better than was expected, and those fried in their own oil with a little salt, as really good; while those fried in butter may become rancid, which is the fault of the butter. He considers that locusts will hardly come into general use for food except where they are actually abundant. Yet the fact remains that they do *make very good food*. When freshly caught in large quantities, the mengled mess presents a not very appetizing appearance and emits a rather strong and not very pleasant odour; but rinsed and scalded, they turn a brownish red, look more inviting and give no disagreable smell.

On 11th July, 1875, Mr. RILEY presented a box with *Caloptenus spretus*, prepared as food preserved in the U.S.A., to the French Entomological Society.

MILLS and PEPPER (1939) arranged a series of experiments, in which four adult volunteers received a meal of 25 beetles, 25 larvae, a few pupae, cast skins, eggs and some excrement of *Tribolium confusum* boiled in two ounces of oatmeal. No deviations from the normal in any individual was noted, either in the analysis of blood and urine or in general condition. Accidental swallowing of flour beetles in cooked cereals thus causes no injury to human beings.

L. O. HOWARD (1916), the grand old man of applied entomology, approached the problem in a typical American way. During the worries of the first world war, he felt that the time was favourable for considering the question of new and cheap food supples. 'Any new cheap food should be especially welcome'. Many insects are eaten in backward countries and the Romans eat certain larvae as delicacies. The historico-geographical literature on edible insects

60

consumed by wild peoples is rather extensive. But practical experiments under modern conditions are almost non-existant. Hence, such experiments are most desirable and most agricultural colleges are in an excellent position to do this work. They should first select those among the common pests which are fit for consumption and taste well, and this first step should be followed by careful analysis of the food value of these species. Locusts or grasshoppers and white grubs, those of *Lachnosterna* in U.S.A., of *Melolontha* in Europe suggest themselves. The former have been eaten by so many peoples that their food value may be accepted as established, though the civilized world needs convincing about the white grubs. And a boy following a plough could pick up a day's ration of the latter sufficient for a family in a short time.

HOWARD then prepared for some experiments with *Lachnosterna*-grubs without further ado. He arranged with Mr. J. J. DAVIS of Madison (Wisc.) that he prepare the grubs by clipping off the extreme anal end, at the same time holding them under running water and pressing the body gently to remove the grit and the intestines, leaving white pieces of flesh. These were put in salty water into an autoclave, and subsequently sent to Washington where they were washed in cold water and generously treated with a French dressing of salt, oil and vinegar, seasoned with white pepper, paprika and salt. Some of the grubs were boiled into a broth, some prepared as salad. The salad was very palatable, although all the guests discarded the tough chitinous skin after chewing. Only one of the ten guests found the taste disagreable. The broth was really appetizing. A week later DAVIS prepared a stew, tasting agreably like lobster, and all were sorry when it was finished. Most of us discovered no especially distinctive flavour, while two found it slightly acid. HOWARD concludes: 'I feel sure that white grubs have a positive food value and I am especially sure that the prejudice against insects as food is perfectly unreasonable'. However, he advocates the need for a throrough sterilization before their consumption on account of the possible pollution of the grubs by the soil from which they came.

This was, however, not the first attempt by L. O. HOWARD to test the edibility of insects. In 1885 (vide Fladung, 1924, p. 8) he and his chief, Prof. RILEY had feasted on cicadas (*Tibicen septem-*

decim) and the experience is described by L. O. HOWARD as follows: 'With the aid of Dr. RILEY's cook, he had prepared a plain stew, a thick milk stew, and a broth. The cicadas were collected just as they emerged from the pupae and were thrown into cold water, in which they remained one night. They were cooked the next morning and served at breakfast time. They imparted a distinct and not unpleasant flavour to the stew, but were not palatable themselves, as they were reduced to nothing but bits of flabby skin and the broth lacked substance. The most palatable method of cooking is to fry in butter, when they remind me of shrimps. They will never prove a delicacy'.

E. B. FLADUNG bases himself (1924) mainly upon the diet of certain N. American Indians of locusts, cicadas, caterpillars, including some of the outstanding insect foods in other regions. He states that the extent to which insects contribute to the diet of mankind is little known, and concludes that insect-eating is widespread only where no agriculture exists and where vegetation does not thrive.

J. H. FABRE (1922, V, pp. 262 ff; 1924, X, pp. 102 ff.), the charming master of modern entomology, was also stimulated by his thirst for knowledge to test for himself the taste of the famous Greco-Roman insect delicacies.

One hot July morning the larvae of *Cicada plebeja* emerged from the soil for the nymphal moult. FABRE with all his family started to collect them. After two hours of intensive search four cicadas were found. Following the advice of ARISTOTLE that they are best before their skin bursts for the moult, they were killed by being submerged in water. These were fried in oil with a pinch of salt and a little onion. The dish was found enjoyable but tasted less like grasshoppers than like shrimps. Its consistency, however, was so thin and to chew it was so much like parchment that FABRE resolved not to accept any other of the dishes recommended by ARISTOTLE.

The village club, consisting of JULIAN, the teacher, GUIGUE, the blind man, and J. H. FABRE decided one day to try the taste of the larvae of *Cerambyx heros*, which they quite properly took for the *cossus* of PLINY. They grilled the grubs over a fire, seasoning them with salt and let them turn to a golden colour. The club approved the dish unanimously, its taste recalling that of roasted almonds and vanilla. They admired the refinement of the old Roman gourmets. Only the skin had to be discarded. Yet PLINY had also recommended fattening the *cossi* in flour. Against his own expectation FABRE found that the larvae, although not getting fatter, did at least remain in excellent condition for a long period.

J. BEQUAERT's 'Insects as food. How they have augmented the food supply of Mankind in early and in recent times' (1921) is an important paper. After mentioning the cicadas of ARISTOPHANES, the *cossus* of PLINY and the locusts of the Bible and at Niniveh, he states: 'Nowadays the use of insects as a diet is practically restricted to native or half-civilized peoples; but even so they form an important item in the food supply of mankind. Although many of those considered are too scarce to furnish more than an occasional dainty morsel, or are reserved for special classes or purposes, other kinds are gathered in great quantities, dried and preserved for a time as part of the staple food supply of the tribe'. Locusts, ants, termites, caterpillars, cicadas, flies, etc. are discussed from the point of view of their importance as basic food. Some of his original observations will be found incorporated into the later chapters. Here it is sufficient to quote his conclusions: 'In spite of the weight of the historical evidence, it is not our purpose to argue the value of insects as food in our own diet. What we eat and what we do not eat is, after all, more a matter of custom and fashion than anything else. Many years ago the learned French physician J. J. VIREY concluded: 'Man may eat insects, as nothing in his anatomical organization or his physiological functions is opposed to it. He should eat insects'. In my opinion the habitual consumption of insects may not be without danger. Most of them have such a heavy, indigestible skeleton of chitin that their continued use may well lead to dyspepsia. In addition, their small size makes it impossible to eliminate from their bodies all organs in which the waste products are accumulated, and which, because of their recognized poisonous properties, are as a rule carefully removed in the case of our meat and fish. Those inclined toward food-reform may be interested in the book of V. M. HOLT, 'Why not eat insects?' They will find there an array of recipes for the preparation of various insects . . . . If the time ever comes when insects are universally used as food, HOLT's book will undoubtedly be greatly treasured by all gastronomers; as BRILLAT-SAVARIN says: 'Anyone inventing a new dish does more for the happiness of his fellow men than all the philosophers, writers, scientists and politicians together'."

A number of other popular compilations are devoted to entomophagy, all of them from the point of view of curiosities. A few of them should be mentioned as they have at least gone over some primary literature or have even added some observations of their own, such as Doctor J. L. SOUBEIRAN (1870), P. BARGAGLI (1877), or E. DAGUIN (1900). G. PORTEVIN (1934) gives a brief survey to which some historically interesting illustrations are added. Other compilations, like those of L. FAILLA TEDALDI (1882), J. DE GAULLE (1873), and many others have no value. Another short paper, which seems difficult to obtain, is that of RAUM (1925). His compilation is entirely based on secondary sources. He has not even taken the trouble to go through the report of ROESEL VON ROSENHOF, which he discusses fully, as is shown by his remark on FRISCH. He completes his poor material by a suggestion, namely that the eating of the honey ants of Mexico and Colorado is not based on the desire for honey, but on the appetizing effect of the formic acid, as is the case with bee-eating in Ceylon for the stimulating effect of their poison glands. He mentions a bouillon from the thorax of may-bugs recommended in old cookery books.

A. HYATT VERRILL (1938 pp. 159 ff.) states that the idea alone of eating insects is disgusting to most people. Yet the phytophagous insects are among the cleanest of creatures and the carnivorous insects also are much cleaner than birds. And yet many scavenging crustaceans are regarded as a delicacy. As a matter of fact, many people still eat insects . . . . Evidently primitive man eats all meats on which he can lay his hands. Our ancestors long ago declared insects as taboos. No reasoning will probably ever succeed in breaking these taboos and let us return to insect diets.

F. NETOLITZKY (1918/9) has written a most informative paper 'Beetles as Food and Medicine'. NETOLITZKY states that entomophagy is more than a mere curiosity, explained by 'De gustibus non est disputandum'. Primitive man could neglect no source of food. Through insect-eating he discovered certain stimulations and side-effects which ensured for a number of insects their place in popular medicine. The weak stimuli of many other insects were attractive, and they became a favourite food or delicacy, including even lice. Yet the consumption of large and fat beetle grubs in the tropics is not based on their taste, but on their protein and fat contents. People with a predominantly vegetable diet are in great need of proteins and fats, of which insects are an important source, especially before the stages of agriculture and husbandry have been reached. The extensive list of beetles used as food or medicine is valuable for the exactness of the determination of the species. This paper, as well as that of J. THEODORIDES (1949), is important for the wide range of original sources consulted.

The latest book on our topic is a thesis 'Les Insectes comestibles' by E. BRYGOO (1946), who classifies the role of insects in human nutrition into four categories:

1. Food collected by primitive food-gatherers.
2. Survival of ancient customs (locusts).
3. Return to primitive food-gathering in times of famine, as is the case with the Mois every year before the harvest.
4. As delicacies (*cossus*, termite queens) or as a perversion of taste, as was the spider-eating habit of the famous astronomer DE LALANDE.

BRYGOO distinguishes between the gathering of every available food, as in the primitive food-gatherers and the selective gathering of certain insects which are an appreciated food. He also points out the nutritive value of insects in human diet.

In some recent books a special chapter is devoted to insects as human food. The more important of these are M. BURR (1939), C. A. EALAND (1915), C. T. BRUES (1946) and S. W. FROST (1942). Many facts from original sources are contained in F. COWAN's 'Curious facts in the History of Insects' (1865). A number of papers based mainly on original observations in various countries which will be fully dealt with in the following chapters are also of great importance. As models of these most important contributions we would like to mention the papers of BRISTOWE (1932) on Siam, of NGUYEN-GONG-TIEU (1928) on Indo-China, of HOFFMANN (1947) on China, of WALLACE (1852/53) on Amazonia, of McKEOWN (1944) on Australia, and of DECARY (1937) on Madagascar.

3. RELICS OF ENTOMOPHAGY IN EUROPE.

In Europe the eating of insects ceased long ago, apart from the collection of wild honey which is still practised in the forests, especially among the Slavic peoples. We have collected at random a few records without any pretention of giving more than a few hints of the occasional survival of the habit.

The great CUVIER (Animal Kingdom, Insecta I, p. 163) points out that the disgusting habit of eating lice is not confined to the Hottentots, the Negroes of Western Africa, monkeys and the American Indians, for it has been observed to prevail among the beggars of Spain and Portugal. CUVIER (ibid. II. p. 205) mentions a report by LATREILLE of how in Southern France the children are very fond of the fleshy thighs of grasshoppers. Also Dr. TROUESSART (vide DAGUIN 1900, p. 27) saw children in France catching *Orthoptera* on the River Loire, pull off their wings and crack their hindlegs with evident pleasure. And RADOSZKOWSKI, a president of the Russian Entomological Society, mentions (vide SIMMONDS 1885, p. 360) that to this day locusts are extensively used as food in S. Russia, either salted or smoked. Similar information has been repeatedly reported of the Tartars of Crimea.

The eating of cock-chafers, especially of *Melolontha*, must have been widespread, particularly in remote regions and in cock-chafer years or in times of famine. Yet the references to this habit in literature are rare. Thus we read that during the famine of 1688 in the county of Galway in Ireland the poorer people had a way of dressing them and living on them as food (in Philos. Transactions, vide COWAN 1865, p. 49). ERASMUS DARWIN (1800) recommended grubs and adults of cock-chafers as food. WESTERMANN (1821, p. 419) reports cock-chafers as an article of food with the mountain inhabitants of Europe. Related species were also eaten. Thus the brothers VILLA relate how the peasants of Lombardy eat the abdomen of *Melolontha aprilina* Duft; similar beetles are also eaten in other provinces of Italy, and in Moldavia and Walachia both cock-chafers and *Rhizotrogus pini* Ol. are eaten (vide BARGAGLI 1877, p. 6). J. DE GAULLE (1873) reports that the pea-peelers in Southern France claim the rejected peas for food, stating: 'Ça engraisse'.

The Reverend GILBERT WHITE in his famous Natural History

66

of Selborne (1854, p. 293) mentions that in his village around 1765 there was an idiot boy who from childhood showed a strong propensity for bees; they were his food, his amusement, his sole interest. In the summer he was perpetually hunting for honey-bees, bumble-bees and wasps. He had no fear of being stung, but would seize the insects with his bare hands and at once disarm them of their weapons and search their bodies for their honey-bags. He had been known to overturn hives for the sake of honey.

EYLMANN (1908) recounts that in the marshes of the Lower Elbe in Hannover the children hunt for bumble-bees to suck the crop which is full of honey.

CONSETT (1789, p. 118), in his travels in Sweden, mentions a young Swede who ate live ants with the greatest relish. He also states that in some parts of Sweden ants are distilled together with rye, to give a flavour to their inferior kinds of brandy. Apparently even today ant-pupae are used there for the production of good gin.

The galls of sage (*Salvia spp.*), which during a certain season are rather juicy like bad apples and crowned with rudiments of leaves, are gathered every year as an article of food by the inhabitants of Crete. OLIVIER (1813 I, p. 139) confirms this statement of TOURNE-FORT and adds that they are esteemed in the Levant for their aromatic and acid flavour, especially when prepared with honey and sugar, and form a considerable article of commerce from Scio to Constantinople, where they are regularly on sale in the market. This is the gall of the Cynipid *Aulacidea levantina* Hed.

The galls of ground-ivy (*Glechoma hederacea*), produced by *Aulax latreillei* Kieff. (= *A. glechomae* Latr.), have been eaten as food in France. They have an agreeable taste, and largely the same smell as the plant on which they grow. RÉAUMUR (1737, vol. III, p. 416), however, is doubtful whether they will ever be as popular as good fruit.

SOME REPORTS ON SPIDER-EATING

While spiders are not strictly insects, they are close relatives and they are usually felt to be more repulsive than insects. Nevertheless, there are a number of reports from all over the world of spider-eating.

Apart from the spider-eating on New Caledonia which was

reported by LABILLARDIÈRE, and a few other notes on spiders, referring mainly to their use in mixed ragouts, especially in the monsoon regions, we wish to mention two additional reports. In Kamchatka spiders are rare, but they are eagerly sought for by sterile women, as their consumption is believed to bring fertility and to ease their labour (KRACHCHENINNIKOW 1764, p. 164). In Brazil certain spiders are believed to be strong aphrodisiacs (DE WALCKENAER 1837, p. 182), and the same quality is ascribed to them in folk medicine throughout the world. WALCKENAER also tells the anecdote about the famous 18th century astronomer DE LALANDE. After his return to France every Saturday he visited the naturalist QUATREMERE D'ISJONVALLE in his villa, and liked nothing more than to eat caterpillars and spiders when they were in season. In the garden LALANDE easily found enough to satisfy his first hunger; but as Mme D'ISJONVALLE liked to do everything well, she collected them in the afternoon and served them for LALANDE immediately he arrived. The astronomer declared the spiders to taste like hazelnut, the caterpillars like stone-fruits.

ROESEL VON ROSENHOF (VI, p. 257) tells of a German who used to spread spiders instead of butter on his bread as a purge. SHAW (vide KIRBY and SPENCE I, p. 343) mentions how Anna Maria SCHURMANN used to eat spiders like nuts, to which she compared their taste. RÉAUMUR (II, p. 342) mentions a young lady who, when she walked in her grounds, never saw a spider which she did not catch and eat on the spot. In certain regions of France the children hunt spiders for eating and compare their taste to that of nuts (BRYGOO 1946, p. 19).

Among primitive people the Bushmen consider spiders a dainty (SPARRMANN I, p. 201). COWAN (1865, p. 354 f.) adds a number of further reports. Spiders are an article of food for American Indians and Australians (LABILLARDIÈRE). The people of Maniana, south of Gambia, are cannibals: they eat spiders, beetles and old men (MOLIEN). In Siam egg-bags of spiders are considered a dainty (TURPIN). The Caribbeans eat spiders, frogs and any kind of worms (PETRUS MARTYRUS). The Guaharibos of the Orinoco and the Piaroa Indians eat tarantulas (WAVRIN). In Madagascar *Epeira nigra* Vins. and *Nephila madagascariensis* Vins. are fried in oil and fat (DECARY). In various places spiders are eaten as an aphrodisiac.

The eating of scorpions is widespread in S. E. Asia. ALEXANDER VON HUMBOLDT (Pers. Narr. II, p. 205 and IV, p. 571) repeatedly observed the eating of scolopenders by South American Indians.

III. AUSTRALIA

1. GENERAL INTRODUCTION

Entomophagy is common all over Australia. The aborigines have now disappeared, or have become 'civilized'. Fortunately science has studied the ethnology and ethology of the Australians before it was too late. Especially in the many fundamental works. of Sir BALDWIN SPENCER we find numerous interesting facts, which go far beyond the mere statement that this or that insect has been eaten. This unique store of insect lore connected with food habits will be described in some detail. This procedure is all the more justified, since these habits, ceremonies and legends shed light on conditions of the stone age, which have elsewhere perished without leaving any trace.

Life was difficult for the aborigines of Central Australia in periods of long-continued drought, when food and water were scarce; but under ordinary conditions, except in the dry desert, where it was never pleasant, their life was by no means miserable or hard. In the steppes kangaroos, wallabies, emus and other game were not scarce, while rats and lizards, etc. were constantly caught without any difficulty by the women, who also collected large quantities of grass seeds and tubers, and, when in season, fruits, such as that of the native plum (SPENCER and GILLEN 1899, p. 7). In the coastal districts fishery and marine evertebrates made life still easier. And in the irrigated areas of the tropics and sub-tropics of North Australia, it was always pleasant. On the Alligator River, for instance, as long as lily seeds and roots abound, fish and wild fowl are available, and the natives remained for months at the same spot. All day long the women and children were in the water, gathering lily 'tuck out', while the men speared fish and caught water-fowl, climbed trees after flying-fox, oppossum and honey-bag, or hunted kangaroos and emus. When they had thinned out the lilies, and honey-bag, and fish and fowl grew rare, they moved to another camp and continued in this way the whole year round (SPENCER 1928, p. 775).

Discussing the Arunta tribe of Central Australia SPENCER (1928, pp. 206, 209) recalls that they had not reached the agricultural stage, neither cultivating cereals, domesticating animals or laying in a store against times of scarcity. They certainly had seed plants that would have lent themselves to cultivation, but the kangaroo was doubtless unfit to produce milk or to become a beast of burden. The Arunta are primitive food-gatherers. They live from hand to mouth, without any thought of the morrow. Nothing comes amiss: acacia seeds, lily roots and stems, yams, honey of the wild bee and honey ant, grubs, kangaroos, emus, snakes, rats, frogs – in fact everything edible was eaten, including some things, such as flies and pounded ant-hill clay, that we would scarcely call by this name. When food is abundant they ate in plenty and were perfectly happy; when it was scarce they accepted the situation without grumbling, tightened their waist-bands and starved philosophically, waiting for something to turn up. There were, of course, times when they were hard pressed, and during a long continuance of drought their life was not happy. In fact they were absolutely at the mercy of their surroundings, Fortunately their very lack of any power to control nature, though they are firmly convinced that they can do so by magic, has been the means of sharpening their powers of observation, and they can obtain water and food in what is for them abundance, in a place where a white man would die of thirst and starvation. There were, however, times when even the aboriginal with all his bush-craft – and this was simply marvellous – was unable to contend against the fierce heat and drought of Central Australia and perished miserably. These seasons and periods of famine were responsible for keeping the native population of Australia at such an extremely low level. Early in the morning, in summer, but not until the sun is well up in winter, the occupants of a camp are astir. Time was no object to them, and if there was no lack of food the men and women all lounged about, while the children laughed and played. If food was required, the women went out, accompanied by the children, carrying digging-sticks and troughs (*pitchis*), and the day was spent out in the bush in search of small burrowing animals: lizards, small marsupials, honey-ants and grubs. The men set off, armed with spears, spear-throwers, boomerangs and shields in search of larger game.

Apart from social insects, insect life in the drier parts of Australia is abundant only in the short period after the onset of the heavy rains. Yet ants, honey and termites are available throughout the year. Their greatest importance as food is in the dry steppes and deserts. Their vital value is shown by the significant place they occupy in the rather restricted gesture language. Thus in the Warramunga language the sign for the famous witchetty grub is: the fingers are loosely pressed on the palm, the thumb lying on the first finger. Or that for 'I have found a honey-bag' is: the first finger and thumb are extended; the second finger is bent half-way over, the third and fourth completely. The hand is held still while the thumb and first and second fingers are flicked (Fig. 5; SPENCER 1928, p. 441).

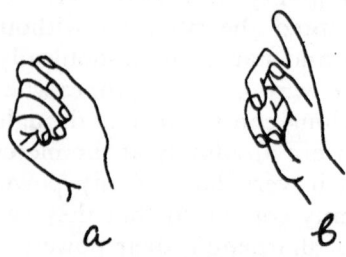

Fig. 5 Gesture language of the Warramunga tribe. a. Wichetty Grub. b. 'I have found a Honey-Bag'. (From SPENCER 1928 I, pp. 439/9).

E. C. STIRLING (pp. 51 ff.) in his report on the animal food of the natives of Central Australia as early as 1896 expresses the opinion: 'There are few living animals that come amiss to the Central Australian aborigines. To mention the names of all that are eaten would be largely to recapitulate the zoology of the district'. He gives the following good survey of their insect food:

Lerp manna: At various localities, particularly in the bed of the Todd River near Alice Springs and, in that of the Hugh near its junction with the Finke, the leaves of *Eucalyptus rostrata* bore the small white conical or tentlike coverings of the larvae of psyllids. Similar structures were seen on the leaves of *E. microtheca*. According to Dr. T. DOBSON (R. Soc. Tasmania 1, 1851, p. 235) the lerp insect is *Psylla eucalypti*, but this is perhaps not the insect found on all species of eucalyptus. These coverings are the result of a glutinous secretion from the bodies of the larvae, which partly takes the form of a fret-work of tubular hairs. Though small, not usually exceeding 4 mm.

Fig. 6. Ants *(Iridomyrmex detectus)* visiting the Psyllid lerp-manna *of Spondyliaspis eucalypti.* Dbs. on an eucalyptus leaf. Curtoisy of Dr. A. J. NICHOLSON, Canberra.

in diameter, they sometimes occur in great numbers, and can then be collected in quantities by the natives. In the localities mentioned entire trees were cut down to obtain them. The substance is definitely sweet, and on analysis the Lerp manna of *E. dumosa* (FLÜCKINGER, WATTS, Dict. of Chem. 1875, 2nd Suppl.) contained both a starch-like substance, the Lerp-amylum, and a sugar.

Witchetty grubs will be discussed later. Other smaller larvae from the roots of acacias are similarly used. Various species of caterpillars, which at certain times appear in great numbers, are collected and eaten.

The honeypots of the Sugar Ants form a favourite article of consumption whenever they can be obtained, and, as in the case of the witchetty grubs, there appears to be a special ceremony to promote the supply. *Melophorus inflatus*, the common *yarumpa* of the natives, and the *ittutunie (M. cowleyi)* are certainly eaten, while this is also probable of the honeypots of *M. midas*. The abdomens, distended by honey, are bitten off and swallowed.

73

Bee-honey or *ultaamba* is eaten, though no honey-making bees were met with by the expedition in Central Australia.

The food of the Pitjendadjara tribe of the Mann range in C. Australia is described by MOUNTFORD (1946), as follows: 'Food not inviting but palatable'. The menu of my companions was certainly varied; in fact, they ate everything that was edible; grubs, lizards, ants, kangaroos, emus, grasses and seeds of many kinds. I have eaten many of these foods with relish. The large white wood grubs, although loathsome in appearance, are particularly palatable, although I must admit it took a lot of determination to eat the first one. They are, indeed, surprisingly similar to roast pork . . . Honey ants are as sweet as any honey . . .' MOUNTFORD adds a photo of a really well fed infant and explains: 'Fat and saucy, a native baby thrives on a diet of mother's milk, white grubs and honey ants. The child's home is in the Mann Range, where previous travellers' reports indicated that the country was too bad to support even aborigines'.

Mr. SCHULTZE mentions as food an insect (*Cicada sp.?*) which comes out of the ground in summer wherever *Eucalyptus robusta* occurs. We have information to the same effect, and we frequently saw split pupa cases projecting out of the ground, but Prof. SPENCER tells me he is unable to identify the species.

A. F. CALVERT (1898, p. 17) in his little monograph on the Aborigines of Western Australia begins with the statement that traditions have taught the children of the bush how to provide for their natural wants, and well-armed intelligent white people will die of hunger in the desert, where the native will find a sufficiency of food. He will never kill his totem- or *kobang*-animal, should he find it asleep, and he only will gather his totem plant (or insect) under certain conditions and at special seasons (ibid p. 19). The natives subsist entirely on hunting and fishing, or on wild roots and fruits, yet every tribe has its proper territory for every purpose, and within the tribal areas private property of families is recognized (ibid p. 20). We wish to add here that in all desert tribes every prey is readily divided with every member of the tribe who desires to take his share. Elder relatives, such as parents and parents-in-law, have the first choice with certain species of game, which indicates that some degree of responsibility for the elders rests on the younger members of the tribe. Fear of magic is obviously here of greater significance in habit formation than purely humanitarian motives.

The sympathies of travellers have been often concentrated upon the supposed scarcity of food. As a rule, however, the natives have an abundance, although they may run a little short in the height of the rainy season, or when they are overcome by torpor in very hot weather. In the journal of Sir GEORGE GRAY we find the following list of food articles of the Western Australian aborigines: six kinds of kangaroo, 29 of fish, one whale, two species of seals, wild dogs, three kinds of turtle, emus, wild turkeys, two opossums, 11 species of frogs, four kinds of fresh water shellfish, every sort of sea shellfish except oysters, four kinds of edible grubs, eggs of birds and lizards, five animals of the rabbit class, eight snakes, seven iguanas, nine species of mice and rats, 29 different roots, seven mushrooms, four species of gum, two kinds of manna, two species of *by-yu* (nut of the zamia palm), two sorts of mesembryanthemum, two of small nuts, four of wild fruits, besides the seeds of several plants. CALVERT says that this can hardly be called a starvation bill of fare (ibid. p. 24). Grubs, which are extremely palatable, are procured from the grass tree and also from an excrescence of the wattle tree. They are eaten raw or roasted, but seem to be greatly improved by cooking. CALVERT was told (1898, p. 28) that they have a nut-like flavour, but he never had the courage to sample them.

And T. G. CAMPBELL (1926, p. 410) comments as follows: Among the Australian aborigines the variety of food-stuffs is considerable, not only from choice, but also owing to the difficulty of obtaining a permanent and regular supply of any one articles of diet . . . . In fact almost anything capable of being chewed is regarded as food. During drought and the absence of large game, the maintenance of an adequate food supply is one of the greatest problems for these mere food-gatherers. Thus it is not surprising that larger-bodied or abundant insects are eagerly collected. Most insects appear in a district for a comparatively short time each season, and the appearance of species needed for food was treated by the natives as a matter of great importance.

The *Bugong*-moth (*Euxoa infusa*) assumed great significance for some New South Wales tribes. This is one of the commonest Austra-

lian Noctuids and the migrating moths were eaten, while the caterpillar causes damage to crops in some year. As a matter of fact,
the full ecology and annual life-cycle of the species is still largely
obscure. The hunt for the 'sugar-bags' of *Trigona*-bees is mainly
restricted to the more tropical and sub-tropical parts of Australia
where the green tree ants (*Oecophylla smaragdina*) are also eaten, as
well as being used as a refreshing drink. The honeypot ants (*Melophorus*, *Camponotus spp.*, and others) are eagerly sought for in Central
and Northern Australia. In Central and Southern Australia the
caterpillars of large Hepialid and Cossid moths, the witchetty grubs
of some ichtiumas, are of great importance. Everywhere the larvae
of beetles, such as the longicorn *Eurynassa australis*, are collected from
living or decaying trees (mainly *Eucalyptus* or *Acacia*). Apart from
these, many other butterflies, grasshoppers, termites, cockroaches,
cicadas, lerp-insects, as well as numerous other insects have been
used as food by the Australian natives.

Insect totems, their intichiuma ceremonies and legends.

We cannot discuss here the most complicated social inter-relations
of tribe, class and totem among the Australian aborigines. For reasons which will be mentioned later, it is rare for all or even two
members of a single family to belong to the same totem. Quite a
number of totem animals of the Australian natives are taken from
insects which are of primary food importance, as are almost all
totem animals. Thus we find as totems in Central Australia a cicada
– (*wutnimmera*), a small grub – (*unchalka*), a longicorn beetle grub –
(*idnimita*), a manna-producing leaf-hopper – one on eucalyptus or
another on acacia (*ilpirla*), the honey ants (*yarumpa*) and the
witchetty caterpillars (*udnirringitta*), etc. The preponderant insect
totems of Northern Australia are taken from various names for
honey-bags.

A few introductory remarks on the totem ceremonials of the
Arunta tribe will be useful. The Alcheringas are the mystical ancestors of the totem, mainly their first generations which themselves
developed from the totem animal into man. These Alcheringas
carried with them on their wanderings one or more sacred stones,
called Churinga, which are often translated as bull roarers or whirlers.
Each Churinga, like any other of the sacred stones or rocks or trees,

76

1—4. Baskets for collecting insects and plants used by the Melville Islanders, N. Australia (SPENCER fig. 438, 4, 553, 2 and 5, plate XIV, 3).

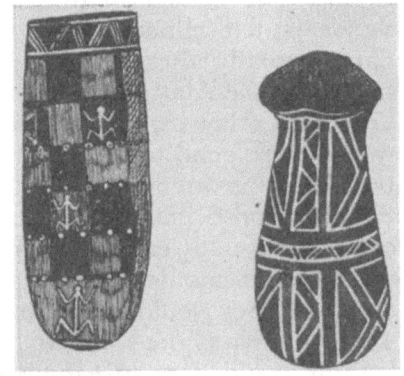

5—7. *Pitchi* troughs of Central Australia serving to collect insects, plants, etc. 1. Bark *pitchi* of Red Gum tree. 2. Hard-wood *pitchi* of the Arunta tribe. 3. Boat shaped *pitchi* of the Warramunga tribe (from SPENCER 1928 I, fig. 115, 117, 126).

8. Simple forms of sticks, used partley for throwing and partly for digging by the natives of Victoria. (from SPENCER 1922, pl. 5).

Fig. 7. Various containers and tools for collecting insects and honey in Australia.

is intimately associated with the spirit of a certain individual ancestor. The Churinga stones are kept in a sacred storehouse, the *ertnatulunga*, or in the totem centre, the place of the sacred *intichiuma* ceremonies, which aim at inducing productivity of the totem animal. While the magic meaning for producing totem abundance is clear, these ceremonies do not involve any appeal to the assistance of supra-natural beings. These fertility-inducing *intichiuma* ceremonies are the most important seasonal events in the camp; yet there is no tradition giving an account of the origin of these peculiar rituals. The *intichiumas* differ from totem to totem. While their season is determined by the approaching season of totem abundance, usually after the heavy rains, their exact time is fixed annually by the Alatunya, the leader of the totem ceremonial.

In most other regions the totem animal is forbidden as food (this would correspond to eating one's own ancestors) but such absolute prohibition is rare in Australia. The point is, however, not clear, as SPENCER and GILLEN (1899, p. 168, etc.) declare that the members of a totem usually eat only sparingly of the totem animal. Yet at the *intichiuma*-ceremonies they are under obligation to eat them as otherwise they would endanger the abundance of the totem animal. The Alatunya of the witchetty grub totem, when asked by SPENCER how he could dare to eat his own ancestors, replied that it was essential for him to eat a small quantity of witchetty grubs, as otherwise he would be unable to perform the ceremony.

The great *intichiuma*-ceremony of the Arunta will be described fully later. Another such ceremony (SPENCER 1928, p. 373 f.) contains some details of interest. The Alatunya of the witchetty grup totem of the Emily Gap was decorated with lines of down, representing a celebrated Alcheringa. He kept bending down over a shield, ornamented on one side with wavy lines which represented the wanderings of the grubs in ancient times (Fig. 8a). On the other side of the shield a number of large circles represented the shrubs on which the grubs fed and smaller ones the eggs laid by the adult insect. The old man was supposed to be trying to induce the insect to lay eggs, so that the natives, who were very fond of them, could have plenty to feed on (Fig. 8b). He himself, being a witchetty grub, would not eat them, except just a little at one special time. The Alatunya of every totem group must, once a year,

Fig. 8. Witchetty totem ceremonies of the Arunta tribe. Above: Part of the ceremony. Centre: Discussion after the end of one of the ceremonies. Below: Old men explain totem rituals to a young man at the end of the ceremony. A decorated shield used in the ceremony is pressed against his stomach by the man sitting opposite to him. From Sir B. SPENCER, Wanderings in Wild Australia. London. 1928. I, Fig. 206/208.

perform the *intichiuma* ceremony to bring about the increase of his totem animal or plant. The natives firmly believe that by this magic they can make animals breed and plants grow. The ceremony is always held shortly before the usual breeding time. If the animals increase, then of course it is the man's magic which has made them do so; if not, his failure is put down to the strong counter-magic of some evilly disposed enemy. When the animal has increased in numbers, men who do not belong to that particular totem bring a supply of it into the camp and offer it to the headman, who eats just a little of it and then hands the rest back to the others, telling them that he has made it for them and that they may now eat it. The headman of each totem must eat a little of his totem animal, so as to have some of it inside him, or else he would lose his power over it and could not make it increase. Thus every totem makes his totem animal for the use of the other totems. When the ceremony was over its meaning was explained to the son of the Alatunga, to whom it will be passed on at the death of his father (Fig. 8).

Apart from the Churinga stones, wooden Churinga are usual in

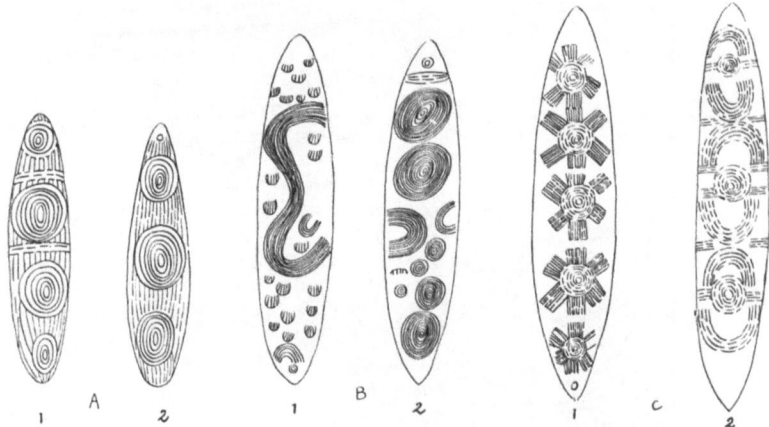

Fig. 9. Some sacred wooden objects or *Churingas* of the Arunta tribe (1 front, 2 back): A. The *Churinga knanja* of the celebrated *Ilatirpa* of the Honey Ant totem at Ilyaba. B. the *Churinga* of a Witchetty Grub man at the Emily Gap storehouse. c. The *Churinga* of a woman of a Small-Fly totem (From SPENCER I, 1928, pp. 300, 301, 305).

the *intichiuma* ceremonies. Fig. 9 represents a number of those from insect totems, with views from both sides. Fig. 9A shows the *Churinga knanja* of the celebrated Yarumpa-ancestor Ilatirpa, which is kept in the storehouse at Ilyaba. The series of circles (a) with the central circle represents the eye, (b) the intestines, (c) the stomach painting and (d) the posterior part of the man. The circles (g) on the reverse represent the intestines of the Alatipa birds. Fig. 9B is the Churinga of a witchetty grub man in the EMILY GAP storehouse: (a) indicates a large grub, (b) a lot of grubs in a hole which has been scooped out in the ground, (c) a man sitting down squeezing the dirt out of the animals before cooking them. On the reverse side the curved lines (d) indicate grubs, (e) the eggs of various sizes and (f) marks on the body of the grub. Fig. 9C is the Churinga of a woman of the Injiparilla, a small-fly totem. The women collect this in great numbers, rub the wings off and grind up the bodies. On one side the five series of concentric circles (a) represent the flies, the radiating lines (b) their tracks as they fly in swarms. On the other side the circles (c) are young women, the curved lines (d) women sitting down, the straight lines (e) their tracks. (SPENCER 1928, pp. 300 ff.).

In some totems we find peculiar associations with certain birds. Thus, around each of the sacred holes of the witchetty grub totem, at which the later part of the *intichiuma* ceremonies is performed there are a number of stones standing on end, representing certain birds, the Chatunga. These birds are looked upon as the mates of the witchetty people, because in the Alcheringa ancestors certain witchetty grubs changed into these birds. The latter abound at the time when the grub is plentiful and are very rarely seen at other times, and they are then supposed to sing joyously and to take a special delight, as they hop about among the *udnirringa*-bushes all day long, in watching the witchetties lay their eggs. The witchetty men will not eat the bird, as this would give them 'bad stomachs', and they speak affectionately of the birds (SPENCER and GILLEN 1899, p. 447 f.). Similarly the *yarumpa* or honey ant people have as mates a little bird called Alatipa which, like the honey ants themselves, only frequents *mulga* scrub country. The Alpirtaka birds, small 'magpies' of the *mulga* scrub, are considered as another mate of this totem. Both birds were once honey ant people, and are occasionally said to have guided the Arunta to the nests of the honey ants, a striking parallel to the honey-guide of Africa.

Many insect totems also occur in the northern tribes of Australia (B. SPENCER 1914, pp. 183 f., 188, 197, 207). The 90 totems of the Kakadu-tribes, which mainly inhabit irrigated, fertile country, contain the following insect totems: *Jailba* (sugar-bag), *Morma* (sugar-bag), *Murkailpu* (sugar-bag), *Pedanitja* (sugar-bag), *Tjailba* (sugar-bag), *Tjinangu* (sugar-bag) and *Tjikali* (wood-grub). While the tribes of Central Australia never make in their totem ceremonies any personal appeal, the Kakadus very conspicuously do so, crying fiercely and insistently: *'Brau! Brau!'* (Give! give!), asking the totem representative to provide them with this animal or with that plant, believing that he is able to control his food supply by means of magic. As in Central Australia, each totem group performs ceremonies for the increase of the totem animal or plant, the *tjutju-*ceremonies, which are the equivalent of the *intichiuma* ceremonies of the Arunta-tribes. When a member of any totem group dies, the respective totem animal is taboo to all other members until a certain ceremony, the Orkban, has been performed, in which the totem animal is cooked. Totems also cause certain restrictions to marriage. Thus, in the Mungarai tribes a Negabullan-man of the sugar-bag totem married a Ngangiella-woman of the crocodile totem and their children are Ngepalieri-men.

A magic amulet deposited into a white-ant hill makes an enemy very hot within three days; he will cry for water and die. Yet pounded termite hills, *mupalangu,* are eaten as a cure for certain diseases, such as fevers. SPENCER (1914, p. 260) saw one woman with a mild malaria fever eating the *mupalangu* and she had perfect faith in the efficiency of the remedy. E. HEGH (1922, p. 670) mentions the eating of the earth of termite hills in certain regions of Central Australia.

2. THE WITCHETTY GRUBS AND GHOST MOTHS

One of the most important insect foods of Central Australia remains unfortunately unidentified – the witchetty grubs, to which an important totem in this area was devoted. STIRLING (1896, p. 53) writes as follows about his experiences: 'Under the name of *wit-*

chetties – a name used by the whites as well as understood by the natives throughout the central districts, but apparently not a word belonging to the language of either of the local tribes – are included certain larvae, the mature form of which is uncertain. One very large white kind, known at Alice Springs as *udnirringita*, believed to be the larva of a big longicorn beetle, reaches a length of 10–12cm. and the thickness of a finger, and is much appreciated. The promotion of the supply of this grub forms the motive of one of the most important food ceremonies of the Arunta natives. They are found principally in the roots of *Eucalyptus rostrata*, which are sometimes completely riddled by them even to the destruction of the life of the tree. They are dug out with yam-sticks, collected in large quantities and eaten after slight roasting. Whites who have tried them speak well of their flavour'. Dr. N. B. TINDALE, perhaps the most competent living authority, thinks that in Central Australia the two most common witchetty grubs are those boring in the roots of *Acacia kempeana* and of *Salsola kali*, both of which are Cossid larvae, which have never been properly identified.

The probable solution is, that the caterpillars of ghost moths (*Hepialidae*), large forms of which abound in Australia, of *Cossidae*, and perhaps other big caterpillars living in abundance upon certain shrubs were all called witchetty grubs. It is even not impossible that some common wood-boring beetle-larvae were included under this name. Yet for this latter conclusion we have only the authority of white travellers, who were not very competent in entomology. The witchetty-totem ceremonies of the Arunta point much more in the direction of various caterpillars. Without any further identification we include in this chapter all the caterpillars eaten in this group.

The witchetty grub or *udnirringita* was an important source of food to the natives, especially in Central Australia. Most writers assume that it is the caterpillar of ghost moths (*Hepialidae*), which produce a fair number of very large moths in that continent. The larvae which grow up to 5 cm. long and as thick as a finger, feed in the timber of growing trees or gouge tunnels in the bark, sapwood or roots, while others live on the roots of grasses and cause great damage to pastures. Their common and widespread occurrence and their dominant role in the diet of many tribes supports this diagnosis. The only disturbing factor is a note by SPENCER and GILLEN (1899) on the food habits of the witchetty grub used for the witchetty grub totem of the Aruntas, which are said by tradition to feed on a certain shrub and on its fruits. Both, ghost moths and their caterpillars, were consumed.

Large numbers of pupae, found in the ground at the foot of gum trees, are dug up in winter, and baked in hot ashes. They are the transitional forms of large green processional caterpillars, which crawl in lines on the stems of trees in search of a place to rest during their pupation. Of this transformation, and of their ultimately becoming moths, the aborigines are well aware. In addition to these there are many delicacies, chiefly collected by the women and children, and cooked in hot ashes, such as grubs, small fish, frogs, lizards, birds' eggs, lizard and tortoise eggs. The grubs are about the size of the little finger, and are cut out of trees and dead timber, and are eaten alive, while the work of chopping is going on, with as much pleasure as a white man eats a living oyster; but with this difference, that caution is necessary to avoid their powerful mandibles, ever ready to bite the lips and tongue. Roasted on embers, they are delicate and nutty in flavour, varying in quality according to the tree into which they bore, and on which they feed. Those found in the trunks of the common wattle are considered the finest and sweetest. Every hunter carries a small hooked wand, to push into the holes of the wood, and draw them out. With an axe and an old grub-eaten tree, an excellent meal is soon procured, and when the women and children hear the sound of chopping, they hasten to partake of the food, which they enjoy above all others. The large fat grubs, to be found in quantities on the bank of marshes, drowned out of their holes in times of floods, are gathered and cooked in hot ashes by the women and children (T. DAWSON, 1881, p. 20).

Dr. BASEDOW (1925, p. 122 ff.) writes of the witchetty grubs as follows: 'The most popular and at the same time most widely distributed article of diet in the insect line is the larva of the big *Cossus*-moth, commonly known as the witchetty (or witchedy) grub. The two varieties which make themselves most conspicuous are, firstly, one living in the roots of such shrubs as the *Cassia* and certain species of *Acacia*, and, secondly, one which bores into the butt of

84

the *Eucalyptus*. The first cannot usually be located by the eye, but its presence is determined by ramming the yam-stick into the ground under the root of the shrub, and testing its resistance to leverage. If the grub is present, the root will readily snap, whereupon the native soon unearths it by digging with the stick and his hands. This variety is smaller than the tree-grub, and is mostly of a yellow colour. The abode of the tree-grub is detected by the native's keen eye, in the small holes the young larva bores into the bark and lives in until it attains the mature moth stage. The larva lives in the butt or in any of the longer limbs of the tree; consequently it may at times be concealed in the bark high above the ground. In the latter case the native will have to climb the butt and effect an ascent, no matter what the shape of the tree happens to be. Various methods are made use of . . . . .

The witchetty grub is extracted from its hiding place by means of a light hooked stick. This implement is 10 to 15 cm. long and is usually cut from a small pronged twig, one arm of which is left the required length, the other cut short and sharpened to form the hook. The stick is inserted into the hole occupied by the witchetty grub, hook foremost and pushed in until the grub is penetrated; then it is withdrawn, the hook bringing the grub with it. As the hole is usually small at its entrance, the bark is first cut away to a small depth with a tomahawk in order to avoid the constriction when the grub is being withdrawn. The witchetty-hook is known throughout central and southern Australia. The Arunta word for it is *ullyinga*. The witchetty grub is prepared like most things already described, namely by throwing it upon the hot ashes for a few moments until is straightens and expands, but does not burst. Although we Europeans have become averse to eating anything in the grub line, there are many bush people who regularly partake of the witchetty; indeed by many the grub is regarded as a very tasty dish. The flavour of the cooked witchetty is like that of scrambled egg, slightly sweetened'.

F. P. DODD (in OBERTHUR 1916, p. 33) reports on the great Australian Cossid *Xyleutes eucalyptii*: Many Australians have eaten these caterpillars and all pronounce them quite palatable. At Townsville, 900 miles farther north, there are still some blacks, but they seem to have become so used to the white man's food that caterpillars, which have to be cut out of though timber, no longer are attractive for them. But here at Kuranda, I have often been disappointed to find that a larva of *X. boisduvali*, which perhaps I had located months before, had been cut out and eaten by some wandering native. These caterpillars are in green trees, and large coleopterous larvae in rotten trees or logs are much sought for and esteemed by the Blacks here, which is 200 miles north of Townsville, and at Heberton we several times met bands of youthful blacks, provided with tomahawks, searching through the bush for these things, principally the caterpillars, which, as a rule, are roasted a little before being devoured.

EYLMANN wrote that while the natives show a definite antipathy to insect food, as is demonstrated by the fact that meat is cleaned of maggots before being prepared, they regard certain insect larvae living in stems of trees as great delicacies, such as the well known witchetty grubs of the interior. It is peculiar that so far the adult insect of these larvae is still unknown. The most common species, called *mematt* by the natives near he Sterling Station, have larvae 8–10 cm. long and 4–5 cm. in circumference. They are yellowish or brownish at both ends, while the other segments are violet and dirty-white. The larva lives mainly in gum-trees, and between 19° and 27° latitude mainly in the red gum (*Eucalyptus rostrata*), or east of Lake Eyre in *E. microtheca*. According to Mr. RÜDIGER of the German mission at Kikalpanina a large moth develops from them. A still bigger larva, called *shabba* by the West-Arunta and *impita* by the Luritjes, measures up to 11.5 × 5.5 cm. This larva is bluish-white but turns dark grey-bluish in the sun. The natives dig it near the red gum trees from the sand of the creeks, where they are found in long tube-like webs, which penetrate vertically into the soil. A third yellowish-white larva is found north of the McDonnel Ranges in the root of acacias. The aborigines of the South coast also eat larvae living in wood. The witchetty grubs are an important part of the food of many tribes. In order to collect them the galleries in the wood or the webs in the soil have to be opened with an axe, a spade or a digging stick. Yet it is not difficult for a man in one of the larger creeks with many eucalyptus trees to collect in half a day as many grubs of the first mentioned species as are needed for a good meal. EYLMANN once met six Diaris near the woolshed of Kilalpanina, who had collected during a trip through the Cooper Creek, which is there eight miles wide and thickly grown with *Eucalyptus microtheca*, about 700 white and violet witchetty grubs. A Lubra-woman collected them in the western Reynold Range, as follows: first, she laid bare part of the external gallery with an axe; then she pulled out the larva with a hard grass-stem. When the galleries were high up in the tree, she cut incisions into the stem with a hatchet in order to gain a foothold for her big toes.

The witchetties are not an unattractive food of little nutritional value. Slightly roasted they taste like boiled eggs. They also give a feeling of satiation, as their body is oily. Many bushrangers of EYLMANN's acquaintance found their taste good. And a white lady living some years ago in one of the ranges of the interior ate the witchetties with such delight, that her boys collected them so intensively that they finally disappeared around that range (EYLMANN, 1908, pp. 284 ff.).

The big Australian ghost moth *Trictena argentata* H.S. (*Hepialidae*) was widely eaten. N. B. TINDALE (1932, pp. 504/5) recounts how one female, after having deposited 29,000 eggs, still had 15,000 fully developed eggs in its abdomen. The moths emerge in the southern districts in the late afternoon after the first autumn rains have fallen. In the arid northern areas of South Australia they emerge at irregular seasons, shortly after or during the heavy rainstorms. They are attracted to light and fire.

G. F. ANGAS (1847 I, p. 57) records that on the banks of the Lower Murray River these 'large ghost moths fluttered into the embers in such quantities that the natives made a capital supper on the scorched and roasted bodies'. Similar incidents have been experienced among the natives of Cooper Creek (F. WOOD-JONES) and of the N. Flinders Range (TINDALE). J. GRAY (1930, p. 6) mentions that the natives of the Wirra Lube at Ororo dug up the grubs and pupae around the big gum trees on the Pekina Creek, and cooked them in ashes. They were known as *barti*.

TINDALE (1938, p. 5–6) completes these statements by further observations of his own on the caterpillars: 'The natives of Warupuyu, where we camped, are exceedingly fond of the grubs. They call them *makotuuta* (i.e. grubs of the red gumtree); these trees. grow on the low sandy banks and on the beds of wide waterless creeks, which only flow after the rare floods caused by summer rains. The natives discover the places where the larvae of *Trictena argentata* are likely to be living by observing minute cracks on the surface of the ground. In such places the earth has fissured and contracted during dry weather, cracking most readily above one or other of the lateral roots of the tree. They dig down along these cracks and find vertical silk-lined tunnels, commencing at a depth of about six inches. They examine the tunnels closely and smell the silken lining. If the material is damp, but has the correct odour, and if they see signs of freshly spun silk, they continue down several feet, using a native digging stick and their hands in excavating them. By persisting in the search, grubs may be discovered, usually 4 to 6 feet deep. The sand at this level is quite damp. Usually the grubs are deep down and much labour must be expended to obtain them, therefore they are considered only as an occasional luxury . . . .'

'On the night when the moths emerge, there is a great feast for the children, for the moths flutter wildly into the numerous fires which are built to attract them'.

'On one occasion when a grub was being dug out, it was injured in the process. The native cooked it by laying it in the hot ashes of his camp fire for about half a minute. When the skin became taut with the warmed juices within it, he raked it out, flicked it with his fingers to remove the adhering dust and offered it to me. It tasted like warm cream or the baked skin on roast pork and was quite delicious'. Considering the difficulties the natives meet in securing this delicacy, it is probable that one could have partaken of few 'rarer' dishes. TINDALE (1932, p. 516), writing about another ghost moth, *Abantiades marcidus* Tind., reports that the Wirrangu of Fowler Bay on the West coast of South Australia dig up the larvae and pupae from around the roots of gum trees and use them as food. The adults fly into the camp fires in great numbers. When this happens they are raked out and eaten. The natives distinguish four stages; *pindi*, the small caterpillar, *valgunda*, the full-grown caterpillars, *tjirgi*, the pupae, and *kunku*, the moths.

The witchetty totem is extensively described by SPENCER and GILLEN (1899), and we follow them in their exploration of this insect totem. Many terms are connected with this insect and its ritual application. *Unchima* or *Churinga unchima* are small round

Fig. 10. Intichiuma ceremonies of the *Udnirringita* or
Witchetty grub totem.
A. Sacred drawings of the totem on the rocks of the Emily
Gap. (fig. 24, p. 171). The three dots in the upper right field
indicate the eggs of the witchetty grub. The same in colours
(red) on fig. 132, p. 632, where another drawing in yellow
represents the chrysalis of the same insect. B. Rubbing the
stomach with the *churinga unchima* during the ceremony.
The men are sitting in one of the *ilthura* (fig. 27, p. 175).
B. Spencer and F. J. Gillen, The Native Tribes of Central
Australia. London. 1899.

stones representing the eggs; the *Umbana* is a long narrow bough shelter used in the *intichiuma* ceremony, representing the pupa of the insect. *Udnirringita* is one of the witchetty larval forms, named from *Udnirringa* bush, the host plant of the insect. *Maegwa* is the name of the adult witchetty insect. *Ilthura* is a shallow cave in the ground, where the sacred stones, mentioned above, are placed. *Alknalinta* is the name of a rock, where in the mythical past the leader of the witchetty insect totem stood, while he pulverised the larvae on which he fed. A number of primitive drawings of the Udnirringita totem have been recorded by SPENCER and GILLEN. Fig. D.D. (p. 148) shows the drawings on both sides of an elongate Churinga stone from a store-house of this totem in the Emily Gap. The meaning of the letters is as follows: a- curved lines representing a large larva, b– a number of larvae in a hole scooped out in the ground, c- a sitting man squeezes the dirt out of the insects prior to cooking them, d- a larva, e- the eggs of various sizes, and f- marks on the body of the larva. Another figure (fig. 132,3) shows a rock-drawing of the witchetty pupa, yellow on grey ground.

The totem of a child is determined by the totem of the spot where its mother believes that it was conceived, and is thus quite independent of the totem of its mother or of its father. The spirits of the totem centres enter the body of the child and take possession of it, being the reincarnation of one of the dead ancestors (pp. 123 ff.)

The most important ceremony of a totem is the *intichiuma*, which aims at securing the increase of the animal or plant of the totem. Most totem animals are not forbidden as food to their totem people. They eat of it and are even obliged during the *intichiuma* ceremonies to eat a portion of it, in order to prevent a failure of its supply and in order to be able to perform the ceremony properly. The local date of the *intichiuma* ceremony is each year determined by the Alatunya, the local totem leader, according to the nature of the season. At Alice Springs the *intichiuma* of the *Udnirringita* totem is performed as follows (pp. 170, ff., fig. 24–28): the men of the totem leave the camp clandestinely, unarmed and without any decoration. The Alatunya leads and the men follow in one broad line. No food is eaten during the ceremony, which begins with a night's rest near a special camping ground. At daylight the men pluck Eucalyptus twigs and follow the Alatunya in single file along the traditional

90

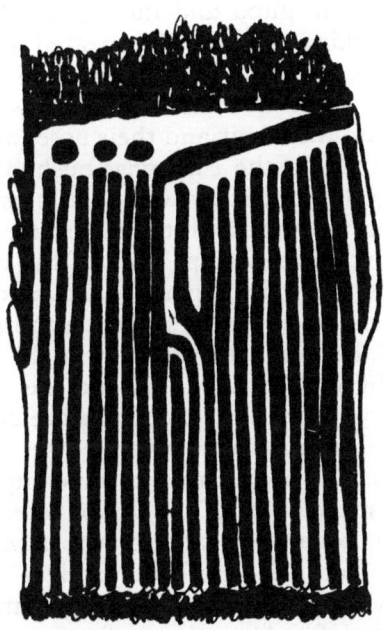

Fig. 11. Churinga Ilkinia of the Udnirringita Totem drawn
on the rocks of the Emily Gorge. Above: Witchetty Grub
in the chrysalis stage. Below: Tracks of Alcheringa Women:
The three spots represent eggs of the *Udnirringita*. From
Fig. 132. B. SPENCER and F. J. GILLEN, The Native Tribes
of Central Australia. London. 1899.

path to the ritual place, the *Ilthura oknira*. This is a shallow cave, within which lies a large block of quartzite, respresenting the adult insect, surrounded by some small rounded stones. The Alatunya begins to chant, inviting the insect to lay eggs, which are later represented by the small stones. The Alatunya then takes up one of the smaller stones and strikes each man in the stomach with it, saying: 'You have eaten much food', then striking the stomach of each man with his forehead. The rock is now stroked by all men with their twigs and they again chant an invitation to the insect to come forth from all directions to lay eggs. The men then march in single file to the nearest *ilthura*, about two and a half km. away There the Alatunya digs out from a hole, about 130 cm. deep, two stones, representing the pupa and the egg of the witchetty insect, and the stomach-striking ceremony is repeated with the pupal stone, the *Churinga uchuqua*, using the same chant. This same ceremony is repeated in nine more *ilthuras*. Turning towards the camp the participants decorate their hair and their noses at a prepared place, where they pluck a twig of the *Udnirringa* bush. The procession then proceeds towards a long, narrow wurley, the *umbana*, representing the pupa, which has been built in the meantime by an elder of the totem who was left in the camp. Men of the other classes of the totem now assemble behind the participants, and still farther away its women. After certain ceremonies all enter the *umbana*, where new ceremonies take place, until the elder from the camp brings food and water. At night a large fire is lit and the people sing around this of the witchetty insect until shortly before daybreak. Then the Alatunya puts out the fire, the singing ceases and all non-participants of the *intichiuma* proper run to the camp. The participants follow later into a special part of the camp, where they strip themselves of all their ornaments and of the *udnirringa* twigs. The totem decorations painted before the arrival at the *mubana* are now rubbed off with red ochre and the normal ornaments are again put on (p. 203 f.). After the performance of this *intichiuma* the witchetty is taboo and must not be eaten until it is abundant and fully grown. Any infringement nullifies the effect of the ceremony and the witchetty supply would become very small. The members of the other classes may eat it at any time, but it must always be brought into camp to be cooked. It must on no account be eaten like other food, out in the bush, or the men of the totem would be angry and the insect would vanish. When, after *intichiuma*, the larva becomes plentiful and fullgrown, the witchetty men, women and children go out daily and collect large supplies, which they bring into camp and cook, so that it becomes dry and brittle, and then they store it away in troughs and pieces of bark. At the same time, those who do not belong to the totem, are out collecting. The supply of the larvae lasts a very short time – the insects appearing after rain – and when they grow less plentiful the store of cooked material is taken to the men's camp, where all the men now assemble by order of the Alatunya. Those who do not belong to the totem place their stores before the men who do; the Alatunya now takes one trough and with the help of other men of the totem grinds up the contents between stones. Then he and the same men all take and eat a little, and hand back what remains to the other people. After this he takes on the trough from his own store and after grinding up the contents, he and the men of the totem once more eat a little, and then pass the bulk of what remains over to those who do not belong to the totem.

After this ceremony the witchetty men and women eat very sparingly of the larvae. They are not rigidly forbidden to eat it, but are even ordered to eat it in small quantities, as in both cases, after over-eating as well as after abstention from the totem insect, the witchetty supply would become scarce.

Pp. 423 ff. describe the traditions about the origin and the early migration of the ancestors of the *Udnirringita* totem. The centre of the totem is near Alice Springs, a territory of about 100 sq. m. About 40 individuals belonged before 1900 to the original totem and to its subsequent traditional branches, and the traditional migrations before arrival are described at length. It is doubtful if local totem groups were usually more numerous.

3. THE BUGONG-MOTH AND OTHER LEPIDOPTERA

The famous *Bugong*-moth of Australia is in fact quite a common looking dark brown to blackish Noctuid moth *Euxoa (Agrotis) infusa* Boisd. (= *A. spina* Gn.). These moths frequently appeared in countless numbers in the southern mountains of the continent, and in dry years along the eastern coasts. They sometimes swarm in Sydney, where they may become a nuisance in houses (TILLYARD 1926, p. 442). This moth was an important article of food for the aborigines, who made a kind of dough or paste from their bodies. Yet its lifecycle and the places where the corresponding mass-development of the caterpillars takes place are still entirely unknown. C. B. WILLIAMS (1930) in his monograph on the migrations of butterflies misses the *Bugong* altogether.

The *Bugong*-moth still exists, but the native tribes which feasted upon them have long ago left the areas where they abound. The best record was given by G. BENNETT (1834, I, pp. 266, 269, 273) who met them swarming by the millions in the Bugong mountains: 'It is named the Bugong Mountain from the circumstances of multitudes of small moths, called *Bugong* by the aborigines, congregating at certain months of the year about the masses of granite on this and other parts of the range. The months of November, December and January are quite a season of festivity among the native blacks, who assemble from for and near to collect the *Bugong*; the bodies of these insects contain a quantity of oil, and they are sought after as a luscious and fattening food. I felt very desirous of investigating the places where these insects were said to congregate in such incredible quantities, and availed myself of the earliest opportunity to do so'.

On the 12th January BENNETT rode to the summit of the mountain: '. . . . This was the first place where, upon the smooth sides or crevices of the granite blocks, the *Bugong*-moths congregated in such incredible multitudes; but from the blacks having recently been here, we found but few of the insects remaining . . . . several of the deserted bark huts of the natives, which they had temporarily erected when engaged in collecting and preparing the *Bugong*, were scattered around.

. . . . At last we arrived at another peculiar group of granite rocks, in enormous masses of varied forms; this place similar to the last, formed the locality where the *Bugong*-moths congregate, and is called '*Warrogong*' by the natives; the remains of recent fires apprised us that the aborigines had only recently left the place for another of similar character a few miles further distant.

1. Female ghost moth (*Leto stacyi* Scott). From K. C. McKeown 1944b, p. 244).

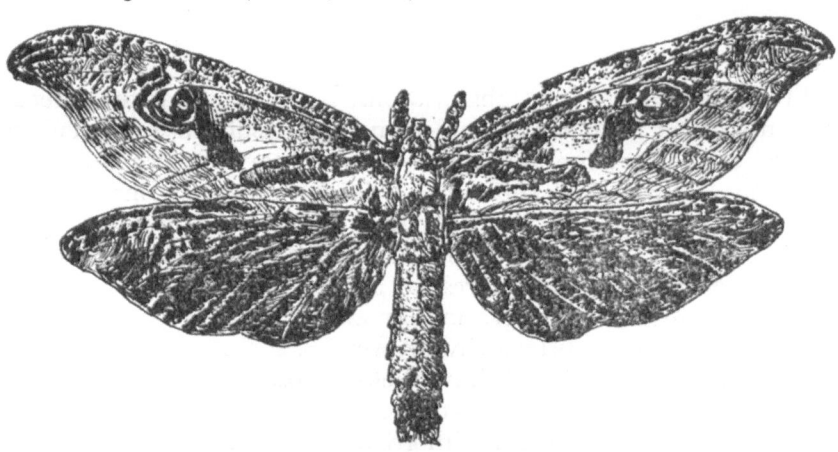

2. Its caterpillar (from Tindale).

3. The Bugong-Moth (*Euxoa (Agrotis) infusa* Bd.). From K. C. McKeown 1944b p. 264.

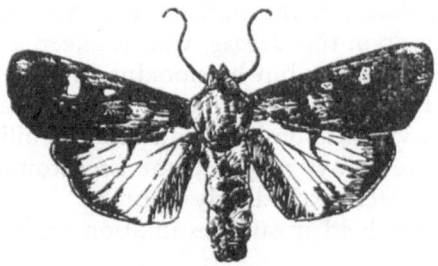

Fig. 12. Ghost Moths and Bugong Moth.

From the result of my observations, it appears that the insects are only found in such multitudes on these insulated and peculiar masses of granite; for about the other solitary granite rocks, so profusely scattered over the range, I did not observe a single moth, or even the remains of one. Why they should be confined only to these particular places, or for what purpose they thus collect together, is not a less curious and interesting subject of inquiry. Whether it be for the purpose of emigrating, or any other cause, our present knowledge cannot satisfactorily answer . . . .

The *Bugong*-moths, as I have before observed, collect on the surfaces and also in the crevices of the masses of granite in incredible quantities; to procure them with greater facility, the natives make smothered fires underneath those rocks about which they are collected, and suffocate them with smoke, at the same time sweeping them off frequently in bushels-full at a time. After they have collected a large quantity, they proceed to prepare them, wich is done in the following manner.

A circular space is cleared upon the ground, of a size proportioned to the number of insects to be prepared; on it a fire is lighted and kept burning until the ground is considered sufficiently heated, when the fire being removed, and the ashes cleared away, the moths are placed upon the heated ground, and stirred about until the down and wings are removed from them; they are then placed on pieces of bark, and winnowed to separate the dust and wings mixed with the bodies; they are then eaten, or placed in a wooden vessel called a *Walbun* or *Culibun*, and pounded by a piece of wood into masses or cakes resembling lumps of fat, and may be compared in colour and consistence to dough made from smutty wheat mixed with fat. The bodies of the moths are large and filled with a yellowish oil, resembling in taste a sweet nut. These masses, with which the *Netbuls* or *Talabats* of the native tribes are loaded, during the season of feasting upon the *Bugong*, will not keep above a week, and seldom even for that time; but by smoking they are able to preserve them for a much longer period. The first time this diet is used by the native tribes, violent vomiting and other debilitating effects are produced; but a few days they become accustomed to its use, and then thrive and fatten exceedingly upon it.

These insects are held in such estimation among the aborigines, that they assemble from all parts of the country to collect them from these mountains. It is not only the native blacks that resort to the *Bugong*, but crows also congregate for the same purpose. The blacks – that is the crows and the aborigines – do not agree about their respective shares, so the stronger decides the point; for when the *arabul*, as the crows are called by the natives, enter the hollows of the rocks to feed upon the insects, the natives stand at the entrance, and kill them as they fly out, and they afford them an excellent meal, being fat from their feeding upon the rich *Bugong*. So eager are these feathered blacks after this food, that they attack it even when it is preparing by the natives; but as the aborigines never consider any increase of food a misfortune, they lay in wait for the *arabuls* with waddies or clubs, kill them in great numbers, and use them as food. The *arabul* is, I believe, not distant from the common crow of the lowlands, the Gungagiar or Worgan. The distinction is, that the 'fat fellers' who fed on the *Bugong* are called *arabul*, the 'poor fellers' the latter name. About February and March the former visit the low land, having become in fine condition from their luxurious feeding. The assemblage of so many distant tribes of natives at this season about the same range, and for similar objects, causes frequent skirmishes to take place between them; and oftentimes this particular place and season is appointed to decide animosities by actual battle, and the conquered party lose their supplies of *Bugong* for the season.

. . . . The quantity of moths which may be collected from one of the granite groups, it is calculated, would amount to at least five or six bushels. The largest specimen I obtained measured 22 mm. with the wings closed, the oily body being 14 mm. long, and of proportionate circumference. The expended wings measured 42mm. across. The coulour of the wings is dark-brown, with two black ocellated spots upon the upper wings; the body is filled with yellow oil and covered with down'*

A. W. SCOTT renders the observations of Robert Vyner, who in 1865 visited the same rocks on the Bugong Mountains with a native,

* The wrong identification of the *Bugong* with the butterfly *Euploe hamata* originating from the Rev. J. G. Wood (Insects abroad), was caused by the misunderstanding of a footnote in G. Bennett's book, where mere reference to the mass migrations of this butterfly (vide Capt. Cook and Capt. King) is made.

as follows: 'The moths were found in vast assemblages sheltered within the deep fissures, and between the huge masses of rocks, which there form recesses, and might almost be considered as caves. On both sides of the chasms the face of the stone was literally covered with these insects, packed closely side by side, over head and under, presenting a dark surface of a scale-like pattern – each moth was resting firmly by its feet on the rock, and not on the back of others, as in a swarm of bees. So numerous were these moths that six bushels of them could easily have been gathered by the party at this one peak; so abundant were the remains of the former occupants that a stick was thrust into the debris on the floor to a depth of four feet.

Mr. VYNER tells me that on this occasion he ate, properly cooked by Old Wellington, about a quart of the moths, and found them exceedingly nice and sweet, with a flavour of walnut, so much so that he desires to have 'another feed'. His clothes, were covered with honey by the moths dashing against them on being disturbed, and smelt strongly of it for several days. At the time these multitudes assembled, the tea-tree and the small stunted-looking white gums were in full blossom, no doubt yielding up their honied treasures to these nocturnal depredators, whose flight, when issuing from their hiding places to the feeding grounds, was graphically described by Old Wellington 'very much like wind, or flock of sheep'. The Tumut blacks report that the moths do not congregate on the high peaks in the spring time, but they first locate the lower mountains, feeding on the blossoms which appear there earlier, and then work their way up to the higher peaks where the plants are later in blossom.

The *Bugong* moths are collected and prepared for food by the aborigines in this wise: A blanket or sheet of bark is spread on the floor; the moths, on being disturbed with a stick, fall down, are gathered up before they have time to crawl or fly away, and thrust into a bag. To cook them a hole is made in a sandy spot, and a smart fire lit on it until the sand is thoroughly heated, when all portions left of the glowing coal are carefully picked out, for fear of scorching the bodies of the insects – as in such a case a violent storm would inevitably arise, according to their superstitious notions. The moths are now poured out of the bag, stirred about in the hot ashes for a short time, and then placed upon a sheet of bark until cold.

The next process is to sift them carefully in a net, by which action the heads fall through, and thus, the wings and legs having been previously singed off, the bodies are obtained properly prepared. In this state they are generally eaten, but sometimes they are ground into a paste by the use of a smooth stone and hollow piece of bark, and made into cakes'.

With regard to the *Bugong* K. C. McKeown (1944, p. p. 68 f.) adds the following observations:

The great hordes of the *Bugong*-moths at times migrate to the coastal regions, and occasionally invade the city. Various theories are advanced, at such times, to account for the presence of these insects in such multitudes. One of the most popular of these is that the moths have 'come in from the sea'. But they are terrestrial creatures and cannot be born there, neither are they capable of so long a flight as would be necessary to bring them from some other land across the ocean. The moths undoubtedly come from inland, aided in their flight by westerly winds.

The number of insects in these visitations is quite beyond calculation. In 1867 the Rev. W. B. Clarke, of St. Leonards, described one of the great invasions of the city and suburbs, and he says that: 'the state of St. Thomas' Church, North Shore, on the 14th September, from the enormous numbers of moths, was such that Divine service could not be held therein'; that 'seven days hard labour in endeavouring to subdue them had been spent in vain'; and that he had 'counted more than 80.000 grouped together upon the windows'. The reverend gentleman certainly had patience. During this period, accounts from Newcastle, seventy miles to the north, Wollongong, fourty miles to the south of Sydney, and other distant parts confirm this statement as to numbers. Within recent years another such visitation descended upon Sydney. They were especially in evidence at a garden party of Government House when every iced cake or tart had a *Bugong*-moth for a 'decoration'.

These great swarms of moths are frequently carried out to sea be westerly winds and drowned; their bodies, like those of the white butterflies, being washed up on the beaches and forming long lines at high-water mark. Scott records a continuous line of these dead moths extending between Newcastle and Redhead, and adds 'probably this exhibition of the fate of these insects in such vast numbers was continued for a considerable distance on either hand'.'

Other Lepidoptera. It is not always easy to decide which of the 'wood-boring grubs' of the earlier travellers really were larvae of beetles and which of moths. Most of them refer apparently to wood-moths (*Cossidae*), the others to longicorn beetles (*Cerambycidae*). References to leaf-eating caterpillars are rare.

The Wongapitcha and other desert tribes do not hesitate to consume quantities of green caterpillars, but do this usually only at the beginning of a good season, when fresh plants are available and the morsel is in consequence claimed to have acquired a sweetish flavour. The only treatment the caterpillars received was to be thrown upon hot ashes until they expand and straighten with the heat. The small hairs covering them were then singed off, but the caterpillars were far from cooked when being eaten (BASEDOW 1925, p. 122).

Several kinds of grubs were eaten in Victoria, namely those taken from the honeysuckle (*tharatum krang*), from the wattle (*martthem krang*) and those from the white gum (*ballook krang*). All the grubs, says Mr. BULMER, are named from the trees from wich they are taken. Some natives preferred to eat the grubs raw, others cooked them by placing them for a short time in the hot ashes of a fire. The common grubs in Victoria are the *Zeuzera citurata* and *Eudoxyle eucalypti* from the wattle and *Eudoxyle nov. sp.* from the gum tree (R. BROUGH SMYTH 1878, p. 207).

The natives of Queensland detected the presence of large grubs by the wood-dust they dislodge, which could easily be seen on the ground or at the entrance of their hole. If the grub was far in, an incision was made in the tree and it was pecked out with a pointed stick. They ate the grubs, either raw or roasted, rejecting the head. These grubs were a delicate food. They have the flavour and consistency of a soft rice pudding enriched with eggs. MATTHEW (1910, p. 90) reports this from experience, having eaten them repeatedly.

The large and fat caterpillars of *Strigops grandis* are eaten in Australia (vide SIMMONDS 1885, p. 355).

4. LONGICORN BEETLE GRUBS AND OTHER BEETLES

Not all the grubs referred to in literature concern beetle grubs. Many of them indicate wood-boring caterpillars. Wherever the references really mean beetles, they mean grubs of longicorns (*Cerambycidae*). Thus, the *hu-hu* larva of the common 5 cm. long *Prionoplus reticularis* Wh. of New Zealand was collected by the Maoris from felled or fallen trees of *Podocarpus* and *Pinus* and esteemed a great delicacy. The same is true for the Australian eucalyptus-borer *Eurynassa odewahni* Pasc. From Western Australia to New South Wales the *bardee*-larva of *Bardistus cibarius* from grass-trees and from *Xanthorroea*-stems were eaten both by natives and white settlers (TILLYARD 1926, p. 232 f.). LUMHOLTZ (1900, p. 114) observed women in Queensland digging out acacia-roots between which they found the highly appreciated larvae of *Paroplites australis* Er., while their husbands hunted for kangaroos. These grubs were immediately roasted in the glowing ashes of the grass-fire. They were consumed on the spot, excepting for those which were set apart for their husbands who would have been angry if they were forgotten.

C. LUMHOLTZ gives an account of the eating of *Eurynassa odewahni* Pasc. in Northern Australia (1890, p. 153 ff.):

'My blacks had found in a large fallen tree some beetle larvae, on which we feasted. There are several varieties of these edible larvae, and all have a different taste. The best one is glittering white, and of the thickness of a finger, and is found in acacia-trees. The others live in the scrubs and are smaller, and not equal to the former in flavour. The blacks are so fond of them that they even eat them while they pick them out of the decayed trunk of a tree, a not very attractive spectacle. The larvae were usually collected in baskets and so taken to the camp. The Australian does not, as a rule, eat raw animal food; the only exception I know of being the beetle larvae.

The large fire crackled lustily in the cave while we sat round it preparing the larvae. We simply placed them in the red-hot ashes, where they at once became brown and crisp, and the fat fairly bubbled in them while they were being thus prepared. After being turned once or twice, they were thrown out from the ashes with a

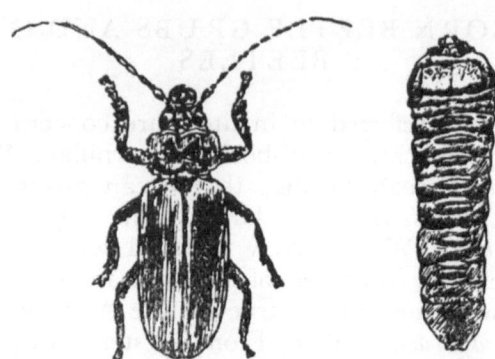

Fig. 13. Adult and larva of the longicorn beetle
Eurynassa odewahni Pasc. (From C. LUMHOLTZ).

stick, and were ready to be eaten. Strange to say, these larvae were
the best food the natives were able to offer me, and the only kind
which I really enjoyed. If such a larva is broken in two, it will be
found to consist of a yellow and tolerably compact mass rather like
an omelette. In taste it resembles an egg, but it seemed to me that
the best kind, namely the acacia-larva, which has the flavour of
nuts, tasted even better than a European omelette. The natives
always consume the entire larva, while I usually bit off the head
and threw aside the skin, but my men always consumed my leavings
with great gusto. They also ate the beetles as greedily as the larvae,
simply removing the hard wings before roasting them. The natives
are also fond of eating the larger species of wood beetles, as well as
of freshwater-shrimps'.

The trunk of the grass-tree (*Xanthorroea arborea*), when beginning
to decay, furnishes large quantities of marrow-like grubs (*Bardistus
cibarius*) which are considered a delicacy by the aborigines of W.
Australia. They have a fragrant, aromatic flavour and form a
favourite food among the natives, either raw or roasted. They call
them *bardi*. The *bardi* are also found in the wattle-tree. The presence of these grubs in the *Xanthorroea* is ascertained in the following
way: if the top of one of these trees is observed to be dead, and it
contains any *bardi*, a few sharp kicks given to it with the foot will
cause it to crack and shake and it is then pushed over and the grubs
are taken out, by breaking the tree to pieces with a hammer. These

bardi are small, but found together in great numbers. Those of the wattle are cream-coloured, as long and thick as a man's finger, and are found singly (cf. COWAN 1865, p. 70).

P. CUNNINGHAM (1827, I, p. 329) wrote: 'Our wood-grub is a long, soft, thick worm, much relished by the natives, who have a wonderful tact in knowing in what part of the tree to dig, and they quickly pull it out and gobble it up with as much relish as we have from an oyster'.

The only other beetle eaten is the Rutelid *Anoplognathus viridi-aeneus* Don., a metallic green beetle. HOPE (1842, p. 132) reports that it is consumed by the New Hollanders (vide W. S. MCLEAY 1819/21). HOPE got confirmation from old Australian settlers who observed the white grubs turn into golden beetles which they identified as that species.

5. HONEY ANTS AND OTHER ANTS

Ants play a very important role in the early food of the Australian aborigines. The pupae or even adults of many species were eaten, or transformed into a refreshing drink. The honey ants, which store honey in the much inflated honeypot individuals, are an important source of food in the deserts of the interior. The black *yarumpa Melophorus inflatus* Lubb., the golden yellow *ittootoonee M. cowleyi* Frogg. and the red honey ant *M. bagoti* Lubb. are the common species. The *Leptomyrmex varians* Em. of Queensland also stores its honey in a similar manner (TILLYARD 1926, pp. 289 ff., MCKEOWN 1944, b, pp. 182 ff.).

Ants are a common food in Australia (CAMPBELL 1926, p. 408 f.). In the arid areas of Central Australia the 'sugar-ants' (*Melophorus inflatus, M. cowlei*) are collected and eaten. During the Horn Scientific Expedition in 1894 places were seen where colonies of these ants had been dug out from the loosy sandy soil. The ants used are food-workers, so gorged with honey that they are unable to move, the 'honeypots'. The distended abdomens are bitten off and swallowed, the remainder thrown away. SPENCER and GILLEN in their 'Northern Tribes of C. Australia' (1899, p. 199) describe a special food ceremony to produce an abundance of honey ants.

The honey ants of the arid and semi-arid parts of Central Australia are a welcome prize for the natives. The honey originates from sweet exudations of certain plant-galls and from the honey-dew of other insects. In their nests many individuals are stored as living honeypots to provide food in seasons and years of need. They are stored one to one and a half meters deep and also occur in other areas.

BASEDOW (1925, p. 146 f.) reports the name *winudtharra* as being used for the honey ants by the Aluridja and Wongapitcha, while the Arunta call them *yarumba*. He mentions how in Central Australia the honey ant (*Melophorus inflatus*) takes the place of bees. 'These remarkable insects live underground, usually in the red sandy loams carrying forests of mulga. Throughout the Macdonnell Ranges and the country north and south-west of them, and in the Musgrave Ranges and the country north and south-west of them, and in the Musgrave Ranges district, they are eagerly looked for by the local tribes. When the entrance to a nest has been discovered, a gin at once sets to by inserting a thin stick as a marker and digging down the course of the hole. This is a somewhat tedious undertaking, and not infrequently she has to dig to so great a depth as to completely bury herself. On several occasions I have unexpectedly come across a woman thus engaged, and neither was she aware of my coming, nor I of her presence, until right opposite her.

The 'honey ant' itself is a modified worker of the colony, which is so overfed by the ordinary workers that its abdomen swells to the size of a marble, about 1 cm. in diameter, in consequence of the liquid honey stored within. With the exception of a few transverse plates (tergites and sternites), the abdominal walls are reduced to an extremely fine membrane, through which the honey can be clearly seen. The insect's viscera are compressed into a small space near the vent. The ant in this condition is naturally unable to move from the spot. It appears that the inflated ants in this extraordinary way provide for the needs of the colony during the barren season of the year, acting as living barrels, which can be tapped as required.

The gin collects numbers of these ants, as she burrows her way downwards, and lays them in her *cooleman*; when the nest has been ransacked she returns with her prize to camp.

When a native wishes to partake of the honey, he grips one of the ants by the head, and placing the swollen abdomen between his lips,

The rock drawings or *Churinga Ilkinia* from the sacred storehouse of a group
of the Honey Ant Totem in the Warramunga Tribe. (From SPENCER and GILLEN.
1899, facing p. 631).

Ant totem ceremony of the Warramunga
tribe. The performer represents a woman
ancestor searching for and gathering ants
on which she feeds. He wears a helmet and
carries leafy twigs attached to each thigh.
From Sir B. SPENCER, Wanderings in Wild
Australia. London. 1928. II, fig. 282.

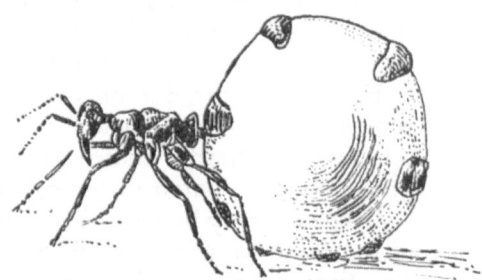

Honeypot ant of the typical Australian species
Melophorus inflatus Lubb. (From K. C. Mc-
KEOWN 1944b, p. 183).

Fig. 14. Australian honeypot ants and their totem.

he squeezes the contents into his mouth and swallows them. As regards the taste, the first reaction the palate receives is a distinct prick of formic acid, which is no doubt due to a secretion produced by the ant in self-defence. But this is both slight and momentary; and the instant the membrane bursts, it is followed by a delicious and rich flavour of pure honey'.

The honey ants or *yarumpa* provide the natives of Central Australia with one of the few sweets at their disposal. The women dig them out with astonishing speed. McKeown (1944, p. 38) describes how the hard ground round the entrance to the nest is loosened with the point of a digging stick, and the loose earth is then scooped out with the hand or a small bowl, the woman alternately breaking up the soil, and throwing it behind her, until a hole is dug large enough to contain her body. She goes on digging deeper and deeper, until she may reach a depth of six feet or even more. Sir B. Spencer describes some of the elaborate ceremonies which were performed by the men in some totem-groups of the Central Australian aborigines, in order to increase and ascertain the supply of the honey ants, to make them prolific and to ensure this source of sweet food for the coming season.

Writing of South Australia Eylmann (1908) reports that ant-maggots are only occasionally eaten by some tribes, but the honey-like matter, which is stored in some workers of Melophorus-species, is considered as a delicacy. The storing crop (in the abdomen) is only slightly larger than a pea. Yet these ants are widely distributed and locally common, thus enabling many groups of the interior to collect this sweet food in sufficient quantities to serve as an important item of their diet. According to Dr. Wetherell this honey is a watery solution of pure fructose.

Honey ants provide only a small proportion of the native food, but as luxury items they are in constant demand and native women seldom work harder than when spurred on by the discovery of an unusually rich patch of honey ant ground (N. B. Tindale).

In the country between Mt. Olga and Ayers Rock, shrub country in the Lake Amadeus basin, Spencer (1928, I, pp. 172 ff.) for the first time encountered the nests of the honey ant (*Melophorus inflatus*), which insect he was very anxious to see in its native habitat. The local Arunta tribe calls them *yarumpa* and are very fond of them. In this miserable part of Central Australia it is one of their chief delicacies. In some places the whole surface of the ground had been turned up, just as if a small army of prospectors had been at work. There is nothing on the surface to indicate the existence of a burrow, except a small opening an inch or less in diameter. From this the central burrow goes straight down. The natives soon found one or two and immediately set to work to dig them out. It was astonishing to watch the speed with which the elder women worked. First of all, the ground round the opening was loosened with the aid of a digging stick, held in the right hand, and, alternately loosening the soil with her stick and then throwing it out over her shoulder, the lubra soon dug a hole just big enough to hold her body. The main burrow went down for between 150 to 180 cm., with horizontal passages going off all round it. A few of the honey ants were found in each of these, but the greater number were in a swollen chamber at the bottom.

In the nests that SPENCER dug up there were only two kinds of ants: the ordinary worker and the honeypots. The latter are a most remarkable instance of the modification of certain members of a social community to serve a special purpose; and similar modifications are found in the dry and arid parts of Mexico and Colorado. Instead of storing up honey in combs, as a reserve supply of food when this is otherwise scarce, these ants utilise the bodies of certain members of the community for this purpose. Exactly how the material for making honey is obtained is not known. It is quite likely that the main source is the exudation of insects, such as Psyllids or Coccids, often met with in the scrub; or the sweet material, afterwards made into honey by the ants, may be derived from the *mulga*-apple, a gall of an acacia. In America the material is obtained from a gall of oaks. The final result is that the honeypot individuals are fed until their crop, in which the honey is stored, becomes so enormously distended that the abdomen has the form of a membranaceous spherical bag with dark little plates, widely separated from one another on the upper and lower surfaces, which represent the whole of the hard rings covering the abdomen in the normal ant. Head and thorax form only a tiny appendage to the abdomen, and this animated honeypot remains quietly, wherever it happens to be when this strange feeding process takes place. When the workers want to eat the honey, they come and tap the sides of the honey-bag with their feet (sic!). In response to this stimulus the honey is passed out in drops from the mouth and is eaten by the others. It is a very strange fact that identically the same modification in ants has occurred in the widely separated areas of North America and Australia. The utilisation of sweet natural substances for the production of honey seems to be restricted among insects to the *Hymenoptera*. Bees store it in combs, ants in the bodies of certain individuals of the community. Both are more sagacious than the Australian savage, who literally takes 'no thought for the morrow' and never thinks of laying in a store of food to help him to tide over bad times, when food is scarce.

It must have been rather a severe trial to the feelings of our savage companions to watch the honey ants that they dug out being transferred to the collecting bottle. They could not understand that they were of any value save as an article of food.

On this occasion we could not find any of the winged sexuals, but later on Mr. Cowle not only secured these, but found a new species of honey ant and a number of specimens of another. Neither of these had the body as swollen as in the case of the commoner kind, and both of them are capable of a certain amount of movement. One of them (*Melophorus cowleyi*) is golden-red in colour. We only came across a single nest of it under a little block of quartzite in a gorge in the Macdonnell Ranges. The nest consisted of irregular branching passages, close to the surface, and in these the ants, the native name for which is *itutuni*, were moving about sluggishly. The other new species is much darker, and though distinctly swollen out, yet the absomen does not become anything like so tensely distended as in the commoner insect. Evidently these two are not so fully specialised as in the *yarumpa* (*M. inflatus*), which is par excellence the honey ant of the arid parts of Australia. Its distribution extends at least from Ayers Rock in the south to Barrow Creek in the north, and far away across the intervening desert into West Australia, from where Sir John Lubbock apparently secured the first specimens.

In Central Australia the *yarumpa* (*Melophorus inflatus*) is of paramount importance and has given the name to a totem. Spencer and Gillen (1899, p. 26) say that the women always have a wooden trough, the *pitchi*, within which they carry food material or a small baby. The only other implement possessed by a woman is a 'yam stick', which is simply a digging stick or a pick. The commonest form consists merely of a straight staff of wood with one or both ends bluntly pointed, and of such a size that it can easily be carried in the hand and used for digging in the ground. When at work, a woman will hold the pick in the right hand close to the lower end, and, alternately digging this into the ground with one hand, while scooping out the loosened earth with the other, will dig down with surprising speed. In parts of the scrub, where the honey ants live, acre after acre of hard sandy soil is seen to have been dug out, solely by the picks of the women in search of the insect, until the place looks exactly like a deserted field where diggers have been at work 'prospecting'. Very often a small *pitchi* will be used as a shovel or scoop to clear away the earth when it gets too deep to by merely thrown up with the hand, as the woman goes on digging deeper and deeper until at last she may reach a depth of 180 cm. or even more. In the scrub a woman will be digging up lizards or honey ants while close by her small child will be at work, with its diminutive pick, taking its first lessons in what, if it be a girl, will be the main employment of her life.

In the *intichiuma* ceremony of the *yarumpa*-totem, as performed at Ilyaba, the majority of the men are Pamunga and Bulthara, only a few Kumara and Purala belonging to the totem (SPENCER and GILLEN 1899, pp. 186 ff.). Early in the morning on the appointed day the men assemble at the men's camp, where they decorate their foreheads, arms and noses with twigs of the *udnirringa*-bush and smear their bodies all over with dry red ochre. Then they march in single file, the Alatunya at the head, to a spot about fifty yards from, and opposite to, the *erlukwirra* or women's camp, where the women and children stand silently. Here the Alatunya, turning his back upon the women, places his hand as if he were shading his eyes and gazes away in the direction of the *intichiuma*-ground, each man as he does so kneeling behind him, so as to form a straight line between the women and the *intichiuma*-ground. In this position they remain for some time, while the Alatunya chants in subdued tones. Thereafter all stand up, and the Alatunya goes to the rear of the column and gives the signal to start. In perfect silence and with measured step, as if something of the greatest importance were about to take place, the men walk in single file, taking a direct course to the ground. Every few yards the Alatunya, who is in the rear, goes out first to one side and then to the other, to see that the men keep a straight line. After half a mile a man is sent by the Alatunya to the *Ertnatulunga* to bring a special stone, Churinga, which is required during the ceremony. The *intichiuma*-ground is situated in a depression in a rocky range, at a considerable elevation above the surrounding plains, and all over the depression are blocks of stone standing up on end and leaning in all directions, each of which is associated with a honey ant man of the Alcheringa. As the party approaches, a messenger arrives who has been sent a long way round, running the whole way.

All the men then group themselves round a pit-like depression in the rocks, which is surrounded with a horseshoe-shaped wall of stone, open at the western end. On the east side is an ancient acacia tree, which is the abode of the spirit of an Alcheringa man, whose duty it was to guard the sacred ground. In the centre of the pit is a stone, which projects for about 45 cm. above the ground, and is the Nanya of an Alcheringa man who originated here and performed the *intichiuma*. On the arrival of the party the Alatunya at once goes down into the pit, and some time is spent in clearing out the débris, while the other men stand round in perfect silence. After a time he beckons to some of the older men to come down and assist him, and they all begin to sing while the sacred stone is disclosed to view and taken out of the earth, together with a smaller, smooth round pebble, which represents a mass of honey collected by the ants and carried about by a man.

When the stone has been taken out, the old men rub it over reverently with their hands, and it is then rubbed over with the smaller stone, after which it is replaced in the ground. This done, the large stone Churinga from the *Ertnatulunga* is brought up, which also represents a mass of honey carried about by a celebrated priest, named Ilatirpa. The Ilatirpa was the leader of the Yarumpa and sent out wandering parties from this spot. Another long, thin stone Churinga from the *Ernatulunga*, pointed at each end and evidently very old, represents the piece of wood which was carried by Ilatirpa for the purpose of digging up the ants on which he fed. The old Alatunya takes up the Churinga, and calling the men up one by

one, each of them walks into the pit, and lies down, partly supported on the knees of two or three of the older men. In this position
the Alatunya, keeping up all the time a low chant, first of all strikes
each man's stomach sharply two or three times with the Churinga,
and then moves it about with a kind of kneading action, while
another old man butts at the stomach with his forehead. When all
have passed through this performance the singing ceases, the Churinga is handed back to the man who brought it, with instructions
to bring it back to the *Ertnatulunga*, and the column forms again and
marches back, taking a different course, again awaited by the women
and children standing in silence.

On the way back a halt is made at a spot in the Ilyaba creek,
where the final act of the ceremony is performed. Certain acacias
and stones are associated with Alcheringa men. On arrival at this
spot all the men sit down, and almost an hour is spent in singing
about the Yarumpa men, of their marchings in the Alcheringa, of
the honey, of the ants' nests, of the great man Ilatirpa, and of those
Yarumpa men who, in the Alcheringa, changed into the little birds
now called *alpirtaka*, which at the present day are the mates of the
honey-ant people, to whom they point out where the ants'-nests
can be found. After some time the decoration of the Alatunya commences, while he leads the singing, which now refers to the men on
the banks, who are supposed, in spirit from, to be watching the
performance from their trees. The decorations on the body of the
performer are intended to represent the chambers in the ants'-nests,
and those on the arms and neck the passages leading to the inner
parts of the nests, where the honey ants are found. The performer
squats on the ground and for some time the other men run round
and round him in the usual way, while he occupies himself with
brushing the ground between his legs with little twigs, pausing every
now and then to quiver. When this is over the decorations are
removed and the party starts back for the men's camp. Fig. 14a
shows a coloured rockdrawing of the Yarumpa near the Barrow
Creek, done in white or black on an artificially ochre background.
Such drawings are sacred, but their exact meaning is unknown.

The tradition on the ancestral wanderings of the Yurumpa is
explained by SPENCER and GILLEN (1899, p. 438 f.), as follows:
'The Yarumpa originated at a place called Ilyaba, in the Mount

Hay country. From this spot, which is the great centre of the totem, they dispersed in various directions. The great priest Abmyaung-wirra started out from Ilyaba to see what the country out northwards was like and, returning after many days, told his people that he had found another mob of honey ants far away in the north-east, and that he intended leading a party to them. After performing many ceremonies which the women were allowed to see, he started off with the party, and after finding two or three honey ant men living in the scrub in various places, they came to Inkalitcha, where the water hole by which they camped was dry. They suffered much from thirst, and so opened veins in their arms and drank blood. Then they went into the ground, and alternately travelling above and below ground, came to a spot on the Burt Plain, where it is related that they made *ilpirla*. This is a drink made by steeping the bodies of the honey ants in water, and then kneading them until the honey is pressed out and mixed with the water. The *ilpirla* was mixed in the hafts of their shields. After drinking some of it the priest left the party, and went on ahead to find the honey ant people whom he had seen before. He found them at Koarpirla making an Engwura, and, returning at once to his party, he led them to that place. The local people were very angry and refused to have anything to do with them, and moreover they opened veins in their arms, making such a flood that all the party was drowned except the priest who returned to Ilyaba, where he finally died. A black hill covered with black stones arose where the wanderers perished, and their Churinga are now in the local store-house.

Another party of honey ant people started out west under the guidance of two priests. At a place called Tabulpma a large *nurtunya* was erected and here one of the leaders remained behind. At Umpira the party met a honey ant woman.

'When she died a large stone arose to mark the spot .... At Lukuria they found a large number of honey ant men and women. On seeing the strangers the women were very angry, and assumed a threatening attitude, stamping and beating the air with their palms extended outwards, shouting: 'Stop, don't advance, we have many Churinga!' The party, frightened, camped a little distance away, where they erected a *nurtunja* and performed *arlitha*. The local honey ants did not come near to them. They travelled on westwards, camping near Mt. Heuglin and on the Dashwood Creek, forming various totem centres. At Amulapirta they stayed for a few hours performing *ariltha*, and all the men had intercourse with a Panunga honey ant woman. On the north side of the Belt Range they camped by the side of a creek, and here they erected their *nurtunya*, and were too tired to take it up again. While travelling on they heard a loud *arri-inkuma* – a special form of shouting – and soon found themselves face to face with a large number of honey ants at a place called Unapuna. The local people resented their coming, and at once drew forth floods of blood from their arms, with the result that all the strangers were drowned, their Churinga remaining behind and giving rise to an important Yarumpa or honey ant totem centre'.

Ants were considered a novel kind of food by J. FRASER (1892, p. 53) and he scarcely believed that they were much eaten; and yet a friend saw a black woman in N. S. Wales put her naked foot on the nest of some red ants, and when they swarmed up her leg, as ants always do when thus disturbed, she scraped them off in handfuls and ate them. Ant-maggots and ant-pupae, dug from the nests are readily eaten.

R. BROUGH SMYTH (1878, p. 207) mentions that the natives of Victoria are very fond of ants. They gather them and place them in a *tarnuk;* they are then mixed with the dry bark of the 'stringy bark' tree, which they tear off the tree and rub in their hands until it turns to powder. When this is thoroughly mixed with the ant-pupae, they take up some in their hands and blow away the loose stuff, and finally get clean pupae to eat. They say they are very good, the taste being something like that of a mixture of butter and sugar. Mr. WILHELMINI mentions the use of troughs by the natives of South Australia for holding the ant-pupae. Their trough, called *yuta*, is about 120 cm. long and 20 cm. broad. The natives open the ant-hills and the pupae are placed in the *yuta*, which is shaken and so manipulated as to retain the pupae and to throw off the dirt and refuse. The season of the ants is in September and October, and during these months the *yuta* is always seen in the hands of the natives. A kangaroo skin, or anything available that will hold the contents of the ants'-nest, is used for shaking and clearing the pupae of dust, etc., when neither *tarnuk* nor *yuta* are to hand. The pupae of the common ant (*Formica consobrina*) are the size of grains of rice;

those of the black and red bull-dog ants (*Myrmecia pyriformis* and *M. sanguinea*) are about 18 mm. long.

The pupae of ants were also a favourite food (N. W. THOMAS 1906, p. 111). Sometimes the pupae were winnowed clear of the ants, but in Queensland the two are eaten together, mixed with salt water. In some parts the ants are allowed by a hungry aboriginal to run up his legs; he then sweeps them off and puts them into his mouth as fast as his hands can serve him. On exceptional occasions ant-maggots are eaten by *some* tribes (EYLMANN 1908, p. 180).

LUMHOLTZ (1890, p. 145) also mentions the eating of small ants and of their larvae in N. Australia. Their 'eggs' have a nutty taste and are agreable and refreshing.

Ants and ant-maggots were readily eaten by the aborigines of N. S. Wales. One method of securing the adult ants was to stand upon the nest so as to disturb the inmates. Thus disturbed the ants would swarm up the legs of the native hunter, who scraped them off in handfuls; they were also collected from the nest, which was broken open by sticks (CAMPBELL, 1926, p. 408).

In N. Australia the adults and larvae of the green tree-ant (*Oecophylla smaragdina*) are eaten by many N. Queensland natives. In the Cooktown district their nests were opened upon a rock or some smooth object, causing the adults to scatter, leaving behind the larvae. These were collected and formed into a ball, by rolling them in the palms of the hands, after which the mass was swallowed. If the contents of one nest failed to satisfy the cravings of hunger, then several were collected and treated in a like manner, the contents being rolled into one big ball, which was usually washed in water before being eaten, the water being afterwards used as a drink (CAMPBELL 1926, p. 408 f.). Often the bodies of the ants were crushed and mixed with water, the liquid so formed being consumed separately, while the residue of the ants' bodies was either eaten or discarded. SAVILLE KENT (1897, p. 253) reports: 'Mashed up in water, after the manner of lemon squash, these green tree-ants form a pleasant acid drink which is held in high favour by the natives of N. Queensland and is even appreciated by many European palates'. A similar liquid gained from the bodies of *Oecophylla* ants and larvae was commonly used by the natives of N. Queensland as a remedy for many ills, such as stomach troubles, headaches, coughs, colds, etc.

6. HONEY-BAGS

All the honey-bees of Australia belong to the genus *Trigona* (*Meliponidae*). The small workers, 4 to 6 mm. long, are all stingless. They nest in trees and are eagerly sought out by the natives, who will follow one of these bees for miles to find the nest. The honey is stored in large quantities, but the flavour is not always very pleasant. The common *Trigona carbonaria* Sm. is found almost everywhere. HOCKINGS gives a good account of it under the name *Karbi*. The European honey-bee (*Apis mellifica* L. and *A.m. ligustica* Spin.) now abound everywhere and appear in many places to have driven away the native flower-hunting bees (TILLYARD 1926, p. 305). In both cases the word 'everywhere' should not be accepted too literally, of course.

H. J. HOCKINGS (1884) observed the biology of the Australian bees of the genus *Trigona*, of which he noted two species in Queensland, and others living to the South: the *Karbi* or *Keelar*, probably *Trigona carbonaria* Sm. and the *Kootchar*, *T. cassiae* Lock. Both nest in hollow trees. Their development takes about three months. The eggs are laid in cells already filled with food and are immediately sealed up. The *karbi* only gather a small quantity of honey. Both species begin to build from the bottom upwards. 4 or 5 lbs. of honey would be a fair yield from these hives, which are not much valued even by the natives. To obtain the honey the comb is usually squeezed and the honey thus becomes largely mixed with pollen, and assumes a sour taste which it does not have in a pure state. These small bees, although being stingless, watch the entrance of their nests and are capable of defending them fiercely. In December, 1882, HOCKINGS observed that two nests swarmed simultaneously, came in contact with each other and most of them died in the fight for the possession of a box. When the honey is removed, the bees attack the intruder in a most determined fashion, biting at his hair, beard and eyelashes, and smearing them with gum. They crawl inside one's clothes and into one's ears, nose and mouth. Very few persons could continue, such operations long without a veil, in consequence of the sticking of the eyelashes and the painful biting of the eyelids. The aborigines usually kill the bees by smoke and then take the honey at their leisure.

The *kootchar* is timid; it emits an ant-like odour which seems to give it a great deal of protection from other insects. Possible intruders are plastered over with small globules of a greenish gum-like matter, apparently extruding from their mouths. Their wax is grey or buff-coloured. When the hive is taken out, they devour the honey greedily at first, then they get much excited, dashing aimlessly about, without attacking the intruder. The *kootchar* are very industrious, and at times fight fiercely among themselves. They also close the brood cell immediately after oviposition. As much as 50 lbs. of honey may occasionally be obtained from one nest. At night they close the entrance hole by gum.

Both species are common. The *kootchar* live only on sandy soil, mainly near the coast, the *karbi* more in the inland scrub. Two nests of *kootchar* can easily be united by simply removing one queen and packing her brood-nest, bees and all, against the brood in the second hive; if any bees return to the old hive they may be shaken in at night and the hive removed. This cannot be done with the *karbi*, as they would fight and kill each other. When either species takes up its abode in a space too large for them, they partition it off by a wall of eucalyptus gum mixed with odd bits and pieces. Their natural enemies include *Achroea grisella* F., small birds and the beetle *Protaetia mandarinea* Wbr.

The importance of honey in the native diet has been described by Dr. BASEDOW as follows (1925, pp. 145 ff.):

'All along the north coast, a welcome addition to the daily fare is wild bees' honey, the 'sugar-bag' of the semi-civilised tribes. The wild bee establishes its hive either in a hollow tree or in a crevice in the ground, and the hunting native – man, woman or child – is ever on the lookout for it. When the exit of the hive has been discovered in the ground, from wich numerous bees are flying, the lucky finder immediately begins to dig down carefully along the narrow channel until he reaches the honeycomb. If the supply is limited, it is usually removed in toto by hand and lifted to his mouth without further ado. If, however, there is a goodly amount available, the whole of the comb is collected and placed in a *colleman* or other food carrier and taken to camp.

When a hive is located in a hollow tree, the native places his ear against the butt and listens; by frequently altering the position of his ear like one undertaking a medical auscultation, he can gauge the exact position of the hive by the murmur and buzz beneath the bark. It is then a simple matter for him to cut into the cover and collect the honeycomb. Some of the experienced hunters can 'smell' their way for a considerable distance to a wild-bee hive.

The Victoria River tribes have invented an ingenious device by means of which they can secure honey from otherwise inaccessible fissures in rocks and hollows in stout-butted trees. A long stick is selected, to one end of which is tied a bundle of vegetable fibre or pounded bark. With the bundle forward, the stick is poked into the cleft leading to the hive, and, when the honeycomb is reached, it is turned around and allowed to absorb some of the honey. Then the stick is quickly removed and the absorbed honey squeezed from the fibres into a receptacle. The process is repeated, time after time, until the greater part of the honey has been obtained.

Wild-bee honey is very liquid, but, nevertheless, quite as sweet and tasty as that of the Ligurian bee. The wild bee, moreover, possesses no sting, and so offers no serious resistance to the enthusiasm of the collector. The bee itself is comparatively small, about the size of an ordinary house fly.

There are no bees in the deserts of Central Australia, but in their stead appears the honey ant'.

Brushes for collecting honey from a wild hive are in use among the S. American Indians, and also among the Victoria River tribes of Australia. These tie a bundle of vegetable fibre of pounded bark to the end of a long stick and force this into the nest. When the honeypots are reached, the stick is turned round and round so that the bushy end becomes thoroughly soaked in honey. Thereupon the stick is withdrawn, the honey adhering to the fibre is squeezed into a receptacle and the process of inserting the stick into the nest is repeated (H. BASEDOW 1925, p. 283 f.; Sir G. H. WILKINS 1928, pp. 246 ff.).

C. LUMHOLTZ (1890, pp. 178–181) describes native honey collecting in N. E. Australia as follows: 'On our return to Herbert Valley we passed one of the Eucalyptus forests which adorn the lower part of that mountain range. They are the favoured place of bees. My native boy who was very fond of honey, which·the natives consume in large quantities, hurried to discover the precious insects. They refuse the maggots, even when they are very hungry. The wax is used as a glue in the manufacture of various tools. It also serves as pomade in the arrangement of their coiffure.

Fig. 15. Honey-hunting in Northern Australia. From C. LUMHOLTZ, 1890, p. 179.

The Australian bee is smaller than ours. It deposits its honey in hollow trees, often at a great height. The native, when passing beneath trees which may grow up to 30 m. in height, in clear weather can still discern these insects, owing to their good eyesight, when these congregate around the small entrance hole of their nest. In open forest the natives never forget to look out for bees'-nests and whenever I met them there, their eyes were always fixed upon the trees. Whilst my eyes are at least twice as strong as normal, I could rarely recognose the bees, even after the native had pointed out their nest to me. Yet one day, guided more by the ear than by the eye, I was able to distinguish a small swarm, about 4 m. up, which caused my men some astonishment . . . . and they loudly commented on the white man who could find honey.

It is an amusing sight to observe the natives gathering honey. One of them will climb the tree and cut a hole large enough to put his arm through, whereupon he takes out one piece after another of the honey-comb, and as a rule does not neglect to put a morsel or two of the sweet food into his mouth. He drops the pieces down to his comrades who stand below and catch them in their hands. At the same time the bees swarm round him like a black cloud, but without annoying him to any great extent, for these bees do not sting, they only bite a little. Most of the honey is consumed on the spot, but part of it is taken to the camp, being transported in baskets specially made for this purpose. These baskets are of the same shape as the other baskets made by the natives, but more solid and smaller in size; they are made of bark, so closely jointed with wax that they will hold water. Sometimes the honey is carried a short distance on a piece of bark, a border of fine chewed grasses being layed round the edges in order to keep it from running off. Sometimes also a palm leaf is used, which is folded and tied at both ends, so that it looks like a trough. It is the same kind of trough as the natives use for carrying water and can be made in a few minutes.

In almost every hive some old honey is to be found which has fermented and become sour, because these bees, which have only rudimentary stings, are not in possession of any poison to preserve it with . . . . Furthermore, the honey of the bees with rudimentary dards causes some inconveniences, such as diarrhoea from which I suffered, together with many natives, whilst I could eat great quantities of European honey without the slightest inconvenience. The old honey which the bees do not eat themselves, looks like soft yellow cheese, and the civilised natives call it old-men-sugar-bag. The natives do not reject it, but mix it with fresh honey and water in the troughs just described. Fresh honey is sometimes mixed with water.

This mixture of honey and water is not drunk, as one would suppose, but it is consumed in a peculiar manner. The natives take a little fine grass and chew it, thus making a tuft which they dip in the trough and from which they suck the honey as from a sponge. While they sit crouching round the trough, and as each one tries to get as much as possible, the contents quickly disappear. Where spoons are lacking this would seem a natural and practical invention, and is surely calculated to secure an equitable division of the honey, as in this way it is difficult for any one person to get more than his share. After the meal the tufts are placed in the basket, where they are carried as long as they are fit for use.

The wild honey of Australia, generally deep brown, cannot compete with European honey of the best quality; its aroma is too strong to be positively agreable. It remains firm and fresh in the tree-trunks, even during the great heat, and forms a healthy and agreable food, which I never regarded as very attractive. I was soon disgusted by it, even when it had to take the place of sugar. Never did we find honey in the vast thickets'.

And further on we read (LUMHOLTZ 1890, p. 253): 'High up in a tree a great quantity of honey was discovered. My men concluded from the nature of the tree that it was hollow from crown to root, and that the honey would drop inwards and be lost, if taken in the usual way. Therefore they borrowed my hatchet. The wood was so hard that the tree fell only after $1^1/_2$ hours. This tenacity is shown by the natives when they are on the track of something promising immediate profit, while otherwise they are indolent or lazy. This day my men were repaid for their trouble, and I was greatly surprised at the enormous quantity of honey within the tree-trunk. It was a delightful, hard and fresh honey, in spite of the heat'.

H. PRIEST (1932, p. 113 f.) describes an experience on the Murray River in North Australia: 'My halting at this beautiful spot was providential; for in the tree at the foot of which I had cast my bundle with such relief, I discovered a hive of bees. The limb in which they had built branched off at no great height, and it was not too wide in girth for my little tomahawk. Unfortunately the foliage hung over the creek, and there was every chance of the whole branch falling into the water. But I was rewarded by securing as fine a 'sugar-bag' as I had ever seen.

The honey of the small native bee is not contained in a honey-comb like that of the common domestic bee. The comb takes the form of clusters of egg-shaped cocoons held together by a wax and pollen cement. Each cocoon contains about a teaspoonful of most delicious honey, which is easily extracted by crushing the whole comb in a cloth and allowing the sweet stuff to filter through. I did not get a great supply, it is true, as the colonies are small and the bees themselves not much larger than houseflies. But I never enjoyed a pot of honey more. The wax, too, was eatable, and very good. I did not learn until then what a valuable food honey is. Its effects upon the system are almost instantaneous. This particular variety of bush honey was remarkable, producing a most agreable sense of physical well-being, The honey is mostly gathered from the blossoms of the various gum trees, the nectar of which is quite innocuous and much sought after by the honeyeaters and many other birds'.

The only report from S. Australia is given by EYLMANN (1908), who declares bees'-honey to be one of the delicacies of the natives which is also appreciated by the white settlers. Missionary SCHULZE observed two species of bees in the area of the Finke River. EYL-MANN observed near Hermannsburg and elsewhere only a small species, similar to our housefly, which nests in high trees. It is common and its nests are easily discovered in the scrub, where trees are scattered and not very high. The collection of the honey, which in some tribes is the privilege of the men, offers no difficulties, as the bee is almost as harmless as the housefly. Before consumption the honey is not separated from the combs and maggots, etc. At the Birdum Creek there was a camp of the Goarangos, where a young man consumed a mixture of honey and water, by alternately dipping a grass-bunch into it and then sucking the grass.

Occasionally the hunter in the bush of Victoria in olden times would see a single busy bee feeding on the flowers near his track. He would adroitly catch this bee and affix to it a particle of down, and follow it until he found its nest (R. BROUGH SMYTH 1878, p. 206).

In New South Wales wild honey is collected by a small stingless bee, not as large as the common housefly. Its honey-nest is generally found at the summit of remarkably high trees. The honey is of delicious flavour, after it has been carefully separated from the comb, the cells of which are generally filled with small flies (sic!). The natives, however, devour it just as they find it, and are very fond even of the refuse comb, with which they make their favourite beverage, the *poull*, and of this they drink until they become quite intoxicated (T. H. BROWN 1846, II, p. 248).

On the north coast of Australia the wild-bee honey is sometimes dissolved in water and drunk. This is nearly always done when the comb which has been obtained in the first place, is mixed with sand and grit, or when the honey is absorbed in the fibres of the collecting bag (BASEDOW 1925, p. 153).

In Queensland there are two varieties of the native bee, both very small. They have no sting and make only a faint hum. One kind, the *killa* of the Kabi is dark; the other, the *kavai* of the Kabi is light greyish and its honey is less esteemed. Their nests are in hollow trees. The natives made a spongy mat out of the inner bark or the best tissue of a tree by beating and chewing it. This they would dip into honey, which it would absorb like a sponge. The method of use was very sociable and economical. Members of the household would take a suck in turn, and after the substance of the honey was exhausted, the flavour would cling to the bark rag and reward the sucker for his exertions (J. MATTHEW 1910, p. 86).

THORET informed M. REVERET-WATTEL (1875, p. 753) that in N. S. Wales the natives actively hunt for the honey of stingless bees, but that by that time wild swarms of the European honey-bee had also spread throughout the forests of Australia. And G. BENNETT (1864) discussing the benefits of animal and plant introduction and acclimatisation of Australia states as his opinion that the honey of the native stingless bees has a more agreable taste than the honey of the introduced *Apis mellifica* L.

The natives near Powell Creek determined the whereabouts of the 'honey-bag' in three ways (B. SPENCER 1928, II, p. 547 f.). The simplest is that of coming by chance across a tree where the bees can be seen flying in and out of the nest. The second is more in-genious. A native will catch a bee and fasten on to its body a small speck of light, white fluff and follow it up as it flies away to its hive. The third, which SPENCER often saw them using, was to place the ear on the trunk or bough of a likely-looking gum-tree, when, if it contained a nest, the low hum of the bees at work could be heard. During the rainy season and in very cold weather the natives say that the bees close the entrance with mud ... When once the 'honey-bag' has been located, it is chopped out. The comb is quite irregular in shape, varying, of course, in different hives. The cells are like little round balls and very much resemble those of the English bumble-bee. However, they are somewhat smaller, the largest being about 0.6 cm. in diameter. Some of them contain honey and some pollen and nectar which has not yet been made into honey. A third kind of cell is much smaller and contains the eggs, maggots and pupae. The whole mass is scooped into a piece of bark. Hundreds of bees get mixed up with the pollen and honey, but the natives do not mind this and eat the whole with relish. The honey itself is excellent.

The importance of the honey-bags for the Kakadu-tribe of N. Australia is easily demonstrated by the number of names given to them. B. SPENCER (1928 II, p. 857) watching the boys hunting for and cutting these honey-bags out of the trees, learned that they have four different names: the *moro* are in the main trunk, some distance up; the *podanerji* right at the foot; the *tjungara* high up on the trunk; and the *tjaina* high up on the boughs, with a short wax funnel projecting from the bough, by which the bees enter.

This importance is further enhanced by the analysis of the bark-drawings made by the Kakadus, which are done in ochre and white on the bark-colour (SPENCER 1928, II, pp. 808 ff., figs. 525, 529, 530, 533). The first illustrates Jerobeni gnomes, the male carrying a basket on his left shoulder in which to collect honey-bags. In the three hunting scenes the animals are always disproportionately larger than the hunting men. In all of them either men or women were obviously hunting for honey, as they carry their *numalka* or dilly bags around their neck as do all honey-hunters, and some of them have their bags more or less filled with *mormo*, i.e. sugar-bags.

In Northern Australia bees are a totem of the Mara tribes.

They have a large heavy stone on the banks of the Barramunda Creek, representing a honeycomb and brought there by the totem ancestors. By ceremonies in which the stone is scraped and the powder blown in all directions, this is turned into bees. (J. G. FRAZER 1910 I, p. 228). Honey is the totem of the Kamilarvi on New South Wales (ibid p. 24).

N. W. THOMAS describes an Australian game based on honey-gathering, in which the children imitate their elders (1906, p. 134): the girls squat on the ground and place their hands, fingers downwards, one above the other, thus representing a tree. The hands are then knocked down from above, imitating the felling of the tree. Then the arm of one girl is made to represent the bough where the honey is located. By the symbolic chopping at her elbow with an axe, the honey is secured and put in a trough of cupped hands, where it is mixed with water. The arm is 'cut off' at the elbow and not at the shoulder, as women are forbidden to obtain honey from the higher parts of the trees.

Sir W. BALDWIN SPENCER, the importance of whose collection of aborigine ethnography in Australia can scarcely be over-emphasized writes about the collecting of honey in North Australia as follows: 'One of the choicest foods of the black fellows is what they call 'sugar-bags', that is the honeycomb of the native bee. This little insect is about 3 mm. long, and when looked at casually is much more like a small fly than a bee. However, the presence of two pairs of well-developed wings at once shows that it is not a fly. It nests in hollow limbs of different kinds of trees – gums, lance woods, etc. – the white ants providing it with plenty of opportunity for building its hives. When taking a spell at any place, such as Karrabobba, we often used to go out wandering in the bush with one or two natives, and they always were on the look-out for opossums and 'sugar-bags' . . . .'

B. SPENCER (The native tribes of Northern Australia, 1914, pp. 301 and 327) also reports the following traditional legend: 'The two brothers Numerji came to a big blood-wood tree on which many bees were feeding on the honey in the gum-tree flowers. The younger brother said, hearing the buzzing: 'What is it, is it bees? Ha! Mormo!' The elder brother was carrying a stone toma-hawk, which he handed to the younger man telling him to go and

Fig. 16. Ghost drawings of northern Australian tribes. All these figures, either the man or the woman, bear a bag for collecting the sugar-bag. The bag is carried: by the men, top left; by the kangaroo hunter, top centre; by the emu hunter, top right. Below: by one of the kangaroo hunters. From Sir B. SPENCER, *Wanderings in Wild Australia*. London. 1928. III, figs. 523, 529, 530, 533.

cut a forked stick. He himself gathered a leaf with a spider cocoon on it, which he shredded out, and having done this, climbed the tree by means of the forked stick which was placed slanting against the trunk. He put a little bit of the web on each bee that he could reach, singing out to them to go home, and at the same time, telling his brother to watch which way they went. They followed them up and put leaves into the holes they entered so that they would know, later on, where the honey-bags were. Some of the bees, on which he put webs, he sent away to distant parts to make honey-bags for the natives there . . . . When he had done this the old man went to the first honey-bag, cut it out with his stone-axe and ate it . . . .'

While this legend indicates the importance of the sugar-bags in the past, as well as some of the ways to find them, the next paragraph describes their importance at the present time. 'The Kakadu-tribe stays in the Alligator Rivers district as long as they can get lily seeds and roots in abundance, fish and wild fowl . . . . All day long the women and children wade up to their necks in the water gathering lily 'tuck-out', while the men spear fish, catch wild fowl or climb the trees after flying-fox and honey-bag. When all these have become rare, they move into another camp, where the same round is gone through day after day and month after month all the year round'.

Other observers also mention how the natives attach a piece of white fluff to a bee in order to trace the forager back to its hive, following it on its flight with their keen eyes, until it reveals the hiding place of the precious store (K. C. McKEOWN 1944, p. 179).

K. L. PARKER (1905, p. 110) in his monograph on the Enahleyi tribe of North-western N. S. Wales reports that ant larvae and frogs, special gifts from some spirit in the stars, are considered to be excellent food by the camp Epicureans. He made many inquiries about honey and bees:

'I had to hear the stingless little native bees humming before I could see them. As to knowing which tree had honey in it, unless I saw the bees, that was quite beyond me, while a mere toddler would point triumphantly to a 'sugar-bag' tree, recognising it as such by the wax on its fork, black before the rain and yellowish afterwards. This honey is good strained but as the blacks get it, it is mixed up with dirty wax and dead bees.

I deplored the sacrifice of the bees one day, but was told it was allright. Whoever had chopped the nest out would take home the waxy stick they had used to help get the honey out. They would throw the stick in the fire, then all the dead bees would go to a paradise in the skies, whence next season they would send Yarragerh Mayrah, the Spring Wind, to blow the flowers open, and then down they would come to earth again'.

R. SEMON (1903, p. 173) mentions that, while exploring along the

Burnett River, he was never certain whether his native scouts were sleeping in the scrub or were looking for 'sugar-bags', the nests of the native stingless bees. The natives lose many working hours through these sugar-bags. Still worse was the occasional discovery of a nest of the European honey-bee. Dr. COLE of Gayndah was an ardent bee-keeper and from his apiary wild European bees had recently spread along the middle Burnett. Conditions there are apparently rather favourable for their existence in a state of nature, while they are unable to maintain themselves without human help in tropical North Queensland. When a tree was discovered containing a nest of wild European bees, often at a height of 10 meters and more above the ground, all the natives joined efforts to cut the giant tree, the work often taking a whole day. In N. E. Australia near Cooktown (ibid p. 297) the beautifully coloured bee-eaters- (*Merops ornatus*) abounded. ASMUS complained that these birds caused many difficulties with bee-keeping. In general, the bees here in the tropics have many more enemies and their honey many more friends, than in Europe or in the temperate regions of Australia. Apart from the many insectivorous birds, among which the bee-eaters excel, the very common spiders and wasps attack them, many lizards hunt them, and the predacious ants rob them. The bee--keeper therefore has always to be on the look-out and has to destroy the nests of ants and of wasps around the apiary, and to shoot the birds, if he is to succeed. ASMUS never saw swarms maintain themselves in a wild state in these tropical areas, as we had observed so commonly along the Burnett.

W. L. PUXLEY (Wanderings in the Queensland Bush, 1923, p. 133), finding *Apis mellifica* in a wild state in N. Queensland, writes as follows: 'Many trees were inhabited by swarms of bees. I was puzzled at first to find that they were of the same species as our English bees, until I heard that numbers of these had 'gone wild' and taken to the bush and had prospered there exceedingly, like so many introduced species. One day I spent with a settler taking some of these bees. He made hives in his spare evenings, and having got several ready and marked the trees, he and a companion would cut down the trees one after the other – no long matter for these hardy men – and would then take the honey and put the bees with a little of their own honey into the hive on the ground.

Sometimes they would try to swarm again, but this he tried to prevent, as it was then necessary to cut down another tree as a rule, and this wasted his time; but I saw him hive three swarms in a day succesfully, and he then left them till the evening, when he moved them to his farm. In this way he is gradually building up a large business, for he sells honey to shops in the towns which has cost him almost nothing in the first instance. I may mention that during the day I spent with him he obtained two large pails full of rich honey excellent in flavour though slightly 'woody', but perfectly clear'.

7. THE LERP-MANNA OF LEAFHOPPERS.

Important sources of sugar for the inhabitants of the Western Desert are the various lerp-insects, which sometimes appear in masses on *Acacia aneura*. The scales of these Psyllid larvae are scraped off and soaked in water to obtain the sugary liquid developedby bruising them. At some seasons the mouths and teeth of the natives are sore and stained with the remains of these insects. On such occasions they have been feeding on them by running the acacia twigs through their lips (N. B. TINDALE).

Apart from bees'-honey and that of the honey-ants, the Australians did not know and use any other sweet substance. Only in various parts of the continent a sweet secretion of various leaf-hoppers was collected from certain trees and either eaten or infused in water and (according to N. W. THOMAS 1906, p. 112) even fermented. The leaf-hoppers or lerp-insects (*Psyllidae*) form an important insect group in Australia, where they take the place of the Aphids. In most of their species the immature stages form a protective scale-like cover of sugary or waxy composition. These lerp-scales are frequently of beautiful and intricate design. The larvae are often found in great and dense colonies on the leaves, stems or twigs of their specific host-plants. The common sugar lerp-insect, the *Spondyliaspis eucalypti* Dob., is widespread in Australia. The scales are white, conical and sugary. The insects often sit in dense colonies on eucalyptus leaves. They fall to the ground beneath the trees and, as 'manna' are much prized by the natives for their sweetness. The natives also used them as food and concocted a sweet drink by steeping them in water (McKEOWN 1944b, p. 106). In years of heavy infestation many eucalyptus trees lose their leaves prematurely and look rather unhealthy, yet real damage is rare (W. W. FROGGATT 1923, p. 9).

In many parts of Central Australia the leaves of the red gums

Fig. 17. Typical Lerp-scales of Psyllid-larvae in Australia. The sugar lerp-insect *Spondyliaspis eucalypti* Dob. (left), *Lasipsylla rotundipennis* Frogg. (centre), and *Eucalyptolyma maideni* Frogg. (right). From K. C. McKEOWN 1944b, p. 105.

(*Eucalyptus rostrata*), growing along the river beds, are covered with lerp-manna – white, conical structures, about the size of a small lentil, which are secreted by the larva of a Psyllid. On account of their sweetish taste large quantities of these cones are collected and eaten. The Arunta refer to manna as *prelja* (BASEDOW 1925, p. 147).

The secretion of Psyllids is found in the interior on twigs and leaves of *Eucalyptus rostrata* and *E. microtheca*. It is shaped like the shell of a *Patella* and is in colour and taste like sugar. When these patella-like protuberances are removed from the food-plant, an coloured insect, 1.5 mm. long, remains. It begins to crawl very actively and upon stimulation secretes a fine white thread, which closely resembles the threads covering the exterior of the patella-like scale. EYLMANN (1908) has no further observations on this insect. Yet BROUGH SMITH in his 'Aborigines of Victoria' reports that in South-eastern Australia a Psyllid-larva produces small, similar formations of white colour on the leaves of eucalyptus. According to Dr. TH. DOBSON from Hobart Town they are called *Psylla eucalypti*. These insects are apparently identical or related to the insects observed by him. This secretion is often found in surprising quantities on the two eucalyptus trees referred to. EYLMANN saw trees of *E. microtheca* where few of the leaves were without it.

This affords many natives of the interior, where these eucalyptus-trees are growing in abundance, the opportunity to collect substantial supplies of this sweet at certain seasons. The scales are small, 1.5 to 2 mm. high, 4 to 5 mm. in diameter, but they are easily removed by teeth and tongue from the leathery leaves. At Kilalpanina in July and August, 1900, the boys of the missionary station often made excursions into Cooper's Creek to delight in the 'manna of the desert'. Probably not only the aborigines of the interior, but also those of the South coast obtain a similar sweet, as Rev. G. TAPLIN in his book on the Narrinyeri mentions that this tribe consumes the manna dropping from the peppermint gum (*Eucalyptus sp.*) in a watery solution.

J. DAWSON (1881, p. 20) writes as follows: 'A sweet substance called *buumbuul* (manna), resembling small pieces of lump sugar, with a fine delicate flavour, which exudes and drops from the leaves and small branches of some kinds of gum trees, is gathered and eaten by the children, or mixed in a wooden vessel with acacia gum dissolved in hot water as a drink. Another kind of manna, also called *buumbuul*, is deposited in considerable quantities by the large dark-coloured cicadas on the stems of white gum trees near the River Hopkins. The natives ascend the trees, and scrape off as much as a bucketful of waxen cells filled with a liquid resembling honey, which they mix with gum dissolved in cold water, and use as a drink. They say that in consequence of the great increase of opossums, caused by the destruction of the wild dog, they never get any *buumbuul* now, as the opossums eat it all'.

LANGLOH PARKER (1905, p. 114) relates one of the legends connected with manna: 'One year the manna just streamed down the Coolabah and Bibbil trees; it ran down like liquid honey, crystallising where it dropped. The old blacks said: 'It is a drought now, but it will be worse. Byamee has sent the manna by the little Dulloorah birds and the black ants, because there will be no flowers for the bees to get honey from, so he has sent the manna. Each time he has done so a great drought has followed'. – And indeed it was followed by one of the worst droughts Australia has ever known. Byamee, it is said, first sent them the manna because their children were crying for honey, of which there was none except in the trees that Byamee, when on earth, had marked for his own. The women murmured that they were not allowed to get this; but the men were firm, and would neither touch it nor let them it, which so pleased Byamee that he sent the manna and he always would when a big drought threatened'.

This tradition about a rich manna-year as fore-runner of a severe drought is widespread in Australian literature. Yet we are unable to decide whether it goes back to one tradition or to several independent ones.

The *intichiuma* ceremony of the Ilpirla or the Manna totem of the Arunta has alo been described by SPENCER and GILLEN (1899, p. 185 f.). The *ilpirna* is a manna, similar to the well known sugar-manna of the gum-trees, but peculiar to the *mulga*-tree (*Acacia aneura*). About five or six miles to the west of Ilyaba is a great boulder rock, curiously marked with black and white seams, where the Ilpirla totem performs its *intichiuma* ceremony. On the top of the boulder, 150 cm. above the ground, there is a similar stone weighing about twenty pounds, together with smaller ones, all of which represent masses of *ilpirla*. The large boulder, on which the others lie, has the same significance and is supposed to have been deposited by an ancestor of the totem.

When the ceremony is performed, a clear space is first of all swept round the base of a stone, and after this the Alatunja digs down into the earth at the base of the boulder, and discloses to view a Churinga which has been buried there from ancestral times, and is supposed to represent a mass of *ilpirla*. Then he climbs on to the top of the boulder and rubs it with the Churinga, after which he takes the smaller stones and with these rubs the same spot, while the other men sitting around sing loudly an invitation to the dust produced by the rubbing of the stones to go out and produce a plentiful supply of *ilpirla* on the mulga-trees. Then with twigs of the mulga he sweeps away the dust which has gathered on the surface of the stone, the idea being to cause it to settle upon the mulga trees and so produce *ilpirla*. When the Alatunya has done this, several of the old men in turn mount the boulder and the same ceremony is repeated. Finally the Churinga is buried at the base in its old position, and with this the ceremony closes.

8. VARIOUS OTHER INSECTS

The insects which we have quoted so far as food of the aborigines of Australia excited the attention of travellers and anthropologists mainly because of their more or less habitual large-scale consumption, and because they were conspicuous over wide areas as an important part of the normal, primitive, native diet at one season or another. There is, however, not the slightest doubt that the total range of insects in their diet was considerably larger than those included in the previous enumeration. Together with all other edible produce of the environment, many more insects were eaten whenever the occasion offered itself. Once many insects were accepted as regular or even delightful food, we have not the slightest reason to doubt that, even if the quantities were small and the supply locally and seasonally limited, their consumption never assumed a sufficient extension to induce their inclusion among the tribal totem animals.

We have found a few further references to other insects in the early travel books and papers which are enormously scattered and have still been only very inadequately utilized. The only astonishing fact is the absence of large-scale and habitual eating of locusts and grasshoppers on the one hand, and of termites on the other. In the case of termites which are found throughout the ever hungry steppes and deserts, we are at a loss to find any explanation at all. With the grasshopper there exists a parallel to the Middle East, which however in the light of the extensive entomophagous habits of the Australians seems not very convincing. In ancient Assyria the big desert locust (*Schistocerca gregaria*) was readily eaten by all and still is in present-day Irak, while the more common local Moroccan locust (*Dociostaurus maroccanus*) is not. It is a smaller species, perhaps with a harder exoskeleton; yet even if they have a mediocre taste, the similar small Australian locusts should still be attractive when appearing in large quantities.

It is appropriate to quote here a deviating opinion of EYLMANN (1908, p. 278 f.), who states expressly that the natives of South Australia eat only a few insects. The main reason is the poverty in edible species of the local insect fauna. In their mainly animal diet, vegetable food was unimportant, being either small in quantity and rare, or unhealthy. During the cyclic droughts vegetable food was very rare and fat game was entirely lacking. Then the natives of the deserts of the interior lived on a diet rich in proteins, but poor in fats and carbohydrates. They lose weight because of the small quantity and the unbalanced composition of their food. The native's diet at this time in general lacks non-proteins, as is illustrated by his constant eagerness to look for fats and carbohydrates. With regard to his animal food the native is by no means selective. He eats every vertebrate and every large mollusc. *Yet many insects are apparently not to his taste.* He even refuses locusts and grasshoppers, which are eaten by so many more civilized and better nourished peoples. That this refusal is not general will be seen from the following references, which, however, do not change the fact that they were obviously not a habitual article of mass-consumption throughout Australia.

SPENCER and GILLEN (1899, p. 771) report that the Maras of the Gulf of Carpentaria ate locusts. LUMHOLTZ (1890, pp. 186, 240) makes a pertinent observation from Queensland: the natives amused themselves by running round and frightening the grasshoppers. The women secured large quantities of them in baskets. In one place a number of natives sat round a fire eating them. First, the contents of the baskets were thrown into the fire in order to burn off the wings and legs, whereupon each grasshopper was roasted individually; they taste like nuts, but there is, of course, very little to eat on them.

Termites and their larvae were eaten, usually only when better food was not available, but were apparently not much appreciated. EYLMANN (1908, p. 179) came more than once to a camp where famine ruled, yet the nearby termite hills were not even touched. He was told by a woman from the Pine Creek that only lubras (old women) occasionally ate termite larvae. The traveller only rarely finds termite hills which have been opened by human hands. McKEOWN (1944, p. 177) also mentions termites as occasional food.

BENNETT (1834) mentions the *galan-galang*, a certain cicada, as being cooked for food in New South Wales, whilst EYLMANN (1908, p. 179) says that near the mission station of Hermannsburg a large cicada is eaten. He also states that the contents of certain walnut-sized galls on eucalyptus-twigs are eagerly eaten in South Australia.

Hyatt Verrill (1938, p. 162) refers to a still more peculiar food in New Zealand, namely the entomophagous fungi developing in large Scarabaeid grubs, which the Maoris eat with great delight.

Lumholtz (1890) is the main source for the eating of lice. He states: 'The natives are not troubled with fleas, but they are full of lice, which are rather large, of a dark colour, and quite different from the common *Pediculus capitis*'. They frequently went astray and came into his quarters, but fortunately did not find there the necessaries of life . . . . 'These insects were also found on the body, and their possessor may constantly be seen hunting them, an occupation which is at the same time a veritable enjoyment for him, for to speak plainly – he eats them. The natives also practice this sport on each other for mutual gratification, and the operation is evidence friendship and politeness (1890, p. 117). If a black man desires to show how glad he is to meet his old friend, he sits down and takes his friend's head into his lap, and begins to look for the countless little animals that annoy the natives, and which they are fond of eating (1890, p. 223). The dingo is an important member of the family. It sleeps in the huts and gets plenty to eat, not only meat, but also fruit. Its master never strikes, but merely threatens it. He caresses it like a child, eats the fleas off it, and then kisses it on the snout'. (1890, p. 178).

9. TASMANIA

The natives of Tasmania or Van Diemen's Land, who were still more primitive than those of the Australian mainland, became extinct shortly after the arrival of the Europeans, so that we know very little about their habits and customs, and the references to insects are scarce.

Davies (1846, vide Noetling 1910, p. 281) stated that the natives of Van Diemen's Land were fond of a large white grub, which most probably is the larva of *Zeuzera eucalypti* (McKeown) found in rotten wood, and that the 'eggs' (read: pupae) of the large ants were considered a delicacy. The latter he identified as probably belonging to *Diamma bicolor*, a mis-identification, as that insect is not an ant, but a Thynnid parasite of mole-crickets (Mc-Keown).

NOETLING (1910, p. 290) identifies three species of ants with the help of A. LEE, namely, *Myrmecia pyriformis*, *Colobopsis grasseri* and *Camponotus consobrinus* and comments: 'The number of ants distinguished by the natives is remarkably large, but the native names are all combinations of the word *tietta* or *taita* with another word. Considering that the pupae of these insects were delicacies, it is hardly surprising that they distinguised such a large number'. The 'white grub', which was also considered a delicacy, was in name distinguished by the Tasmanians from the ordinary *Lepidoptera*. Among the latter they ate the large white caterpillar, about five cm. long, found in rotten wood and in the *Banksia*, and also the 'eggs' of the large ant (DAVIES 1846, p. 414).

The French naturalist LABILLARDIÈRE, during the stay of the 'Recherche' at Tasmania in 1792/3, observed a lady devouring parasites from her own head: 'We observed with disgust, that like most natives, she crunched these filthy insects between her teeth and then swallowed them'. (vide BONWICK 1898, p. 17).

10 NEW GUINEA AND POLYNESIA

VON MIKLUCHO-MACLAY (1875), B. HAGEN (1899) and VAN DER SANDE (1907) agree that the Papuans of New Guinea are practically omnivorous, eating, in addition to vegetable food, every type of animal which is edible and harmless. No aversion appears to derive from prejudices or outward appearance. Animal food, at least that of larger mammals and birds, is a luxury for the common people. BIRO (1899, p. 96) reports, for example, on the consumption of hornets' nests as a delicacy. MIKLUCHO-MACLAY (1875, p. 70) states expressly: 'All insects without exception, especially big beetles, are eaten raw or cooked by the Papuans'.

LABILLARDIÈRE (An VIII vol. II, p. 239 f.) observed the mass-eating of the *Nougi*-spider (*Epeira edulis*) on New Caledonia by children.

G. GARNIER in his Voyage in New Caledonia (vide FAILLA TE-DALDI, 1882) describes how the Kanaks of the island collect roots, 'worms' and grubs of beetles to complete their basic diet. Natives of New Caledonia eat grubs of *Mallodon costata* (vide SIMMONDS, 1885, p. 355).

On Dutch New Guinea big brown beetles, a still undetermined Dynastid, are eaten by the Papuans. LEEFMANS, to whom we owe this information, also heard there about the eating of large cater- pillars. SCHELTEMA (p. 382), referring to the Papuans of Dutch New Guinea, mentions that plants, snakes, and in general everything which runs, flies, creeps or swarms is acceptable as food. D'ALBERTIS (New Guinea I, p. 261) tells how disgusted he was to see a Papuan take off his coronet of cassowary feathers to hunt for lice and devour them.

The polyphagy of the Polynesians is confirmed by H. A. BERNAT- ZIK's studies on Owa Raha (1936, p. 71, 73, 91), one of the British Solomon Islands. The natives still live in the stone age. Fishery and marine products are the main source of their diet. The collection of shellfish and insects is the task of the women and children. Animals eaten are: all edible, non-poisonous fishes, including sharks and eels, cuttle-fish, all kinds of shrimps, prawns and crabs, *Birgo latro* being especially esteemed, marine worms of all sizes, turtles and their eggs, sea-urchins, oysters and other shellfish, domestic fowl, the eggs of which are reserved for the children, wild fowl, especially the large *Megapodus* and their eggs, some pigeons and ducks, cranes and cormorants, pigs, flying fox, dogs, dugong, opossum, whales and dolphins, crocodiles and their eggs, waranes, snails, the grubs of some beetles, especially attractive being those of the big rhinoceros beetles (*Oryctes rhinoceros* L.). The following are not eaten: birds of prey, parrots, cats, bats, lizards, geckoes, snakes, spiders, adult beetles, cicadas, praying mantis and butterflies. The main vegetable food is yam and taro. Fish and insects are roasted on the open fire before consumption.

IV. AFRICA

1. GENERAL INTRODUCTION

A most instructive picture on the food and diet of the primitive savages of Africa has been given by M. BRIAULT (1943): 'We have to imagine that primitive Africa, before the penetration of Arabs and white men, was a continent almost bare of basic foods, especially those of vegetable origin. The only endemic vegetable foods are banana, citronella, gourd, beans and peas, sorghum, millet, durrha, and perhaps taro. The natives of the primeval forest, including the Negrillos, have no plantations at all and live exclusively by hunting and fishing' (p. 82). 'Meat, fish and pastes of caterpillars or big palmworms are wrapped into a large leaf, made supple whilst passing them over the fire. Some salt, some spice, some drops of wild lemon, and a pleasant meal is taken from the hot ashes, but it never lasts long' (Fig. 25, p. 87). 'When game is insufficient or lacking, the worries of revictualling pass to the women. They go out to explore the rivulets, the caves, the ponds, the shrub and the bogs, and return with rats, lizards, even with snakes, occasionally even with a small monkey or porcupine, or with supplies of caterpillars, palmworms, tadpoles or young fish. These dishes of the lean days are stewed with salt, spice and lemon, which are an a priori accompaniment of every native dish. Nevertheless, famines are not rare'. (p. 90).

The famous Swedish traveller A. SPARRMANN, the friend of DE GEER, made many pertinent observations in the three volumes of his Travels to the Cape of Good Hope (1778). In the first volume (pp. 263 ff.) we read: 'As unaccustomed to agriculture as the monkeys, the Bushmen, like them, are obliged to search in hills and mountains for wild roots, seeds and plants to be eaten raw in order to maintain their miserable existence. Occasionally, however, their table contains some other dishes, such as insect larvae, caterpillars,

termites, locusts, snakes and certain spiders. Despite this rich selec-
selection of food they often lack the essentials of life. Famine dries
them out into mere skeletons. Yet in a few weeks one of these
starving Bushmen can grow fat again. Their iron stomachs never
feel over-filled; they absorb everything, and when occasionally they
can no longer retain the food they have devoured, they do not
hesitate to begin immediately to feed again.

S. S. DORNAN (1925, pp. 114 and 128) describes the Kalahari
Bushmen as practically omnivorous, ignoring nothing which is
edible. Hunger compels them to use everything eatable, both animal
and vegetable. They catch and devour mammals, herbivorous and
carnivorous; birds, snakes, iguanas, fish and insects. Practically
every living thing is devoured. All kinds of buck, large and small;
zebras, lions, leopards, hyenas, jackals, porcupines, hares, rats, mice,
birds of all kinds including nestlings, snakes, poisonous and non-
poisonous; tortoises, iguanas, frogs, barbel, locusts, grasshoppers,
flying ants and wild honey, and often the lice of their persons are
eaten. In common with many savage tribes some of them seem to
have a preference for these little parasites. The same had been
noticed in connection with the Hottentots by the travellers at the
Cape . . . . The Bushmen sometimes suffer great privations during
bad seasons. In a time of drought the animals migrate, and they
must follow them or starve. They are then reduced to great straits,
emaciated to a dreadful degree, so that it is a wonder they survive.
Their bodies are fearfully shrunken, so that they look and are
literally bags of bones. Only the toughest and strongest survive, and
there must be great mortality amongst the young and infirm, yet
in such times they are very kind to each other, contrary to what is
commonly believed about them.

The Bushmen live on the honey of the wild bees, which they
smoke out of hollow trees, rocks or holes in the ground. Grasshoppers, crickets and locusts are abundant . . . . The natives turn out
when a swarm alights and do their best to destroy and capture as
many as possible. They dry them in the sun, and use them as food.
The legs and wings are pulled off and they are fried in fat and kept
in bags, and are not at all bad eating. . . . The well-known Bushman's
rice is a species of ants with long bodies and black heads.'

G. TESSMANN in his monograph on the Pangwe of the Cameroons

mentions a number of insect foods. The wild honey of *Apis adansoni* is widely collected and highly esteemed (I, p. 108). A number of taboos exist for the uninitiated and for pregnant women. For the former the larva of the beetle *Angosoma centaurus* F. is forbidden. Both groups are forbidden to eat edible caterpillars from webs, the grubs of *Rhynchophorus phoenicis* F., and termites because of their womb. Free-living caterpillars and palmworms are not forbidden. The larvae of *Odonata* are eaten, yet cause the children to urinate heavily. All other lower animals are taboo. Head-lice are often killed by specially prepared oils, yet often the lice are just picked by hand, crunched and eaten (II, p. 187, 190).

O. BAUMANN (1887, p. 163), referring to certain Bakongo tribes of the Congo, mentions that they eat with great pleasure thick white larvae which live in the decaying leaf-roofs, as well as termites.

DE LISLE (1944) in a study on the beetles of the Cameroons notes that in various regions the natives eat many insects of all orders, to compensate for the lack of proteins in their other food. The eating of termites and locusts is well known. On the markets in the Dschang region he has seen baskets full of *Popillia*-beetles for sale. These gourmets are easily trained into valuable insect collectors (p. 57). He mentions especially *Popillia femoralis* (p. 64).

Prof. TH. MONOD collected a personal observation of A. AYOS on the palmworms, a delicacy of the natives, who distinguish two kinds: the former is found in living palm trees and they are collected when the palms are felled, the other lives in decaying galleries in felled palm trees and these are more highly prized.

The same worms of the raphia palms are the delight of the Badjoués of the Cameroons. H. KOCH (1944) mentions one of their, proverbs: A child which has its father does not eat decaying worms.

A valuable description of many insects eaten *regularly* as food by various tribes in Tanganyika territory comes from W. V. HARRIS (1940 pp. 45 ff.). They form a not unimportant seasonal change in diet and there is little doubt that they are a source of animal protein too often lacking in the native menu. Many also contain fats. As many are eaten raw, they do not lose their vitamins.

Orthoptera. Locusts, such as *Nomadacris septemfasciata*, *Locusta migratoria* and *Schistocerca gregaria*, are most widely eaten in Tanganyika and opportunities for this are unfortunately too frequent. The collecting energy of the natives bears a definite relation to the frequency of locust invasions in a particular district. Each of the three locusts is eaten in turn, adults and hoppers. The wings and hindlegs of the adults are removed before roasting. Fried in butter, they have a mild and not objectionable, flavour, reminiscent of shrimps. Flying locusts and even more hoppers are sun-dried for storage and eaten as a flavouring with porridge. To the west and south of Lake Victoria the green grasshopper *Homorocoryphus vicinus* is eaten in large numbers. Small shepherd boys catch them and impale them on skewers, usually grass-stems. They are eaten fresh or dried. The fat, yellow, giant cricket *Brachytrypes membranaceus*, a common garden pest of roots, is dug out of the ground, where it is not uncommon, and is roasted and eaten as a relish.

Isoptera. Termites (*Acanthotermes spp., Macrotermes spp., Termes spp.*) are eaten in Tanganyika by most living creatures. The larger species are eaten casually, only *Acanthotermes* being widely sought after. In the western parts of the territory dried termites are offered for sale in the local markets during the season. The low mounds of *Acanthotermes* or their emergence holes are covered with a light frame of branches supporting banana leaves, bask cloth or blankets, airtight except at one end with an opening close to the soil about one foot in diameter. On the threshold of this opening a bowl is sunk into the ground. When the termites emerge, usually around 2 p.m., they make for the brillantly lit opening. Bumping each other, few manage to fly streight out. Many drop their wings and fall into the basin, from which the collector removes them to his gourd or petrol tin. They are eaten raw, often alive, and also dried for storing. The bodies of the sexuals are plump with reserve food-stores, to feed them for the next weeks when all their energies are devoted to colony foundation. The queen termite is also sought as food, especially those of *Macrotermes*, or less frequently that of *Termes*. They are sometimes 12 to 13 cm. long, pallid sausage-like egg-producing machines. They are roasted and highly appreciated by most natives.

Lepidoptera. The caterpillars of the wild silk-moth *Anaphe infracta* are sought and eaten in several parts of Tanganyika. They are gregarious and build communal nests of strong yellow silk on the branches of *Bridelia micrantha*. Each nest contains many larvae, which are cooked fresh, or dried and powdered for storage. The

140

large, ugly black-velvet coated Saturniid larva of *Bunaea caffraria*, 10 cm. long with 12 mm. long spines, is searched for by the Matengos near Lake Nyasa each season and collected in large numbers. When fully grown they are eaten after roasting.

Coleoptera. Beetle grubs are not popular. Those of the big coconut beetle are ignored. Only the tree-boring maggots of the weevil *Sipalus aloysii-sabaudiae*, common in abandoned ceara-rubber plantations, is regularly eaten. The old ceara-trees are split open at the proper season and yield a good harvest of grubs which are boiled or roasted.

Hymenoptera. Honey of *Apis adansoni* is in considerable demand, but most natives also have a taste for the grubs in the brood combs. These are usually eaten raw, being chewed up with the comb they share with the honey. At other times they are shaken out and added with honey to the stiff meal porridge, which forms the basis of a native meal.

Diptera. The Lakefly *Chaoboris edulis* of Lakes Victoria and Nyasa is eaten. Clouds of these gnats drift in masses like smoke-screens over the lakes and the natives rush into the clouds, whirling about their heads hemispherical baskets on long handles. As a basket is filled, the contents are squashed by hand into solid masses of dead gnats, which are then moulded into suitably sized cakes, and dried in the sun. *The lakefly is an important food* in the limited areas in which it occurs.

In April when the termites swarm in the Bantu country (Buchner, vide Simmonds, 1885, p. 351), the natives cover their conical hills with a dense matting of banana-leaves, while within this cover vessels with funnel-shaped entrances are placed. In these vessels a large number of white ants, males and females, are caught and roasted on the spot. Buchner found them a great delicacy. Also a big cricket, and a beetle grub from hollow trees are sought for and roasted. But especially a large caterpillar, the *Ugoungoo*, is harvested by the natives like a field crop. It is about 5 cm. long, black with yellow rings, lives in the savanna and is perhaps a *Crepis sp.* Whenever it appears in great numbers the Negroes march out in full force from their villages, camping out for weeks in the wilderness to gather and cure the crop. After the intestines have been pressed out, the caterpillars are dried before the fire and rolled up in packages of fresh leaves. To a civilized taste they are most digusting, the smell reminding one of that of our cabbage *Pieris*. Yet the Bantu refuse to eat reptiles and amphibians even in times of starvation!

BRYDEN (1936, pp. 215, 218) informs us that the deadly arrow poison of the Masarwa Bushmen of the Northern Kalahari desert is composed of the juice of a toxic bulb, snake poison, and the entrails of a noxious caterpillar known as the *n'gwa*. When the long desert drought prevails and game is scarce the lot of the Masarwa becomes much harder. They must, perforce, camp by some permanent waterpit, where their sphere of operations is limited. In these times they subsist upon almost anything they can pick up – snakes, lizards, tortoises, bulbs, roots, caterpillars, and so forth.

Wherever the natives of S. Africa have the opportunity, they boil, roast or grill locusts, which they dry in huge quantities for sale on the markets. Termites cooked in butter, caterpillars grilled or roasted and even spiders are likewise consumed (MOFFAT, SPARR-MANN).

On insects as food in South Africa there is an informative, but not easily accessible early paper by H. VON P. BERENSBURG (1907): 'The natives of South Africa consider the locusts a very welcome food and eat them roasted or dried. The Hottentots and Bushmen welcome the arrival of a swarm, which gives variety to their menu, which, in the arid districts of South-West Africa, is rather a simple one. They prepare also a fat brown soup from their eggs. The larva of a giant click beetle (*Tetralobus flabellicornis*) is considered a delicacy by the Negroes of Central Africa; as is the larva of a South African longicorn (*Malodon downesii*) by the Kaffir, the palmworm by the natives of Madagascar and India . . . . Of moths the fat Saturniid caterpillars of *Gynanica maia*, *Nudaurelia belina* and *Bunaea caffra* are emptied, stuck on small sticks, and roasted over the fire by the natives of Natal.

SPARRMANN counts these caterpillars among the delicacies of a Bushman's diet . . . . Termites afford an abundant supply of food to some African tribes. The Hottentots eat them boiled and raw, as also do some tribes of Central Africa. The Indians in Natal collect the flying insects and devour them after having removed the wings. ANDERSON, on his journey to Lake Ngami, saw the natives make two holes in a termite hill and drive the termites out by smoke from strong-smelling materials. In this way thousands of insects are captured and a paste made of them. Sometimes some flour is added. Honey and honey-combs full of young and old maggots are also eaten by African tribes, and are preferred to the pure honey . . . . When LIVINGSTONE and his party visited Lake Nyassa they were attacked in certain seasons by great swarms of small gnats, which sometimes became so abundant that they congregated inches thick on the bottom of the boats, were captured by the natives, boiled and then baked into cakes about one inch thick, of a dark brown colour and of a taste like caviar'.

'The first cause that led human beings to take to insects as food may have been utter starvation. This was doubtless the case in S.W. Africa, where water is scarce and the soil yields very poor products, and in some parts none at all. The starved creatures were thus compelled to seek sustenance in the form of the insects they came across, and subsequently became fond of this form of nourishment, as affording a change in their daily menu'. We have shown in the introductory chapter that this over-simplified approach by a present-day European is devoid of any real foundation.

This article by BERENSBURG was almost unobtainable, even in London, until Prof. J. C. FAURE of Pretoria, who had recently published some information on bugs eaten in S. Africa, was kind enough to copy the paper for me. He added the following most welcome personal remarks on insects still eaten there in our times. Apart from the large Pentatomid bugs eaten in Eastern Transvaal, FAURE mentions termites. In Namaqualand, in sandy country, nests of *Microhodotermes viator* Latr., are dug out, and the flyer-nymphs are eaten. These look like rice-grains when cooked, and this is supposed to be the origin of the Afrikaans name for termite, namely *rysmier* (rice-ant). In the Transvaal the winged sexuals of large fungus-growing termites like *Macrotermes swaziae* Full. are regularly collected and eaten.

The large Buprestid beetle *Sternocera orissa* Bug. is collected and eaten in the adult stage in the Transvaal. The big Saturniid caterpillars of *Gonimbrasina belina* Westw. and related species are collected and eaten. The ant *Carebara vidua* Sm. with its large females and males has minute workers. The females, possibly also the males, are considered a delicacy by the natives in Transvaal. 'I nearly forgot to mention that waggon-loads of adult migratory locusts are collected, especially if the females are full of eggs, and eaten by natives in South Africa. This applies especially to *Locustana pardalina* Walk. and *Nomadacris septemfasciata* Serv., but I have no doubt *Locusta* and *Schistocerca* would also be eaten'.

2. TERMITES

Throughout practically the whole of tropical Africa termites are such an important addition to the regular diet of the natives that most travellers comment there on (BEQUAERT 1921, p. 194). So anxious are the Azande and Mangbetu of the Uele district to secure these so-called ants, that termite hills are considered by them private property, and during the harvest of these insects fights, often with a fatal outcome, occur between rival claimants. BEQUAERT learned from Mr. LANG, the leader of the American Museum Congo expedition, of an ingenious automatic device by means of which the natives of certain regions visited collect the winged sexuals of the termites their nuptial flights. They tightly enfold the termite mound in several layers of the broad leaves of a Marantaceous wood reed, the interstices soon being closed by the termites, which usually join the inner leaves to the nest. A projecting pocket, built on one side of the leaf cover, serves as a trap; for when the winged termites begin to swarm, they find no egress and finally drop in masses into the pocket from which they are scooped out by the watching negroes. In other instances the nests themselves are dug up to obtain the workers, soldiers and huge, fat queens, which form a dainty tit-bit when broiled over the fire. At Banalia along the Aruwimi River in December 1913, BEQUAERT was rather surprised to find, among many strange articles of food offered for sale by the natives at the weekly market, baskets of dried soldier termites. JUNKER, one of the first white men to reach the Azande country, relates how Chief Ndoruma sought to win his favour by sending him twenty large baskets of termites, each load so heavy that it was all a porter could carry. In this instance the contents made such an excellent oil that a chicken cooked in it tasted as delicious as if fried in butter.

'Most fugitive Hottentots have a thick stick with a pierced stone as its head, weighing two pounds or more, giving the stick more force, when they use it to root out roots or bulbs, or to open the termite hills, *as termites form a great part of their food*. Often I have seen how these fugitive old men exhausted their remaining strength in opening one of these hard hills, often finding them empty in the possession of another animal, which had managed to enter the nest and to devour all the termites and their stores. The Hottentots who collect the swarms of sexuals of *Termes capensis* De Geer within a short time became fat and in good health. They boil the termites in their mud pots or sometimes eat them raw. When I saw that the only son of my host tasted them, I did likewise and I discovered no taste beyond a freshening of the palate' (SPARRMANN, 1778, II, p. 22, 98).

The astonishing account of *Termes bellicosus* given in the Philosophical Transactions of the Royal Society (1781, p. 167 f.) by HENRY SMEATHMANN, based mainly on observations on the Banana Island, includes a good account of termite consumption. He states that not only all kinds of ants, birds and carnivorous reptiles, as well as insects, are on the hunt for them, but the inhabitants of many countries, and particularly of that part of Africa where he was, eat them. 'I have not found the Africans so ingenious in procuring or dressing them (as König describes from India). They are content with a very small part of those which at the time of swarming, or rather of migration, fall into the neighbouring waters, which they skim off with calabashes, bring large kettles full of them to their habitation, and dry them in iron pots over a gentle fire, stirring them about as is usually done in roasting coffee. In that state, without sauce or any other addition, they serve them as delicious food. And they put them by handfuls into their mouths, as we do comfits. I have eaten them dressed this way several times, and think them both delicate, nourishing and wholesome; they are something sweeter, but not so fat and cloying as the maggot of the *Rhynchophorus palmarum*, which is served up at all the luxurious tables of West-India. SPARRMANN says that the Hottentots also eat them . . . I have discoursed with several gentlemen upon the taste of the white ants; and on comparing notes we have always agreed that they are most delicious and delicate eating. One gentleman com-

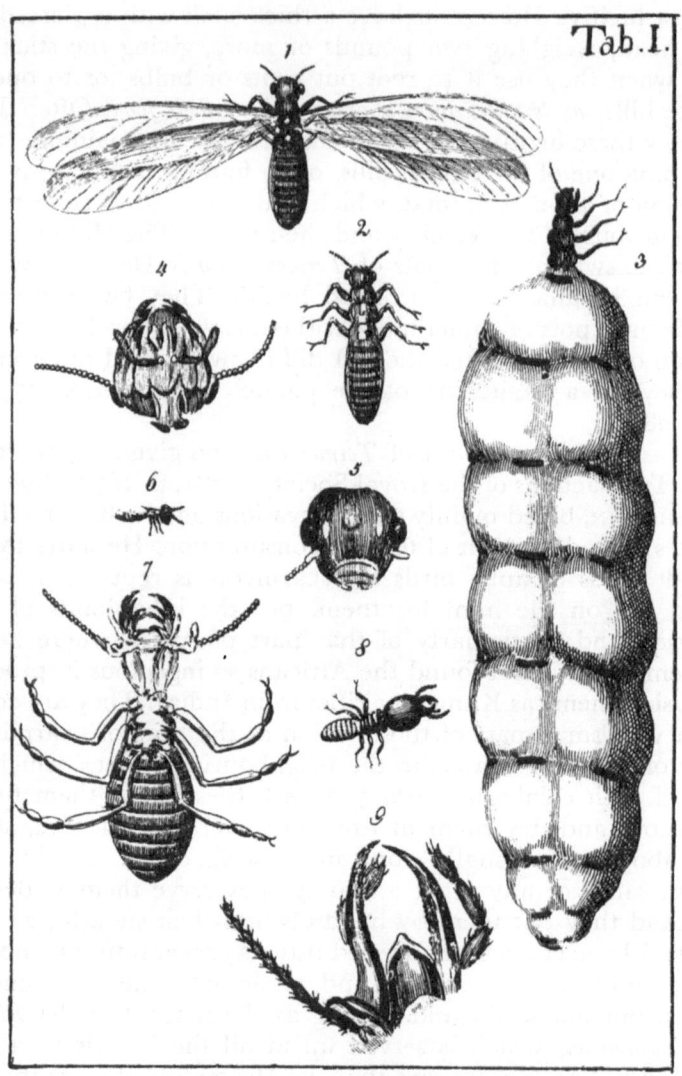

Fig. 18. Stages and castes of *Bellicositermes sp.* from the original paper of HENRY SMEATHMANN on the biology of the African termites.

pared them to sugared marrow, another to sugared cream, and a paste of sweet almonds'.

There are innumerable later records of termites as food in tropical and in South Africa. The most elaborate observations are contained in the monograph of E. HEGH (1922, pp. 669–678) 'Les Termites'. HEGH discusses the utilization of these insects for food, mainly based on his own experience in the Belgian Congo. Termites are esteemed as a delicacy everywhere. Some natives prefer the winged sexuals, others the workers; some decapitate them and eat the abdomen only, others consume the entire insect. Some devour them raw, yet more often they are roasted on heated stones or in the coffee boiler. Their taste, when roasted, is not disagreable to the European palate. The winged termites are caught in water vessels which are put near the termite hill and the sexuals drop into these after their nuptial flight and die in the water. Workers and larvae are chased by fumes from their nest and captured in jars put at the entrance holes. Fig. 19 (p. 671) illustrates a hill of *Acanthotermes spiniger* Sjöst. near Stanleyville, showing how the natives cover the hill with large leaves to prevent the insects flying away. This cover of leaves on the hilltop and its side pocket where the termites accumulate and are collected by the natives, were removed just before the photo was taken.

The following notes are the answers to a questionnaire sent by HEGH to the administration of the various districts of the Belgian Congo. Answers referring to the use of the earth of termite hills for building huts or the use of the hills as stoves have been omitted.

Lower Congo. At Lempa Mpese termites are much eaten. Women prefer the larvae, destroy the fungoid nests and eat their earth (R. P. DEVOS). At Nsonna Mbala all natives eat termites, roasted and hot. The remainder are sold in the markets, where one cup costs 10 centimes, one liter 1 franc. The winged ones are caught towards evening. A brush is held over the exit hole, and they are emptied into a vessel, as soon as they are covered with the insects, which are then roasted. The termites of the fungoid hill nests are collected. In Matadi the winged sexuals are caught by hand only. In the Congo da Lemba termites are rarely eaten.

Central Congo. In the Central Congo and Central Kwitu the winged '*lunsu*' are eaten by everyone; the big queens (*Kake le Suwa*) by the women. The people are strongly geophagous and inciden-

Fig. 19. A hill of *Acanthotermes spiniger* Sjöst. near Stanleyville showing how it is covered by the natives with large leaves to prevent the swarming sexuals from leaving. The cover on the hill-top and the lateral pocket where the termites congregate and are collected by the natives have been removed before the photograph was taken. From HEGH 1921 (Fig. 452).

tally eat also the earth of termite hills. The Lufimi eat termites and catch the sexuals after they have shed their wings.

Lake Leopold II area. On the shores of this lake, as well as at Eala, no tribe eats termites. Yet at Bangala the natives hunt them as food with passion. Before the flights begin the natives go into the forest on a day of heavy rain to clean one or two places at the foot of each termite hill and to dig a groove 55 cm. deep and 30 to 35 cm. broad. When they observe at night that the winged termites open their exit holes, they fix a torch of copal or resin at an angle of 45° to light up the capture groove. Then they proceed to another hill. The swarming sexuals are attracted by the light and drop easily into the groove, from where they are later collected into baskets.

148

For drying they are spread out in a thin layer in the sun on clean soil and later collected for storage.

In some regions of the Central Congo termites are much esteemed and collected at certain seasons. They are a luxury reserved for the chiefs only, who have territorial rights over the termite hills in their area.

The natives of the Lake Leopold II district eat winged sexuals, after having dried them in the sun. They cook them and add pepper; they also extract from them a very good oil tasting like hazelnut. They catch the termites in grooves, 40 cm. deep and 30 cm. wide. When the termites fly by day their catch is nil, yet at night their crop is good. The natives then light a fire of herbs to attract the insects, which drop into the grooves and shed their wings. One termite hill yields up to 4 or 5 baskets full of termites. At Lulonga one species only, called *Mankenena*, is eaten.

Kasai Sankuru. In Sankuru-Kasai the natives build towards the swarming season a small hut of twigs or rushes above the swarming exits of the hills. The insects thus cannot fly away and congregate in the prepared spaces, from where they are collected by the natives. When the termites swarm before these huts are built, the natives assemble and catch them as they come out of their nests. The wings are removed and the animals are then eaten raw, roasted on a fire or dried in the sun. The soldiers also are eagerly eaten. For this purpose a hill is opened and women and boys collect them. Children also introduce a palm leaf into a hole and cry: *'hou! hou!'* as they stir the leaf, whereupon plenty of soldiers seize it and are withdrawn with the leaf from the nest. Termite-hunting is one of the main resources of the Balubas. They have developed an important trade with them which reaches even to the coast. The sundried, winged termites are wrapped in dry *Maranta*-leaves in parcels of 60 to 80 × 15 to 30 cm. which are sold for 1 to 3 francs. In the swarming season the Balubas assemble to sing legendary songs, thus attracting the attention of the inhabitants to the approach of this important season. Similar songs accompany their work during the termites hunt and capture. They also prepare fetishes in hill shape from the earth of the termite hills.

In the districts of Luebo, Luluabourg, etc., the termites are caught in grooves 20 cm. deep, 40 cm. long and 30 cm. wide.

During the evening and night a fire is lit in the hole. The insects come out and their wings are burned. All other parts of the hill are covered with branches or sticks, only one exit hole remaining uncovered. *In December the populations of entire villages sleep by day and are busy by night with the termite hunt.* In certain districts there are proprietory rights over the termite hills. They are sold for one franc the kilo. When smoked, they may be preserved for a long time. They are also eaten raw or mixed with cooked bananas. The natives catch the termites at every season by driving a pole, 175 cm. long, into the foot of the hills. When this hole is one meter deep, a small broom of grass-roots is introduced into it. The native then bends over the hole and produces with his tongue a particular noise, as he says, to imitate the noise of falling rain. He turns the broom slightly and the termites bite into it; then he draws it back and collects the insects one by one to eat them raw or to throw them into a vessel with water.

Low and High Uelé. In the High Uelé termites are much sought after by almost all tribes. Certain kinds of termites are reserved only for the chieftains. They are smoked or well prepared with palm-oil. Some are caught in grooves, 50 cm. deep, dug near a hill and in which a fire is lit during the swarming season. The winged termites are attracted to the fire, which burns their wings. Otherwise the whole hill is covered with branches and plants, and the insects are trapped within this cage. They are collected a little later by hand, after they have shed their wings. Sometimes two pieces of wood are beaten for a long time against each other on the soil of the termite hill. At Uelé Equator the people are very fond of termites fried in palm oil. The natives also suck the earth of the termite hills, which they say contains salt (Antoine P. E.).

Stanleyville. In Stanleyville the termite hills are covered with foliage before the swarming season and the winged insects are collected in a tunnel. At Maniema swarming occurs mainly in the early hours of the day or by night during full moon in August/October, usually after the first rains. The natives are very fond of termites because of the fat they contain at the swarming season and they hunt eagerly for them. For this purpose they build very large cages with sticks and leaves which cover almost the entire hill. Grooves, 40–50 cm. deep, 20–30 cm. wide, are dug between the hill and the cage. The insects leaving the hill push against the walls of the cage and fall in great numbers into the grooves. When these are full, the natives collect them in the morning and roast them slightly before eating them.

In Tanganyika-Moero termites are sometimes caught in their hills. Galleries are driven into the centre of the hill. Then a stick holding at its end a pad of cotton soaked in palm oil is introduced, and on its withdrawal the insects sticking to the pad are collected. The termites are placed in rotang baskets which are severely shaken to deprive them of their wings.

Right Luelaba. The winged termites are eaten at swarming time. The hills are sometimes surrounded before swarming by a layer built of fine sticks, which indicates a possession claim. During the swarming pockets of banana leaves are formed at all exits, in which the insects are collected (R. MAYNE).

Katanga-Lulua. In Munama winged sexuals are collected. The natives are satisfied to collect them from the ground after they have shed their wings. In Kambove termites are eaten raw or fried in palm oil and seasoned with pepper. HEGH tasted these and found them very good. In the swarming season the natives surround the peak of the termite hills with a mat leaving only one opening of about 5 cm. diameter, through which the termites come out. Above this opening they put a bottle-shaped basket, within which the insects are caught. The basket is emptied every two or three minutes during the swarming. To prevent escape the basket is severely shaken, to make the termites dazed and to break their wings. This is a rather primitive procedure and many termites escape. If it were perfected, the results would probably be excellent (CHARLIERS).

In certain regions of Central Africa the earth of termite hills is eaten (E. HEGH 1922, p. 670).

The Meshonos of S. Rhodesia gather swarming termites, mainly *Macrotermes goliath*, at the beginning of the rainy season (N. C. E. MILLER).

F. BRYK (1927) has given us a full description of the termite season at Mt. Elgon, where they are caught for consumption in almost unbelievable quantities. The swarming sexuals form an important food of the natives. BRYK did not share the esteem in which this dish was held, either roasted or raw, as they tasted to him insipid; but the white boys of that region hunt them no less than the Negroes. At 17.30 hours about eight Kitosh boys surround the reddish platform of a nest of *Odontotermes:* some make music, yet most of them are busy with some ceramic work. From loam and water they form a termite trap (Fig. 20). It looks like a retort and is composed of a connecting tube and a vessel, the trap proper. A slightly curved tube, about 15 cm. long, leads from an exit hole in the termite nest to the vessel. At the entrance of the vessel two large elastic leaves are arranged, leaving a small opening for the termites to enter, but not to leave. Only a small portion of the exit holes are closed in this way, and the swarms leaving them are soon the prey of many birds. Throughout this time the musicians

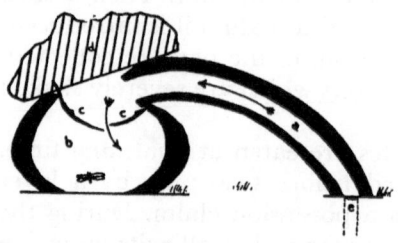

Fig. 20. Schematic cross-section of a primitive termite-trap used in the Mt. Elgon region. e: exit of termite-nest; a: tube of the trap; b: container of the trap; c: leaves between a and b; d: stone covering the trap .(From F. BRYK 1927, p. 1).

continue to beat two sticks against another larger one, thus attempting to imitate the noise of dropping rain. And actually the winged and wingless termites soon begin to leave the nest. During the entire hunt the other boys repair the traps and loosen the covering stone to let some light into the vessel, which they enlarge from time to time. Water is also poured into the exit holes to strengthen the impression of rain. The crop is collected in large jars and bags, for consumption at home. Yet the smallest boys in their impatience put the crawling insects into their mouths and enjoy them. It is night, yet the beating does not yet cease.

At the beginning of the rainy season with the onset of the swarming of the termite sexuals the big termite hunt begins. They are caught in many ways: here water vessels are put near the exit holes of the swarming sexuals, which are dropping into them in masses; the wingless forms are fumigated and gathered in quantities near the exits. There, the termite hills are surrounded by slender sprigs which are set on fire. The termites roast in their own juice, and after half an hour they look like creme caramel. Elsewhere the swarming termites are collected and roasted over fire in pots like coffee (Sir SAMUEL BAKER, vide BRYGOO 1946, p. 25).

An African manna, that is the name quite appropriately given by A. VILLIERS (1947) to the termites. He reports mainly on his observations on the common *Bellicositermes* of the Man region on the Ivory Coast. There the country is sprinkled everywhere with their high hills, often some meters high. Swarming took place in late September through exit holes of about 5 mm. diameter, on a hot and stormy afternoon, when simultaneously a multitude of columns of winged termites rose from all over the forest and united at a height of between 30 to 50 m. into huge whirling clouds of termites, which occasionally obscured the sun. The emigration of these sexuals in dense sequence lasted for a long time, while soldiers and workers patrolled on the ground to protect them against the attacks of ants and other enemies. The sexuals clumsily climb up stones, grasses and shrubs, before taking flight. The Yafobas of Man put conical brooms over the exit holes, which are called *niye* and are made from palm leaves or from low plants. As the sexuals leave their exit holes, they climb slowly up these brooms and are collected from them by handfuls in spite of the bites of the few soldiers who accompany the sexuals up the *niyé*. This termite crop, called *glo* in Yafoba, is gathered by the women and girls. Apart from a few crawling termites which are eaten as they are, the crop is collected in earthern jars containing some water, or in great leaf bags which are sprinkled from time to time with water. In both cases the water is intended to wet the wings, which are shed or remain pasted together and prevent the sexuals from rising for a new flight. Every collector has about six such *niyé*, each of which is put above one exit hole, and she then passes from one *niyé* to another, gathering the available sexuals into the vessels at her disposal. In the evening the termites are consumed in the village raw and alive, or fried. The fried insects taste agreably like shrimp paste. The sexuals which started their flight are heavily preyed upon in the air by swarms of various birds. Yet, says VILLIERS, in spite of all these decimations many new colonies of *Bellicositermes* are established annually, and Africa will remain a continent where termites are a dominant element, to the great pleasure of entomologists, the native gourmets and these birds.

Fig. 21. Catching the swarming sexuals of *Bellicositermes sp.* (Man, Ivory Coast). The raphia-trap is put over the exit holes of the termite hill. As the insects leave, they climb up and when sufficient termites are on the raphia, a woman brushes them off, before they have time to fly away. Courtesy of I.F.A.N.

The Hottentots, when their corn is consumed and when they are reduced to the necessity, eat both the flying white ants and their 'pupae'. These they call 'rice' as they resemble the rice grains, and are usually washed and cooked with a little water. Thus they are quite palatable. When the people find them in abundance, they soon become fat upon them, even when previously much reduced by hunger. A large nest will sometimes yield a bushel of 'pupae' (J. BACKHOUSE, 1844, p. 584).

D. LIVINGSTONE (1857, p. 464) gives an interesting account of white ants: 'In spring at Kolobeng (Angola) white ants swarm in the evenings by thousands. A stream of them is seen to rush out of a hole, and after flying one or two hundred yards they descend; and if they light upon a piece of soil proper for the commencement of a new colony, they bend up their tails, unhook their wings, and, leaving them on the surface, quickly begin their mining operations .... Nothing can exceed the eagerness with wich at the proper time they rush out from their birth-place. Occasionally this occurs in a house, and then in order to prevent every corner from being filled with them. I have seen a fire placed over the orifice; but they hesitate not even to pass through the fire. While swarming they appear like snow-flakes floating about in the air, and dogs, cats, hawks and almost every bird may be seen busily devouring them. The natives too profit by the occasion, and actively collect them for food, they being about 12 mm. long, as thick as a crowquill, and very fat. When roasted they are said to be good, and somewhat resemble grains of boiled rice. An idea may be formed of this dish by what once occurred on the banks of the Zonga. The Bayeiye chief Palani visiting us while eating, I gave him a piece of bread and preserved apricots; and as he seemed to relish it much, I asked him if he had any food equal to that in his country. 'Ah – said he – did you ever taste white ants'? As I never had, he replied 'Well if you had, you never could have desired to eat anything better'. The general way of catching them is to dig into their hill, and wait till all the builders come forth to repair the damage. Then brush them off quickly into a vessel, as the ant-eater does into his mouth.'

JUNKER (1891, p. 338) writes as follows of termites: '25th February. One morning on leaving my hut, I saw a somewhat yellow-

Fig. 22. Collecting winged termites at Man, Ivory Coast. Above: Girl returning from termite gathering. Below: The termite crop is put into a basket and then into a container with water, where they die and shed their wings. Courtoisy of I.F.A.N.

ish-white mass thrown up like little mole-hills in many places that had been cleared of the grass. It resembled greatly fresh weycheese passed through a sieve, but contained some hard round white grains like tapioca. This was the product of some termite . . . . My servants enclosed these places with bits of stick and covered them with foliage, assuring me that something good to eat would soon appear. Then followed the most singular phase: The upper surface of the little heaps, which had been steadily enlarged by accumulations from within, appeared in a few hours overgrown with tiny white mushrooms, distinctly formed, though scarcelʎ 1 mm. in size. These little fungi now began to shoot up on graceful slendeo stems, and next day were already 2 to 5 cm. high and were then eaten as dainty morsels by the natives, ever on the lookout for eatables. I had a dish of them prepared for myself and found them excellent. After their removal the little heaps began to shrink, and at last crumbled away to dust.

With the commencement of the rainy season the natives also begin to gather certain termites. Weeks before the 'harvest' the people mark off those nests which seem most suitable for their purpose. Here they dig a round hole a foot wide and several feet deep, whereby the place is at the same time set apart for a certain person and left untouched by the others. Like everybody else, my servants had done this, and also prepared a quantity of long bundles of dry grass, which here take the place of the resinous torches elsewhere in use. Rainy days and excessive moisture are unfavourable conditions for the appearance of the termites, which may be safely expected on fine evenings following sunny days. Then the people may everywhere be seen with their flaming brands squatting down each at the hole, which he had dug at the foot of the hill reserved for him. The female termites creeping out go straight to the fire without actually rising on the wing; others soar into the air, but also partly wheel round towards the light, while the rest fly away. Those approaching the hole are all swept in with tufts of foliage. Many lose their wings, and most of them are in a dazed state, so that they are afterwards easily transferred to baskets, sacks or pots. The termites for the most part take wing during several successive days, or else at intervals in bad weather'.

H. Noyes (1937, p. 228 f.) compares the welcome of the rainy

Fig. 23. Termite harvest in Central Africa. From
JUNKER 1891.

season with its termite flights in Central Africa to the hailing of the
advent of the oyster season by the British gourmets. The Baganda
like the winged sexuals, alive. When these leave the nests, a contin-
gency which has long been foreseen, they often rise only to collide
with a sheet of bark-cloth spread over the summit of the termite hill
by the natives. The impact breaks off their wings at the sutures
and they fall to the ground within the curtain in white, struggling
masses; their wings are swept aside by human hands, when they
are sifted out from the cloth. Men and women scoop them up in
handfuls, eating a few occasionally, savouring the flavour; naked
children, shrieking with delight, vie with all the birds of the neigh-

bourhood, wild or tame, in chasing and collecting stragglers, munching as they run, stuffing themselves to repletion, heedless of the acute diarrhoea which will presently disorganise their interiors.

CASATI was told by the natives of the Equatorial province of the Sudan, that the alates emerge from their homes on hearing the song: *'Anyeku me kotu'* (Come out in numbers like raindrops). The beating together of sticks and the stamping of feet near likely outlets were also recommended as incentives to urge the insects to come and be killed. An English-speaking Baganda mission-boy informed NOYES that by means of a watering-pot or a perforated petrol-tin, he could persuade the alates to leave their nests at any time of the year. When NOYES pointed out that they could not fly without wings, he protested his superiority. The Baganda consider a compost of termites mixed with maize or millet flour to be an appetising dish worthy of any king. Native epicures, however, since the necessary pans were introduced into the country, prefer them fried.

H. BARTH (1857 III, p. 4) observed for the first time in August, 1851, at Kukawa, rainfall unaccompanied by a storm. The watery element disturbed the luxurious existence of the *'kanam galgalma'*, the large termites. One day later they all of a sudden disappeared from the ground, and filled the air as short-lived winged creatures, the *tsutsu*, and when fried, are used as food.

JOYEUX and SICÉ, in their 'Précis de Médecine Coloniale', report having seen earth-eaters in tropical Africa eating the soil of termite nests, especially that from the deep galeries near the royal chamber. Balls of that earth have a sweetish, not very agreable flavour. H. PETENOSTRA (1927) mentions that the peoples of Djallon water the entrance holes of the termite hills to induce these insects to come out before the onset of the rains. H. JAQUES-RÉLIX (1948) observed that the Bamilékés of the Cameroons cover the termite hills with a roof of stubble in order to collect them when the swarming begins. The insects are worked into the sauce *'na'*, which seasons the basic yam meal.

CAMERON (Across Africa, vide SIMMONDS 1885, p. 369) states that dried white ants are eaten by the natives with porridge as a relish, on account of the scarcity of animal food. They are caught by building a framework of twigs above the hills, covered with leaves cleverly fastened together by stitching the mid-rib of each into the one above it. A small entrance is left at the bottom and beneath it a groove is dug, 60 cm. deep, 30 cm. broad. The termites thus caught are collected in the morning by the natives.

E. DAGUIN (1900, p. 18) remarks quite properly that most travellers in Africa mention the taste of the Negroes for termites. Major S. PINTO comments on a true passion of the Bihenos for these insects, who eat them raw after having destroyed their nests. Sir S. W. BAKER notes from Central Africa that, fried in butter, they are considered a very delicate meal. He himself found that they have rather a good taste, with a light flavour of burned plums.

FLADUNG (1924, p. 7) writes: 'The natives of the Central Lakes region of Africa, when short of tobacco, chew as a substitute the clay of termite hills, which they call 'sweet earth'. The queens are regarded as very fortifying'.

J. D'AGUILAR (1941, p. 116) comments: 'The natives of the Black continent cook them until they turn brown and eat the termites by the handful without any seasoning'.

E. BRYGOO (1946, p. 27) concludes: 'Thus it is well demonstrated that at least in Africa the alimentary role of the termites surpasses by far that of a mere curiosity. Even if one cannot speak of a civilisation of termites, they play a very important role in the life of many tribes. They are the origin of many strictly codified customs and may even determine a rhythm of life of these tribes'.

3. LOCUSTS

While locust invasions were and are a plague for all agricultural peoples, they are a 'manna' for primitive food-gatherers all over Africa. In this continent locusts may swarm almost anywhere at any time, appearing occasionally more locally, at other times over wide areas. The most important species both as a locust and as human food is the desert locust (*Schistocerca gregaria Forsk*), which has its home in the low savannas throughout the Sudan. The desert locust periodically swarms from there all over Africa. The tropical locust (*Locusta migratoria migratoroides* R. and L.) breeds mainly in swampy areas in the forest regions. In the thirties of this century a small, flooded area near the bank of the Niger was the focus for an out-break which within less than two years filled all Africa south of the Sahara with huge swarms of that species. It is also endemic in Madagascar. In South Africa two additional locusts breed: the brown locust (*Locustana pardalina* Wlk.) and the red locust (*Nomad-acris septemfasciata* Serv.). All these locusts formed and form a staple food of all the food-gathering tribes, and were eaten by the primitive nomads and agriculturists of tropical and S. Africa. If reports are most abundant about South Africa, this is mainly due to the early settlement of Europeans and especially of missionaries, who lived among the natives.

SPARRMANN, an early Swedish traveller (1778 II, p. 99) writes as follows: Locusts are another occasional gift of providence to the savage Hottentots, even in the most distant districts. They may appear after an interval of 8, 10, 15, 20 years or longer in innumer-able swarms, arriving from the North and migrating to the South. No obstacle can hold up their migrations, which always follow this direction. They always perish in the Sea, whenever they wish to cross it. The female locusts, which the Hottentots prefer to eat, are less fit for these migrations. They appear rather fat. Some people confirm these facts and mention that the arrival of locust swarms made the Hottentots extremely happy. The locusts devour many plants, yet the natives take ample revenge for this destruction of their vegetables. They devour them in such quantities, that they grow fat within a few days. The Hottentots explain locust years as follows: A master magician in the far north lifts a stone covering a deep trench. From this trench the locust swarms come out to become their, the Hottentots', food. SPARRMANN, however, adds that it is difficult to believe that nature has created the locusts only as food for the Hottentots, yet he is unable to solve the question of their true purpose. 'May they be needed to clean, like fire, the fields from the bad weeds?'

It is not easy to imagine that ugly insect like the locusts may serve as human food, although this is nevertheless observed among various peoples. They even use different ways to prepare them for consump-tion. Some pound them and boil them with milk, others merely roast them over coals and find them excellent. De gustibus non est dis-putandum! But ADANSON (1757, p. 88 f.) gladly left the greatest swarms of locusts to the Negroes of Gambia in exchange for the smallest of their fish.

LIVINGSTONE (vide SIMMONDS 1885, p. 357) claims that locusts are a real blessing to the country. When reduced to flour and mingled with a little salt, they afford palatable food, which keeps unimpaired for months. Boiled, he found locusts disagreable, but roasted they had a vegetable flavour, and on the whole he preferred them to shrimps. A. COLE' (e.l. p. 351) is one of the few with something unfavourable to report. He saw a whole kraal of Kaffirs once die after having consumed an unusual quantity of locusts.

J. BEQUAERT (1921, p. 193) confirms that in many poor regions of S. Africa flights of locusts are such a blessing, that the medicine man sometimes promises to bring them, instead of rain, by his incantations. He quotes Dr. SPARRMANN's account of how greatly the Hottentots rejoice at the arrival of locust swarms. The grateful natives collect and consume this provision so appreciatively, that within a few days they grow visibly fatter and appear to be in a much better state of health. Mainly the female locusts are eaten, especially just before their migratory flight, at a time when their wings are short and their bodies heavy and distended with eggs.

H. SALT (1816 I, p. 222; II, p. 371; pl. 32) describes the great damage caused in Abyssinia by locusts. He states that among the nomads, Dankali as well as Yemenite Arabs, they are a ccmmon food. W. G. BROWNE (1800 II, p. 32) makes similar statements for Dar-Fur.

R. G. CUMMING (1850, p. 69) one April in S. Africa passed through a swarm of locusts, resting for the night on the grass and bushes. They lay so thick that the waggons could have been filled with them in a very short time and were covering the large bushes just as a swarm of young bees covers the branch on which it settles. Locusts afford fattening and wholesome food to men, birds and all sorts of beasts; cows and horses, lions, jackals, hyenas, antelopes, elephants, etc. devour them. He met a party of Batlapis carrying heavy loads of them on their backs. His hungry dogs made a fine feast upon them. The locusts were roasted as food for his men as well as for the dogs.

H. BARTH (1857 II, p. 30) mentions that on the afternoon of the 17th January 1851 he strolled a long time about the market in the Sudan and was not a little astonished to see whole calabashes filled with roasted locusts. *They occasionally form a considerable part of the food* of the natives, particularly if their grain has been destroyed by this plague, as they can then enjoy not only the agreable flavour of the dish, but also take a pleasant revenge on the ravagers of their fields.

The Hill Damaras collect locusts by lighting fires in the direct path of the devouring swarms. In roasting, the wings and legs crisp up and are separated; the bodies are then eaten fresh or dried in hot ashes, and put away for future use (P. L. SIMMONDS 1885, p. 361).

F. LE VAILLANT (1931) writes on the 2nd of January, 1782, from South Africa: 'Joy showed itself suddenly on all faces when a cloud of advancing locusts was sighted, composed of millions of these insects. They passed not much above our heads on a front of almost 1000 m. continuing for over an hour in such a dense stream that they did fall like hail upon us. Those of my men who were accustomed to wild life, enjoyed them and boasted so much about the excellence of this manna, that I ceded to the temptation to eat them. My prejudices were certainly stronger than any real cause for aversion, as I could not detect any disagreable flavour, and they actually taste like the yellow of a boiled egg (I, pp. 202,8). The Bushmen also dried locusts on mats, removing the wings and legs. As these locust had begun to ferment they had a very bad smell. The Namaquis had as their sole provisions some pieces of dried meat and a sack of dried locusts' (II, p. 100).

E. BRYGOO (1946, p. 36, vide ANDERSON) mentions that the Bushmen light big fires when locust swarms are passing and collect the insects, the wings of which are burned. These are arranged on a stalk of straw and fried, or boiled and dried in the sun, and then winnowed to separate the legs and wings, or worked into a paste which is eaten cold. They are also stored as flour. The Hottentots boil a soup from the locust eggs. CAMERON describes how the natives fell even good-sized trees to gather the locusts sitting on them. All early travellers mention the locust-eating habit of the natives of South Africa, many of them devouring these insects in such quantity when the opportunity offers itself, that they fatten conspicuously within a few days. And R. MOFFAT (1843) states expressly that without the locusts hundreds of families would die of hunger. He also refers to a species with red wings which is regarded as inedible. G. KUNCKEL d'Herculais (1893) reports that in Togoland in 1892

King Kuma was obliged to forbid the locust hunt, as it caused the natives to neglect their fields.

PRINGLE (1851) in his song of the wild Bushman, lets him sing:

'Yea, even the wasting locust-swarm,
Which mighty nations dread,
To me nor terror brings nor harm;
I make of them my bread.'

The missionary R. MOFFAT (1842, p. 448 f.) says that the natives of S. Africa seize every opportunity of gathering locusts, when this can be done at night. Whenever a swarm alights at a place not very distant from the village, the inhabitants turn out with sacks and often with pack-oxen and gather loads, returning next day with millions. The locusts are then prepared for eating by simple boiling, or rather steaming, as they are put into a large pot with a little water, and tightly covered. After boiling for a short time, they are taken out and spread on mats in the sun to dry, when they are winnowed, somewhat like corn, to clear them of their legs and wings; and when perfectly dry are put into sacks, or laid on the floor of the house in a heap. The natives eat them whole, adding a little salt when they can obtain it, or pound them in a wooden mortar, and when they have reduced them to something like flour, they mix them with a little water and make a cold stir-about. When locusts abound, the natives become quite fat, and would even reward any old lady who would say that she had coaxed them to alight within their reach. MOFFAT thinks the locusts not bad food and, when well fed, almost as good as shrimps.

The Korannas and Bushmen of the Cape save the locusts in great quantities, and grind them between two stones into a kind of meal, which they mix with fat and grease and bake in cakes. Upon this fare they live for months together, and chatter with the greatest joy as soon as the locusts are approaching (FLEMING 1853, p. 80).

D. and C. LIVINGSTONE (Le Zambesi et ses affluents, vide DAGUIN 1900, p. 25) compare the taste of locusts to that of caviar.

Locust gathering by night is occasionally attended with danger. Thus ANDERSON (1856, p. 284) relates how on such occasions people have been bitten by venemous snakes.

164

Fig. 24. Collecting locusts for food in Central Africa.
From JUNKER 1891.

W. JUNKER writes about locusts as follows (1891, p. 278): 'My people were glad to join the natives of an evening, when they went with lighted torches locust gathering. This was done (near Ndoruma, in Central Africa) not through any absolute want of food, but because of their liking both for locusts and termites, which however do not by any means form a common article of their diet. In C. Africa I only once saw a large harvest, though in Tunis it was of frequent occurrence. I soon overcame my repugnance to such fare, which in fact I found very palatable. The insects were very fat, and when roasted without wings and legs looked like little fish or shrimps' (p. 277 fig. Locust gathering).

4. HONEY-HUNTING AND HIVE-KEEPING

Honey-hunting is widespread throughout Africa, being most important for the food-gatherers; it still plays an important role with the hunters and nomads, less so with the primitive agriculturists. The Massai-warriors took with them on their long expeditions no other food for days except honey. The riches in bees and honey of the African savanna are mentioned by many early and modern travellers, as are also those of Abyssinia and the Kilimandjaro-region. Apart from honey-hunting, we find that hive-keeping is widespread throughout Africa; this is charcterized by the offering of artificial hives to swarming bees. In most cases little care was taken to leave a number of combs for the maintenance of the bee-population. Usually, when the season for the honey-crop approached, all combs were removed from the hive. The empty hive was then offered in the coming season to other swarms. It is most interesting that almost everywhere the honey-hunt and the putting out of empty hives to attract the bee-swarms was restricted to the men. The hunter and the food-gatherer found in the pure sugar of the honey a food which was very adequate to their great physical strain.

The following wild bees are found in Africa:

A. The genus *Apis* is represented by one species only, the African bee, *Apis unicolor* Latr., the main varieties of which are:

Apis unicolor unicolor Latr.　On Madagascar, Bourbon, Mauritius.
Apis unicolor adansoni Latr.　Over the mainland of Africa south of the Sahara.
　Most other names are apparently synonyms of this form.
Apis unicolor fasciata Latr.　In Egyt and S. E. Arabia.

B. Stingless bees of the genera *Trigona* and *Melipona*, usually of the size of a housefly, dark brown to blackish. They nest in the soil, among rocks, in trees and in artificial hives. When nesting on the ground, deserted burrows of termites, wart-hogs or ant-bears are the most favoured sites.

To the pygmies honey is of paramount importance as food, and the Negroes love it. The usual tools of the honey-hunters are ropes, baskets or bags and often a chopping axe.

C. SEYFFERT (1930) has devoted an extensive study to The Honey-bee and Honey in the life of the Africans, where much additional literature will be found. The following survey is based on his study.

Special wooden implements are used by the Hottentots and the Bushmen, especially hooks for removing combs from trees. The honey-hunters make long journeys and even some special 'honey-hunter paths' are mentioned. In the Cameroons wild bees are used as divine judgement, by which people who have sworn a false oath

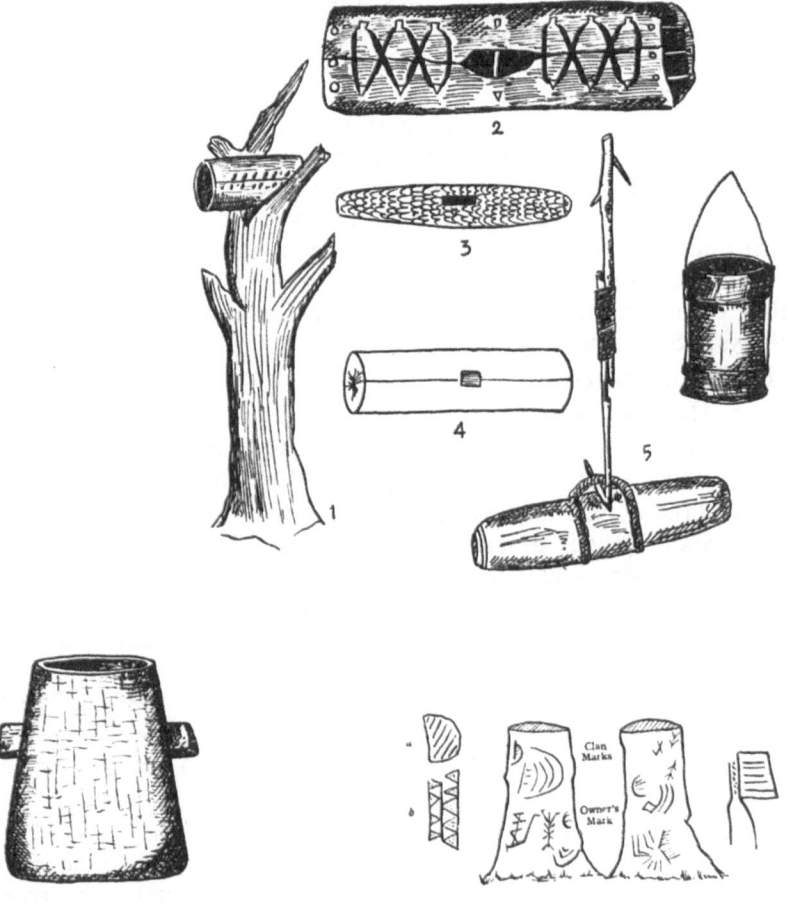

Fig. 25. Various African hives. 1. Bark hive cylinder from Uganda (p. 55, fig. 12). 2. Bark cylinder-hive with pegs and band ornamentation of the Luchatse (p. 53, fig. 11). 3. Cigar-shaped plated bamboo-hive of the Mittu and Bongo (p. 60, fig. 14). 4. Specially constructed, cylindrical trunk-hive from the coast of East Africa. 5. Wanderobbo-hive with special arrangement for suspension. (SEYFFERT, 1930). Below, at left: Vessel for the preparation of honey-beer (mead) 1/10 nat. size. From M. MERKER, Die Masai, Berlin, 1910, Fig. 7, p. 35; Fig. 11, p. 38. Below, at right: Kikuyu bee-hive marks on trees. The pictographs indicate: a. A bee-warm settling on a tree. b. Honey. c. A bee-hive maker. In the centre proprietary marks (from C. W. HOBLEY).

are killed (p. 30). The pygmies take the honey only by night, usually without taking care to leave some combs for the maintenance of the bees. In many regions of Africa the honey season is rather short and although there are many bees'-nests, no flying bees are seen for much of the time. This may be partly a consequence of such maltreatment. It is reported that the Bushmen and the Massai make property marks on trees containing bees' nests by scratching a sign into the bark or by driving in wooden pegs.

In addition to honey-hunting, they also have certain methods of keeping semi-wild bees. This is not yet bee-keeping proper, but often comes rather close to it. The artificial hives in use are, as follows: (SEYFFERT, 1930).

I. Wooden hives are not rare.

    1. Hollow tree-trunks with a bee's nest are cut from the tree and taken to the village (Abyssinia, etc.). Also nests in hollow bamboo-reeds are taken in the same way.

    2. Tree-trunks are artificially hollowed for use as hives. GUTMANN (1922) describes them from the Djaggas as being 1 m. long with walls 30 cm. in diameter and 2–3 cm. thick, and with two round covers with each containing an entrance opening. This is the size for *Apis*, whilst the hives for *Melipona* measure 50 × 20 cm. The application of rituals is needed (conjurations, sacrifices in front of the tree) before the hive is made. The preparation of similar tree-hives by the Wanderobbo (p. 47, fig. 6) 100–130 × 30 cm. is described by MERKER.

    3. Divided tree-hives.

        a. Composed of the upper half of the cut trunk (p. 50, fig. 9) 150 × 40 cm. in East Africa (H. KRAUSS). The tree is first cut longitudinally, each half is excavated and both are then bound together.

        b. Barrels composed of planks. Cylindrical, often packed into hay for protection by the Gallas (CECCHI). Here as well as in other types there is a special arrangement to separate the hive transversly into two halves by a special wooden movable plank so that the honey can be removed.

        c. Box-hives made of *Ferula*-stems in Morocco by the Berbers.

II. Bark-hives. Mainly in the area of the south-equatorial watershed, between the sources of the Congo and the Zambesi. Cylinders 150 cm. in length and 40 cm. in diameter are in common use (LIVINGSTONE 1857, p. 284 and pp. 53, 55, fig. 11, 12).

III. Clay hives.
    1. Jars: mainly in Togo, also in Abyssinia. Relatively restricted in their use.
    2. Pipes: mainly in Upper Egypt and in the Atlas (p. 57 and fig. 13); 200 × 20 cm.

IV. Walled up hives. Very rare bee-towers at Habé, Bambuk, etc.

V. Dung-hives are very rare (Abyssinia, etc.).

VI. Plaited hives. Rare but widespread, usually hung up, often cigar-shaped.

These hives are usually put into the branchings of trees or are hung from tree-branches by a rope, hanging free in the air, high up and as far distant from the trunk as possible, baobab-trees being preferred. In S. Tunisia and in Nigeria pipes are built into walls, and ALVAREZ (1676) reports this from Abyssinia. The position of the hive is almost invariable horizontal. This suggests that the hollow branch rather than the hollow trunk served as model. Where trees are rare, the hives may be hung from high poles (Adamaua) or on a special roof (Ruindi). Occasionally the hives are merely protected against the sun by roofs or covers of hay, which also give protection against rains and heavy dew (RÜPPELL).

The catching of a swarm is only practised as an exception. The number of swarms is so great, that hives offered are usually speedily occupied. A few tribes smear the empty hives with honey, aromatic herbs, etc. in order to attract free swarms. Only in Abyssinia, among the Gallas and a few other tribes, swarms are actively hunted and beaten into the new hives. Conjurations often help in the alluring of free swarms.

SEYFFERT (1930, pp. 71 ff.) discusses in detail the quality, colour and taste of the various varieties of honey, and very extensively the various uses of honey. Honey drinks are apparently the oldest alcoholic drinks of early man. The mead is prepared by the addition of water, the mixture being left for fermentation. Such a preparation may easily have occurred in an almost empty honey jar, into which rain water entered, provided the vessel remained standing for some time. The African honey drinks are:

1. Freshly prepared honey-water, probably derived from the Arabic sherbet.
2. Mead or hydromel was offered to Egyptian sailors by the King of Punt.

SEYFFERT regards it as a Hamitic invention, the *tetsh*. The *tetsh* is consumed by the nobles after a meal, it is highly intoxicating and is usually prepared in the homes. It can be preserved for years, or by distillation may be transformed into brandy.

Honey is also added to reinforce other alcoholic drinks.

Honey and wax are old forms of tribute or gifts, Merker reports from ancient Massai tradition that 25 pots of honey and 40 bee-hives formed the essential part of the nuptial price.

Poisonous honey is repeatedly observed; for example, in Nubia, the honey and mead from the big Euphorbias. Often the honey of the stingless bees is slightly poisonous (pp. 77 ff.).

One of the most terrible punishments of the ancient Germans was to fetter the naked criminal, to smear him with honey, or to dig him into the soil up to his neck, to smear his head with honey, and to expose him to the burning sun and to the stings of bees, until he died (cf. Hittite Code). In Africa the same punishment was found in the south (Latuka, Bornu), while in the S. Sennar and with the Hammedz criminals are roasted in honey (SEYFFERT 1930, p. 167).

SPARRMANN (1778 III, pp. 59–78) writes as follows of bees in Africa: 'Bees often make their nests in the earths of various small mammals. The ratel, their natural enemy and an important visitor to their habitations, has a peculiar manner for discovering and attacking them. Its long claws, which it uses to burrow into the soil help it also to undermine the bees'-nests. The evening is the most convenient time for this purpose. It sits down, holding one hand before its eyes to protect them against strong light, and to see better. When it thus looks around, it sees some bees flying and knows that these bees are returning to their nest, and follows them. In addition, the ratel, in common with the Hottentots, the Kaffirs and the peasants of Africa, has the wisdom to follow a small bird, flying from place to place and calling *cherr, cherr, cherr*, leading those who follow it to a bees'-nest. This small traitor who for his personal interest delivers the bees to their enemies in this way, and thanks to whom I myself found honey in the desert, is the *Indicator indicator*. Bees'-nests on trees have nothing to fear from the ratel. However, when this bird bites a man following it in his foot, it is a sure sign that bees are in the tree. Everybody told the same story.

The honey guide is neither large not beautiful, looking at first sight like a sparrow. Being very fond of honey, this bird guides man and ratel to the bees'-nests. Everytime a nest is taken out, a little honey is left for its benefit, as recompense for its services. Its appetite is awakened mainly mornings and evenings, when it attracts the attention of its allies by its *cherr, cherr, cherr*. It is rare if neither one nor the other appears on the spot where the bird calls. Without interrupting its calling the honey-guide then flies slowly from spot to spot towards the bees'-nest. The followers must neither make much noise nor be in large numbers. One of my Bushmen answered its call by a very sweet whistling in acknowledgement. When the nests are distant, the flights are longer and the bird waits at intervals for its follower and encourages him by new calls. With the approach to the nest the intervals grow shorter and the calls more frequent and stronger. Sometimes the bird even returns and calls impatiently, when it is followed too slowly. When it has arrived at the nest, in a tree or in the ground, it hovers above its entrance for a few seconds, as I have seen twice, and then sits in silence in a tree or shrub nearby, awaiting its bounty. When the bird stops calling, this is a sure sign that a nest is in the neighbourhood. When the Hottentots have plundered a nest with the help of this bird, they usually leave a fair portion with eggs and maggots which the bird appreciates, while they themselves do not. A professional honey-hunter will not be too liberal towards the bird and will leave it just a portion for one meal, thus stimulating the bird to call him to other nests in the neighbourhood'.

Fig. 26. Indicator indicator. KOLBEN (1719 p. 183 f.) mentions that this bird indicates to the Hottentots the way to wild bees'-nests (from KOLBEN, Ex: SEYFFERT 1930, p. 24, fig. 2).

Many wild bees live in the Cape Colony, but the honey-guide is absent; its nest is like that of some finches, being formed from small strips of bark mixed and woven into the form of a bottle, with the neck downward (p. 74: description of bird). LOBO (Voyage en Abyssinie, 1728) was the first to mention the honey-guide from Abyssinia, but only from hear-say. Many Hottentots and even peasants also eat the eggs, maggots and combs of the bees, which they declare to be a delicacy. The honey is sweet and tasty, and even relatively free from wax.

SPARRMANN has never heard about bee-keeping in Africa, except near Constance, where a young son of a farmer left empty boxes outside in places where he was sure that within two or three days a wild swarm would settle. But he did not give them much time and soon devoured them.

'Mosquitos and bees are another inconvenience of the travel to Podor or to Galam, the bees even more so than the former. Everyday towards noon I was attacked by one, two or more swarms of them as they came to settle in the cabin of the ship; and they were difficult to chase away. These bees do not differ from those of Europe except by their small size. They have a particular honey, which never grows hard, but remains always liquid, like a brown syrup. This honey is infinitely superior as regards delicacy and taste to ours in S. France' (ADANSON, 1757, p. 82 f.).

D. LIVINGSTONE in his Missionary travels and researches in South Africa (London. 1857) has much to say about honey and wax: 'In these forests of Angola we first encountered the artificial bee-hives so common in that country. They consist of about 150 cm. of the bark of a tree 37 to 45 cm. in diameter. Two incisions are made right round the tree at points 150 cm. apart, then one longitudinal slit from one of these to the other. The workman next lifts up the bark on each side of this slit, and detaches it from the trunk, taking care not to break it, until the whole comes from the tree. The elasticity of the bark makes it assume the form it had before. The slit is sewed or pegged up with wooden pins, and ends made of coiled grass-rope are inserted, one of which has a hole for the ingress of the bees in the centre, and the hive is complete. These hives are placed in a horizontal position on high trees in different parts of the forest, and in this way all the wax exported from Benguela and Loanda is collected. It is all the produce of free labour. A 'piece of medicine' is tied round the trunk of the tree, and proves sufficient protection against thieves' (p. 284).

In E. Angola he found plenty of honey in the woods, and saw where the Balonda dry their meat, when they come down to hunt and gather the produce of the wild hives (p. 264).

'Whilst at Chihune (Angola), the men of a village brought wax for sale, and, on finding that we wished honey, went off and soon brought a hive. All the bees in the country are in possession of the natives, for they place hives sufficient for them all. After having ascertained this, we never attended the call of the honey-guide, for we were sure it would only lead us to a hive which we had no right to touch. The bird continues its habit of inviting attention to the honey, though its services in this district are never actually needed' (p. 344).

'We remained near a small hill, called Maundo (in Port. East Africa), where we began to be frequently invited by the honey-guide (*Indicator indicator*). Wishing to ascertain the truth of the native assertion that this bird is a deceiver, and by its call sometimes leads to a wild beast and not to honey, I inquired, if any of my men had ever been led by this friendly little bird to anything else, than what its name implies. Only one of the 114 could say he had been led to an elephant instead of a hive, like myself with the black rhinoceros mentioned before. I am quite convinced that the majority of people who commit themselves to its guidance are led to honey, and to it alone' (p. 547).

'The honey-guides were very assiduous in their friendly offices, and enabled my men to get a large quantity of honey; but though bees abound, the wax of these parts forms no article of trade. In Londa it may be said to be fully cared for, as you find hives placed upon trees in the most lonesome forests. We often met strings of carriers laden with large blocks of wax, each 80 or 100 lbs. in weight, and pieces were offered to us for sale at every village; but here we never saw a single artificial hive. The bees were always found in the natural cavities of *mopane*-trees. It is probable that the good market for wax afforded to Angola by the churches of Brazil, led to the gradual development of that branch of commerce there. I saw even on the banks of the Quango as much as sixpence paid for a pound. In many parts of the Batoka country bees exist in vast numbers; and the tribute due to Sekeletu is often paid in large jars of honey; but having no market nor use for the wax, it is thrown away . . . .' (p. 614).

Le Vaillant (1931 I, p. 208) repeatedly states that the natives in S. Africa were very discontented when he wanted to kill one of the honey-guide birds. Azande chiefs in the Congo formerly cut off the ear of any man who dared to kill a honey-guide (Chapin 1939, p. 550).

Many of the 'rees near the Tana River falls in Kenya had hollow boles placed in their forks or hanging from their branches (E. Cobbold 1935, pp. 36 ff.). These cylindrical barrels are the hives in which the natives collect the honey of wild bees. They blow smoke in through the many crevices of the boles and the bees go away stupefied; they then seize the honey, leaving no provision for the bees. The careless natives often set fire to the trees while smoking out the bees, and the resulting damage is very serious in this dry country. Sometimes the bees build their combs in the living camphor trees, which in their old age are hollow as an ancient oak in England. These combs would never be discovered if it were not for the cunning of the honey guide . . . . The bees are soon bemused with the smoke arising from the green leaves and fly away, sometimes halfheartedly landing a sting on their aggressors. As soon as the whole nest is exposed, the comb full of delicious wild honey mixed up with leaves, grubs and all manner of mess is scooped into a cooking pot brought for the purpose. The honey is eaten in the dirty condition in which it happens to be obtained, often black with age and smoke. The honey-guide is more fastidious, and removes carefully the leaves and grubs. Lady Cobbold also reports how the Germans in the First World War in East Africa utilized the bee-craft of their Askaris, to lure a British transport convoy into an ambush of some hundreds of bees'-nests, many men and beasts of transport being killed by the enraged bees. The honey of heliotrope is too fierce for the human palate, burning like mustard (ibid p. 54).

Captain Merker (1910, p. 34 f.) informs us in his remarkable monograph on the Massai, that this people eat honey (*en aisho*) pure and also use it as an admixture to various medicines. Honey-beer (*en aisho namga*) is prepared by dissolving honey in water, pieces of the root of the steppe aloe or of the fruit of *Kigelia aethiopica* being added, in order to improve the taste as well as to speed up fermentation. This mixture stays for 3 to 5 days in a warm spot to ferment. This mead is fairly intoxicating and a banquet usually ends with general drunkenness. The wooden trough for mead fermentation as well as small honey-bowls (*en gira*) and honey-pots (*ol olul*) are normal implements in a Massai household (p. 34 f.). The banning of mead for the younger men is very strict, even up to warrior age. The decalogue of the Wanderobbo (MERKER 1910, p. 280) contains the following fourth commandment: You shall hold peace and not quarrel amongst yourselves. Only old men are allowed to drink mead, as the mead intoxicates younger men, makes them excited and leads them to brawls and affrays. The Gamassia include mead among the nuptial gifts.

The Massai name of *Apis* is *ol oderok*, for the stingless bee *os salihoi*. Their myths describe the discovery of the honey early after the creation of men (p. 272) as follows. At the time when God gave the cattle to man, Naiterogob (i.e. Eve) gave birth to her first son Sindillo, who later helped his father in keeping the cattle. One day, when looking at a bee swarm in a hollow tree, he discovered the honey and brought it into his father's kraal. Since then honey is food for the Massai.

With the Wanderobbo the production and keeping of the artificial hives is the main work of the old men. Many of these hives are distributed around the camp up to many hours' distance away. They are hollow trunks of 120 cm. in length and 30 cm. outer, 25 cm. inner diameter. They are cut with much pain with very primitive tools from a piece of trunk of similar length. The diameter narrows towards each end (fig. 25,5). The outside is made smooth with a knife. The hollowing is done with two tools, a small two-bevelled chisel and a simple chisel, the latter serving to smooth the inner surface. Both openings are closed in a prop-like manner by flat wooden covers, which are fitted into them. Two small holes on each cover serve as entrance holes for the bees. These hives are hung up to branches of the higher trees by lianas, wooden hooks and sticks. An owner's mark is made on the lower side of the hive, identical with that on his arrow-tips. Before the honey is taken out, the bees are driven away by the smoke of a fire beneath the tree. The Wadjagga of the Kilimanjaro stupify the bees by a fungus (*Lycoperdon*). Then one man climbs up to the hive, removes it from its hook and lowers it slowly on a rope to the ground. The bees are expelled or stupified by the additional effects of the smoke. The honey as a result has a smoky taste and contains many dead bees and maggots. But the fierceness of the bees' sting forces the poorly clothed honey-collectors to protect themselves by a thick cloud of smoke. It is not rare for a bee-swarm to attack a caravan, stinging some of the carriers and donkeys to unconsciousness and even to death. Tobacco smoke or loud singing or crying apparently irritates the bees. When the honey is taken out, about a quarter of every comb is left, in order to prevent the bees from leaving the hive. The catching of wild bee-swarms is unknown. The honey is put into leather-bags or woven pots. When bees'-nests are found in a hollow tree, an empty hive, the inner surface of which is smeared with the aromatic honey of *Trigona*-bees, is hung up near this tree. In order to remove the honey from bees-nests in hollow trees, the bees are expelled by a man pushing in a burning branch (p. 234).

The young men did not go to war with the El Dinet. They fought with the bees, which abounded in their country. Openings were seen in all trees and in many places in the hard red soil, all inhabited by bees. Each of these openings had a carved owner's mark (p. 293). The land of the El Garguresh was full of hives, which were twice as broad and half as long as those of the Wanderobbo (p. 290).

Wild honey is an important food on the long marches of the Massai warriors. They are led to the bees'-nests by the *Indicator*-bird (*en johorai*). When this bird sees men, it calls with a jarring sound and slowly flies to the nearest place where honey is found. The warriors follow the bird and collect the honey. When the quantity of honey does not satisfy the warriors, they bury the masticated combs. Then the bird after some time leads them to another 'honey place'. Whenever sufficient honey has been obtained, they leave the rest for the bird (p. 104).

Minor body-mutilations may be expiated with the Massai by payment of some hives. One finger is valued at one, one hand at eight hives. Among some Wanderobbo one finger or a hand is expiated by a fat-tailed sheep (p. 262). This estimate is interesting in connection with the punishments in the Hittite Code of Bogazköy.

The cliff-dwellers of the Elgeyo reserve of Kenya have been described by A. MASSAM (1927 p. 122 f.). In some areas, particularly in the south of the reserve, the natives use honey boxes fairly extensively. There are no restrictions as to the number of these boxes a man may keep, but no one is permitted to put his box in a tree already appropriated by someone else. The 'hives' used are made out of tree trunks. A log of suitable length is split lengthwise down the middle. Each half is then hollowed out, except at the end which are left solid. At this stage it looks like a trough. The two pieces are then strapped together, thus forming a hollow log, which serves as the hive. In the under half a few holes are burnt to allow the bees ingress and egress. A thick strip of bark is fitted over the box to keep out rain. Then the hive is the ready to be placed in a tree, either in the forks of branches, or, when monkeys are troublesome, suspended by ropes made of strong creepers.

Boxes which contain bees are opened twice a year, only half the honey in the box being removed each time. Some of the natives do not care for the job of removing the honey, and employ less timid friends, who are rewarded with half the honey taken. The man who undertakes such work smokes out the bees, using wood which gives off plenty of smoke. He takes no notice of stings, though he takes care that no bees go near his eyes, which are not protected by any covering. Honey is stored in special cylindrical cases made of wood, more or less covered with hide, and fitted with a strap and with a leather lid which fits tightly over the top of the case. Honey is considered a good food, and it is especially esteemed when eaten with pounded termites.

Termites and honey are distinctly luxury foods to the natives, who live principally on the grains they grow (p. 126). The payment for work done for the elders, is made in the form of honey. (p. 134). The older boys also receive instruction in the fashioning of beehives, and make many small models until they become proficient in the craft (p. 100).

The Bushmen of the Kalahari (DORNAU 1925, p. 28) live on the honey of wild bees which they smoke out of hollow trees, rocks, or holes in the ground (p. 110). Their sight is so good that they can follow the flight of bees high up in the air, and so find the direction of the nest and rob the bees of their honey. I have heard it asserted that they can follow the bees to their nests by noticing the drops of honey on stones. The Bushmen will tell you that the bees fly lower and more slowly when they are returning to their nests loaden with honey.

There is an interesting report on folklore and the cunning of the honeybird: 'The honeybird is a very small bird but he is very cunning. He flies about watching where the wild bees make their nests, and when he has found one of their nests he comes and leads the Bushmen to the place. The Bushmen are very fond of honey, and when the honeybird has led them to a nest, they always give it a bit of the comb for its share after they have removed the honey. One day a Bushman was out hunting. He had very bad luck and the buck were so wild that he could kill nothing. Tired out with running in the sun, and very hungry, he lay down to sleep under a tree. A honeybird came and woke him by calling out: 'Come on, and I will show you where there is a bee's-nest'. The Bushman was much pleased and, being hungry, he got up and followed the honeybird at once. It took him a long way, and he was beginning to get tired of running after it, when they came to the bottom of a big cliff. It stopped there and the Bushman began looking up to find the hole where the bees had their nest. The honeybird was sitting on a bush beside him calling out: 'You will give me some of the honey!' Now there was a lion lying asleep behind a bush, at the bottom of the cliff, and the Bushman did not know that . . . . The Bushman was going forward to climb up the cliff, and the lion gathered himself ready to spring upon him. The Bushman turned and ran . . . Just as the lion was about to seize him, he threw away his loin-skin, . . . so the Bushman escaped. He never went with a honeybird after that, and when one came after him he used to curse it and throw stones at it. So that is why we Bushmen are very careful when following a honeybird. We stop when it stops, and stand back a bit and have a good look round before we go forward, lest the bird may have led us up to a lion. So the honeybird is sometimes a very wicked bird' (DORNAU 1925, p. 180).

At Dar-Fur (W. G. BROWNE 1800: II, p. 31) honey-bees are common, but they have no hives. The wild honey is very brown and of disagreable taste.

The province Wojjerat in Abyssinia is famous for its white honey (H. SALT 1816: II, p. 191) which is brought in great quantity to

Antalo. When the natives find a wild bees'-nest they suspend near by a box of wood, called *muggit*, which is rubbed with old honey and which is accessible only through a tiny hole in one of its sides. The bees are thus attracted and collect in great numbers within the box. The proprietor comes by night, stoppers the entrance hole and transports the box to his farm. The bees accustom themselves readily to the new environment and build their cells in square compartments which are prepared for them in the walls, which in a great part of the country are built from mud.

The Zulus (H. SCHOMBURGK, 1910, p. 6) also know the honey-bird (*Indicator indicator*), a small bird, which attracts the attention of man by its twittering. Once it has been observed and is being followed, it flies on and finally settles on a tree, within which bees have made a nest. When the honey is removed, a part must be left, otherwise, so the Zulus say, the next time the bird will guide the man following it not to honey, but, in order to take revenge, to a leopard or to a black mamba, the most dangerous poisonous snake of Africa. 'In no other region have I observed that such a gift is left for the honey-bird', writes SCHOMBURGK. 'The wild bees mainly build their nests in hollow branches. In many places, the natives are very clever in making bee-hives. In Zululand I never saw such hives.

The Kalahari are great hunters of game, but primarily honey-hunters. Everywhere in the forest one finds many of their primitive hives. These are made by joining up with a series of pegs a piece of bark one metre in length, which has been removed from a tree, into the form of a barrel; both the open ends are closed by smaller pieces of bark. A hole is pierced on the lower side for the bees to enter. The hives are hung up in the Bush within the territory of the village and nobody would think of touching a hive not belonging to them' (H. SCHOMBURGK 1910, p. 115).

Wild bees (*Apis unicolor*) are rare in the C. Cacao fields of S. W. Africa. They have become so rare owing to the inconsiderate destruction of every nest by the natives. The honey-guide leads them to the nests. The small black, stingless *Ozomburumburu* (*Trigona sp.*), which greatly molest the Hereros, are very common but they provide no honey. They build separate nests for breeding in small urns, and from these a spoonful can be collected only by long labour. Termites are devoured by the pounds by the Hereros (STEINHARDT, 1922, pp. 113 ff., 207).

In the Kilimanjaro region bee-cylinders are placed in the trees to attract *Trigona erythra togoensis* Stad., the honey of which is greatly coveted. The Botocudo who cannot ordinarily be roused to action would climb the highest trees eagerly and swiftly after honey. The Camacans even domesticated stingless bees, and the bees themselves were placed with wax in water to obtain an intoxicating brew. In the forests of the Amazon adjoining human settlements, trees can everywhere be seen from which an oblong section of the trunk is missing, indicating the spot where a stingless bees'-nest had been discovered (SCHULZ 1905, p. 142).

Hive-keeping in the central and southern Sudan was studied by H. H. KING (1920), who tried to develop an improvement of the native hive, mainly to increase the wax-production of the country. The wild bees of *Apis adansoni* make their nests in holes in trees and fallen logs, in holes of termite hills and in banks, in crevices in rocks, etc. The natives prepare primitive hives, which they place in likely situations at the beginning of the swarming season. These hives consist sometimes of a hollow palm log placed in a slanting position with its butt in an empty earthenware *burma*, and sometimes of basket work, made apparently in imitation of a hollow palm log, and fastened in the branches of a tree. Probably other patterns are in use, but the palm log type seems to be the most common one. In this hive the queen has ready access to any part and the honey-comb cannot be removed unless the colony of bees is first driven out and destroyed. These hives are usually about two meters long and about 20 cm. in diameter. When in use the ends are closed with wads of grass and the bees enter by a rectangular opening midway between the ends. When the honey is taken out, the bees are destroyed by burning and the entire hive is robbed.

The Kaffirs of S. Africa (MARAIS 1912, p. 790 ff.) destroy enormous numbers of hives each year, actively supported by the white settlers.

In N. W. Ethiopia (CHEESMAN 1936, p. 64) all hollow trees have either a swarm of wild stingless bees or a square hole cut where the *adze* has enlarged the entrance in order to admit the hand of the honey-gatherer. MARAIS mentions two kinds of stingless bees in northern Transvaal, one of which (*Trigona cf. clypeata* Friese) enjoys protection against men and ratel, living in very hard, virtually impenetrable ground. Even the honey-guide, partial to wax, does not lead his allies to the ground or arboreal nests of wild bees. Thus he ignores (HUGHES 1933, p. 355) the nests of the tiny black bees, which are generally found in the hollow knot of a tree, and easily spotted by the thick black streak trailing down beneath the opening. The best records of the behaviour of *Indicator indicator* Sparm. have been given by CHAPIN (1924, p. 333; 1939, pp. 549 ff.).

To induce a swarm of *Apis adansoni* in British East Africa to occupy a new hive, the prospective home of the colony is fumigated by burning an aromatic wood, the odour of which is thought to be pleasing to the bees (LINDBLOM 1920, p. 495). As a lure (GUTMANN 1909, p. 206) the Wadschaggas place ripe bananas in the tubular entrance of the hive, so that the wood may become impregnated with the pleasing odour. Some *Trigona* bees are partial to bananas, and perhaps this is true of *Apis adansoni*, of which GUTMANN speaks. This method would seem at least as effective as the chants which are addressed to the insects, in the hope that they will induce them to enter the hive.

About 30 different species of stingless bees (*Trigona spp.*) live in East, South and West Africa. The natives easily distinguish them from the true honey-bee (*Apis unicolor*) and they usually prefer the highly aromatic honey of the former. H. MORSTATT (1921) studied them in the Tanganyika Territory, mainly in connection with their wax production. The biology of the stingless bees is little known, as the total honey production of most species is very small, and the detection of their nests in most cases difficult and only accidental. This is true, for instance, of *Trigona gribodoi* Magr., *T. lendliana* Fr., *T. schmidti* Stad. and most other species. When in certain regions the stingless bees are well known to the natives, this is mainly due to two species, which they detect and utilize quite freely. One of these is *T. clypeata* Fr. (including *var. zebra* Fr.), which had already been studied by MARAIS (1912) as the big *moka*-bee. It burrows a cylindrical tube 60 to 150 cm. into very hard soil with a storage chamber at the end of the gallery, where honey and pollen are stored separately from the brood combs. The honey is filled into small, thimble-sized wax-cells which are grouped together into bunches the size of an orange. The entire bunch is again enclosed by a layer of wax. As stated above, their aromatic honey is preferred to the *Apis*-honey. Its total quantity within one nest ranges, depending upon its size and season, from a few cups to some fifteen pints; in a related smaller form it usually amounts to two water-glasses full. This latter species nests in trees. The same *T. clypeata* often builds its nests in abandoned parts of termite nests which are still inhabited. The storage chamber is about 15 cm. in diameter. The pollen cells are much larger than the honey cells.

The other common species is remarkable, as it is the only African *Trigona* which is adapted to hive-keeping. In the Kilimanjaro region the *buwa* or *T. togoensis* Stad. (including *var. junodi* Fr.) is clearly distinguished by the natives from the *nyuki* (*Apis unicolor*). Its wax is not traded, as the honey can be obtained without destroying the combs, thus leading to true bee-keeping. The *buwa*, like the *nyuki*, readily build their nests in the long honeypipes, the *mzingas*, which the natives suspend everywhere in high trees. They say that the *buwa* draw twigs and leaves into the *mzinga*, and that these leaves are formed into globular receivers for the storage of honey at the sides of the combs. The honey-hunters pierce through these balloons and the outflowing honey is collected into vessels, without disturbing the combs. Their honey is also highly appreciated as medicine. The yield of one active family is estimated at 5 to 6 pints of honey in a year. Like all *Trigona*-honey it is highly aromatic and of a thin fluid consistency. The most important tribe which have reached this stage of true bee-keeping and of domestication of a native bee are the Wadshaggas, and bee- and honey-ceremonies play a great role in their life.

Many African bee rituals have been collected by H. RANSOM (1937, p. 292 ff.). HOLLIS (1909, p. 7) mentions that once his porters were attacked by wild bees in the Nandi country. One Nandi volunteered to pacify the bees, as he belonged to the bee-totem. Naked as he was, he returned, whistling loudly all the time. The bees swarmed on him, but he took no notice of them, and soon came back. The bees were now quiet and HOLLIS' porters could collect the baggage undisturbed. The Bakoanda of N. Rhodesia, who make a honey-beer, belong to the bee-totem (MELLAND 1923, p. 123). In other tribes of the same country the Benangonyi (birds) family totem are paired with the Balembas (bees). The 'birds' tease the 'bees' by saying that one of their tribe, the honey-guide, devours bees and honey (ibid, p. 252).

C. W. Hobley (1922, p. 251 ff.) reports that the Akamba, a Bantu tribe in Kenya Colony, were great hive-keepers. Before the arrival of the white man they hollowed logs of wood and hung them in trees for bees to hive in. They periodically collected the honey, brewed mead, and threw away the combs. In Kikui when a man makes his first bee-hive, he does not hang it on a tree himself, but gets his uncle to do so. The owner of a hive cannot co-habit with his wife until he sees that a swarm of bees has settled in the hive and is building there. Two nights after he is satisfied that this is the case, he may resume marital relations. If on the first visit he finds the hive occupied, he brews beer and pours some on the ground as a libation to the *aiimu*, the ancestral spirits. In a season when there is a dearth of honey the owners of the hives go to the woods, where their hives are, and sacrifice a goat; the meat is eaten and the blood mixed with beer is poured on the ground as a propitiatory libation to the *aiimu* to secure a good honey crop. When a man has hollowed out the log of wood for a bee-hive, he gives the village a ceremonial meal, and at the first honey crop he has to present his mother with some honey.

When the first brew is ready, the father of the hive owner buys it for a goat, which may not be killed. On the second night after the purchase, the parents of the hive owner must co-habit. The Akamba believe that the consumption of the beer in the succeeding ceremony will ensure a good supply of honey; it is actually a magic fertility ceremony. Hobley quotes similar rituals as still existing in Kenya. These tribes also mark their hives with marks of the clan and of the owner, and also the trees used for the manufacture of hives while they are still standing (Fig. 25, below), to claim property rights (Hobley 1941, p. 412).

The Bakunta tribe of Uganda (Roscoe 1924, p. 171) make hives from hollowed-out logs, and more rarely from plaited cylinders of papyrus, 120 cm. long and 30 cm. in diameter. In addition to the honey, they eat the bee-grubs, which they consider a great delicacy. When a man is going to take honey, he must keep apart from his wife for a night. The next day he smokes the bees, driving them to one end of the long hive before taking the honey.

Many honey taboos exist among the Thonga of South Africa. During betrothal the suitor must not eat honey during his first visit to his future father-in-law, because honey slips away like a fish, and the girl he wishes to marry might slip through his hands (Junod 1912 II, p. 107). When the man is married, he must not eat honey with his wife for a year, till she has had a child. The bride is allowed to eat honey at her father's house, but must not eat it in her new home or with her husband. The husband may eat it in the Bush, but he must wash his hands before coming back, lest his wife should notice it; she would run home if she did. When the bees have eaten honey, they fly away from the hive, and the wife may do so too. After a year this misfortune is not to be feared (Junod 1912 I, p. 239). The Thonga also have a magic horn into which honey is poured, from bees which nest in the ground. It is used for divination: when war is imminent the honey ferments and comes through the hole of the horn, and then people prepare for war (Junod 1912 I, p. 363). These Thonga eat all the honey and give all the wax to the honey-guide which is just what the birds wish for (Junod 1912 II, p. 319).

The Lango say that if you eat honey on the way to a hunt you will kill no game; and among one of their tribes, the Joatekit, the women do not eat honey (Driburg 1923, p. 267).

The Bushman perform a dance with a sort of rattle, the *goin-goin*, in order that bees may become abundant, chanting: 'The people beat the *goin-goin* in order that the bees may become abundant for the people, in order that the bees may go into other peoples' places, so that the people may eat honey. Therefore the people beat the *goin-goin* when they dance, so that the peoples' bees may go into other peoples' places, so that the people may eat honey, and go that they may put away honey into bags. And the people carry honey. And the people carrying honey bring the honey home. For the women are dying of hunger at home. Therefore the men take honey to the people at home; that the women may go to eat, for they feel that the women have been hungry at home, while they wish that the women make them a drum, so that they may dance when the women are satisfied with food. For they do not frolic when they are hungry' (Bleek and Lloyd 1911, p. 353).

Many rituals exist restricting the consumption of honey-beer on many occasions and for many classes (Bibliography in Ransom 1937, p. 298).

The active survival of all these habits is clearly documented by a recent report from East Africa. Pater Othmar Morger of the

184

Swiss Mission at Peramiho (1950) accompanied one of his clients, PANGANI, on his honey-hunt in hollow trees. From time to time he blows into a small calabash with two openings one of which he half closes and alternately releases, thus producing something like a mournful song. This tune is to attract his friend *kihegu*, the honey-guide. After half an hour the inconspicuous *kihegu* approached with his call: *chako-chak-chako*. Occasional shrill whistles and deep calls of hi! now replace the earlier mourning call, and develop almost into a dialogue with the bird. PANGANI asks, the honey-guide answers, until suddenly the bird's call voice almost breaks. Now PANGANI collects some dry herbs and lights them with a glowing thread, thick smoke arises from his hand, and grasping his hatchet he approaches a hollow tree and begins to cut a hole, i.e. to enlarge the small entrance hole of the hive. The angry inhabitants swarm around PANGANI's head, who continues his work without paying attention to them. Soon the hole is big enough to permit him to insert his arm inside and, one after another, to extract five combs, each almost one meter long, full of honey and brood. Almost twenty stings are in the hunter's arm but he pays no attention to them. He and his friend ODILO eagerly began to eat the brood combs, and could scarcely understand the Pater's refusal. 'The happiness on their faces showed me what a delicacy I had refused from mere prejudice. They warned me against consuming the honey, as this would cause an unquenchable thirst. The most attractive dish was made by pressing some of the honey over the brood combs. Their appetite was enormous. The wax was later sold. Enough remnants were left to satisfy the appetite of the honey-guide'. Regretting the total destruction of the hive, he obtained the answer: 'If I would leave anything, my neighbour would take it tomorrow'.

Another part of the Pater's story of PANGANI is less clear. He was told that PANGANI was in the possession of more than a hundred hives made from pieces of bark, and these were distributed over a wide area in the bush. The honey-season is in October, when he takes from each hive up to 15 pints of honey, worth the monthly salary of a worker, and in addition the much more expensive wax. A poor relative recently asked PANGANI for a cloth, and received the wax from three of these hives to buy cloths in the town. In spite of the great dispersal of these hives, up to ten hours from his hut, it is extremely rare for any honey or wax to be stolen. Last year one of PANGANI's hives was robbed, but upon his threat to bewitch the entire village if the honey and wax were not restored, it was promptly returned. It seems that if hives really are permanently maintained in this region by leaving sufficient combs to replenish the plundered hive – which is by no means clear from the description – such a procedure must be of very recent origin, deriving from the influence of European bee-keepers and schools. It would be desirable for such improved conditions to be carefully described and for a detailed enquiry to be made into their origin and development in every case.

5. VARIOUS INSECTS AS FOOD

Of the large number of beetles which are eaten, usually as roasted larvae, only a limited number has so far been identified. The main reason is that most beetles and especially their larvae are not available in large quantities. We may mention the following:

Scarabaeidae. Angosoma spp. (larvae; GHESQUIÈRE, 1947).

Goliathus sp. (larvae delicacy in Congo; BEQUAERT 1921).

Oryctes boas F., *O. owariensis* Beauv. and *O. monoceros* Ol. (larvae and occasionally adults eaten in South Africa).

Platygenia spp. are sold living or fried in oil on almost all native markets of tropical Africa (GHESQUIÈRE, 1947).

Cerambycidae. Ancylonotus tribulus F. (larvae eaten in Senegal and Gaboon).

Macrotoma edulis Karsch (larva fried in palm oil on St. Thomé Island, Gulf of Guinea; NETOLITZKY, 1920).

Omacantha gigas F. (ditto in Senegal; NETOLITZKY, 1920).

Curculionidae. Rhynchophorus phoenicis F. This is the palmworm of tropical Africa, about which GHESQUIÈRE (1947) wrote recently: 'The natives of the Belgian Congo greatly prize the larvae of *Rhynchophorus* and search for them by putting their ears to the palm stems. They recognize the appropriate moment for gathering the larvae by the noise which these make in nibbling at the fibres of the host. At Mozambiques, at the Cape, in Sierra Leone and at Fernando Po these larvae have likewise since olden times been sought for by the natives. GHESQUIÈRE states from his own experience that throughout

Fig. 27. Market of roasted caterpillars at Man, Ivory Coast.
Courtesy of I.F.A.N.

the native markets of Africa the pupae or plump larvae of *Rhyn-chophorus* or *Rhinostomi* may be found, either alive or fried in oil, only the body being eaten, not the head.

BURR (1939, p. 210) was offered a huge beetle-grub, bigger than his finger and gently squirming, by a native in Angola. Upon BURR's refusal, the Negro wrapped it carefully in his loin-cloth, to keep it as a treat for supper. His porters as part of their normal routine prepared stews of two kinds of big caterpillars. One was pale green, fat and juicy and certainly had a nutritious air, while the other was black, ringed with yellow.

LIVINGSTONE (vide PORTEVIN, 1934) observed in Angola on the-Quango River that the natives dug for big beetle larvae.

BEQUAERT (1921, p. 197) has collected some observations on caterpillars as food, especially among the forest tribes in Africa. Thus, TESSMANN observed that the Pangwe of S. Cameroon eat not less than 21 different species of caterpillars. They distinguish a number of edible species by name and know also their particular host plants. The caterpillars of the silk-weaving *Anaphe sp.*, common Notodontids of Equatorial Africa, congregate when fully grown in groups of 12 and more. They are eaten and their silky nests offered for sale (*Photo* p. 197). SWINNERTON writes about *Anaphe* in Gazaland: 'This is hardly of special interest in itself, for many other moth-larvae are also eaten by them, but what is perhaps of some slight interest is their alleged differential effect on particular individuals eating them. I was first informed of this by a native skinner and collector in my employ, whose statements I have in general found to be reliable; and he specially remarked that even brothers, eating from the same dish larvae that had been captured and prepared together, differed thus in their reaction: one brother suffering no ill effects whatsoever, the other being always completely prostrated for as much as 2 or 3 days in the more serious cases. This statement has been completely corroborated by such natives as I have since spoken to on the subject. All have further agreed in saying that the larvae are much liked, and that their inability to eat them is felt as a misfortune by those whom they affect unpleasantly'.

BEQUAERT continues that the Medje of the Ituri forest diligently collect in the proper season, in addition to *Anaphe*, various other caterpillars. Those called *ebbo* (*Phot. p.* 198) are especially sought

188

Fig. 28. Market of roasted caterpillars (above: also of roasted termites) at Man, Ivory Coast. Courtesy of I.F.A.N.

after; dried and smoked they can be preserved for many months. The most common species collected by the Mu-Lang (*phot. p.* 198) is the caterpillar of the Ceratocampid *Micragona herilla* Pack. Heavy spines cover the body, which are scraped off before cooking. Two other caterpillars belong to the same family, and another delicacy of the Medje is the larva of the Psychid *Clania moddermanni* (*Phot. p.* 199).

According to TESSMANN the Pangwe also hunt for the aquatic larvae of dragonflies, to which they attribute diuretic properties.

LIVINGSTONE and his company during their travels in S. Africa sometimes suffered much from want of meat, but not of food. He says that to show their sympathy the natives gave the children, who suffered most, a large kind of caterpillar, which they seemed to relish and which they themselves devoured in large quantities (1858, p. 48).

Major S. PINTO (1881, p. 235), while crossing Africa, was offered among other victuals a big basket of hawk-moth caterpillars, which are collected in abundance from plants. The Ganguelas like them a great deal, but his men did not touch them.

N. C. E. MILLER (London) has informed us that in Southern Rhodesia the Bushmen in certain districts collect whole branches full of caterpillars of a Lasiocampid moth (*Brachiostegia sp.*), which they slightly roast and eat.

MJELE, a native commissioner, reported in 1934 (p. 37 f.) on the *Harugwa*-bugs (*Eucosternum* (*Haplosterna*) *delagorguei* Spin.) in the native Bikita reservation of S. Rhodesia, which assemble in great swarms on the leaves of loquat trees. The natives say that they arrive from the south. Their breeding grounds are unknown. Their arrival in dense swarms, resembling a locust swarm, usually indicates the end of the rainy season. The headman Nerumedzo of the reservation is vested with the *Harugwa* monopoly. When the bugs arrive, Nerumedzo sends a sack to the neighbouring chiefs and to the native commissioner at Bikita. This act symbolises the opening of the *Harugwa* season. The natives then negotiate with Nerumedzo to buy permission to gather them. The bugs are first killed in hot water. The thorax is then pressed to remove a certain unpleasant secretion after which they are roasted on an open fire and placed on a granite out-crop to dry out. Usually they are eaten with other food, and only gourmands eat them as a separate dish.

The *Harugwa* have settled in the last twenty years with one exception in the same spot, and in the exceptional year in its neighbourhood. The weather of the coming season is foretold from the number of *Harugwa*: many indicate heave rains, few scarce rains to come. The *Harugwa* stay from April to September, when with the approach of the rains they fly southwards, apparently to their breeding grounds.

The entomologist A. CUTHBERTSON (ibid) adds the remark that ten species of Pentatomid-bugs have the habit of congregating in enormous numbers ln S. Rhodesia. Outstanding amongst these is the *Harugwa*, which appears in swarms on certain trees in March–April in the Belingwe and Bikita districts. It is the cause of serious quarrels among the natives of neighbouring kraals. Despite the nauseous odour caused by a fluid which they excrete, the Bikita natives gather and devour the bugs as soon as they appear.

More recent observations are added by J. C. FAURE (1944 pp. 110 ff.). The natives of the eastern slopes of the Drakensberg in E. Transvaal regard the green stinkbug *Eucosternum delagorguei* Spin. as just as great a delicacy as those of S. Rhodesia. They are believed to be strong flyers and swarm at high temperatures in the air. The bugs settle on a number of shrubs and trees, such as *Acacia ataxacantha australis* B. D. There seems to be only one annual generation. The bugs are present every winter, but sometimes are rare, especially after a wet summer. The bugs seem to congregate in protected hollows on the eastern slopes of the Drakensberg at 700 to 1400 m. altitude. Although occurring in swarms they are never abundant enough to bend down or to break the branches of the trees. The Mapulana call it *Thsonono* (= he farts, he is fat) from their stink and their fatness at the collecting sesoan. The bugs are collected at sundown or early in the morning, or better still on misty, cloudy days. Natives often trespass on Govt. property to obtain this delicacy and they do so much damage to trees in collecting them that they have to be prosecuted. The *Thsonono* are prepared for eating, raw or cooked, as follows: the head is removed by rubbing it across a rough stone, a tree or some other hard object. Then the body is taken between the thumb and two fingers, and squeezed from the posterior end forwards, in order to remove the 'poison' which is supposed to flow out through the neck. Some liquid does flow out, but it probably consists mainly of blood, and possibly of some of the contents of the alimentary canal. Apparently not much of the stinking gland secretion is removed, as the insects are merely gently squeezed. Sometimes the bugs are put beneath a stone under 5 cm. of water to dilute the 'poison', which is said to produce severe pain in a cut and to float on water like oil. Usually the bugs are cooked before they are eaten, yet they are also eaten raw. This tribe also eats locusts, winged termites and large Saturniid caterpillars. But the *Thsonono* are more of a luxury than the locusts, as they cannot be collected in such large quantities, the process of preparation is slower, and as they are too rich to be taken as a piece de résistance. Practically all members of the tribe are fond of the bugs, young and old, either with porridge or alone. The origin and distribution of this custom is unknown. FAURE saw five natives eat several of these bugs in quick succession with evident relish, immediately after removing the head and giving a perfunctory squeeze or two.

'The *mopane*-tree (*Bauhinia*) is remarkable for the little shade its leaves afford . . . . On these leaves the small larvae of a winged

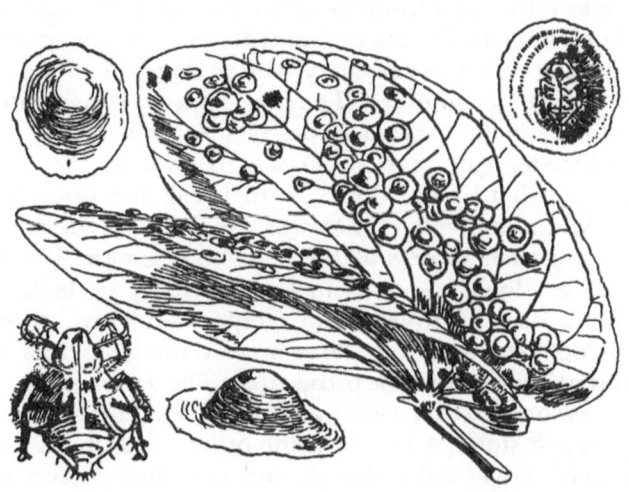

Fig. 29. Psyllid larvae and their edible secretions on the leaves of *Bauhinia*. From D. LIVINGSTONE. London. 1857, p. 164.

insect appear covered over with a sweet gummy substance. The people collect this in great quantities (in the Bushmen country near Maila), and use it as food' (LIVINGSTONE, 1857, p. 164).

Fig. and Note by J. O. WESTWOOD: 'A species of *Psylla*, of which genus a similar secretion is reported from Eucalyptus in New Holland, the *Wo-me-la*, is scraped off the leaves and eaten as a saccharine dainty. The insects found beneath the secretion, brought home by Dr. LIVINGSTONE, are the pupae, flattened, with large scales at the sides of the body, enclosing the future wings. The body is pale yellowish, with dark-brown spots. The secretion is flat and circular, apparently deposited in concentric rings, gradually increasing in size till the patches are 6–8 mm. in diameter'.

'And the *lopane*, large caterpillars 75 mm. long, which feed on the leaves of the *Bauhinia*, and are seen strung together, share the same fate (i.e. to be eaten) (p. 361). The people of the Quango district (Angola) seem to feel the craving for animal food as much as we did, for they spend much energy in digging large white larvae out of the damp soil adjacent to their streams, and use them as a relish for their vegetable diet' (p. 389).

In tropical Africa the large winged queens of *Carebara vidua* F. Smth. at certain seasons emerge in great numbers from termite mounds. They are highly prized as delicacies and eagerly gathered for their swollen gaster which is eaten raw or roasted (J. BEQUAERT 1913, p. 429; 1922, p. 329). Mr. LANG observed the nuptial flight of *Carebara vidua* at Stanleyville in March, 1915, and confirms that only the abdomen is eaten by the natives, sometimes raw, sometimes fried, also crushed.

The Pabouins of Gaboon collect an ant, the *ntchongou*, with spoons and throw them by the thousands into big calabashes filled with very hot water. Boiled and then pounded with a herb similar in taste to chicory, they are an esteemed dish of this tribe (DE COMPIÈGNE, vide BRYGOO, p. 52).

In Uganda mole crickets (*Gryllotalpa africana*) are kept for their chirping and as food (FLADUNG 1924, p. 6).

In Nyassaland *kungu*, a paste composed of mayflies (*Caenis kungu*) and mosquitoes (*Culicidae*), is eaten (FLADUNG 1924, p. 8).

In Nigeria H. BARTH (vide BRYGOO 1946, p. 22) noticed at a market plates full of roasted dragonflies, *fara*, which form an im-

Fig. 30. Basket for collecting lake-fly, Ukara Island,
Tanganyika, from HARRISON.

portant part of the native food in times of famine. The insects
measure about 5 cm. and make a rather appetizing meal.

Some African Negroes bury meat to breed fly-maggots, of which
they are very fond (vide HYATT VERRILL 1938, p. 162).

D. and C. LIVINGSTONE also observed the consumption of the
kungus on the northern shores of Lake Nyassa. The natives cook
these immense swarms of gnats of the summer nights and form
cakes which they like very much, and each of which contains millions of such gnats. He was offered such cakes, one inch thick and
of the dimensions of a blue bonnet of a Scotch worker. They were
brown and tasted somehow like caviar or salted locusts (vide
DAGUIN 1900, p. 28).

P. KOLBEN (1738 II, p. 179) says the Hottentots eat the largest
of the lice with which they swarm; if asked how they can devour
such detestable vermin, they claim it is no shame to eat those who
would eat them: 'They suck our blood, and we devour them in
revenge'.

STEEDMAN (1835 I, p. 266) says of the Kaffirs, that 'except an occasional plunge in a river, they never wash themselves, and consequently their bodies are covered with vermin. On a fine day their karosses are spread out in the sun, and as their tormentors creep forth they are doomed to destruction. It often happens that one Kaffir performs for another the kind office of collecting these insects, in which case he preserves the entomological specimens, carefully delivering them to the person to whom they originally appertained, supposing that as they derived their support from the blood of the man from whom they were taken, should they be killed by another, the blood of his neighbour would be in his possession'.

6. THE PYGMIES OF THE CONGO FORESTS

SCHEBESTA (1938, pp. 67–71) tells us about the food of the Bambuti-pygmies of the Ituri forest. This tropical forest is much poorer in animals, especially in big game, than generally assumed. Animals, such as leopards, elephants, okapi, small antelopes and monkeys, are rare. The Ituri has even less birds, and these are hunted more because of their coloured feathers than because of their meat. Those found include rhinoceros birds, parroquets and guinea fowl. Snakes, including poisonous species, are rare. Smaller animals are more common and these are of fundamental importance for the existence of the Bambuti. Molluscs, worms, insects, etc., form the bulk of their animal food. Pride of place goes to the honey-bee, the produce of which is greatly prized. Everywhere the Bambuti try to find nests of bees. When en route he looks for them on the trees; he immediately follows the call of the honeyguide bird; and he traces their subterranean nests in the soil. SCHEBESTA collected seven different bees in the Ituri, most of them stingless. The termite is only little less important than the honey-bee, and it is doubtful which of the two is preferred by the pygmies. The swarming season of the termites always causes a minor revolution in the small camps. Groups which lived in harmony together for months suddenly separate; every family takes position near a termite hill, in order not to lose this delicacy. The Ituri as a whole is rich in termite hills. Yet certain tracts, such as some near the Oruendu, are without them, and this is regarded as a serious loss of basic food. The pygmies have not the slightest idea of the damage caused by these insects.

Caterpillars, grubs and all kinds of larvae and worms are not uncommon in the forest. Snails, mussels, fish fry, crabs, etc. are also eaten. Butterflies are rarer in the Ituri forest than in the Malayan forest. Yet occasionally enormous nests of caterpillars are met with on tree-trunks and these are carried as a welcome prey into the camp. Not all caterpillars are edible. Rotting trees are beaten with poles and searched with hatchets for the fat larvae which they contain. We may just mention the interesting suggestion of SCHEBESTA that flies and other vermin are not the ultimate reason for the many changes the pygmies make in their camps.

In the Ituri forest the pigmies, when not tracking animals, seem always to be hunting for honey. CHRISTY (1924, p. 40) watched them, after enlarging the entrance to a bees'-nest in a hole in a tree with their spears, put in their arm time after time and bring out pieces of honeycomb. This they would stuff into their mouths regardless of whether it contained grubs, honey or only pollen and throw away the empty comb. All the while a cloud of bees buzzed round them, settled on them, ran about them, but for some reason or other they were not angry bees, and apparently the robbers did not get stung. Their absolute indifference to the bees, and to their stings if they did receive any, seemed unaccountable. CHRISTY occasionally had to make a meal of this wild honey and honeycomb in the forest, and very good it seemed to him; but either in the coarse comb or in the honey there is something which subsequently produces a painful irritation of the throat.

The Bambuti pygmies of the Congo are all more accomplished hunters and collectors than the Guayaki of S. America (SCHEBESTA 1940, p. 47). They live in larger groups, in order to be able to help one or more sick persons, invalids and orphans, which would be impossible within the framework of small families. Motherhood lasts throughout life, suckling lasting up to two years (pp. 51 ff.). In Ituri and at Gaboon also, the women kill crippled children and the more feeble of twins, while the Batwa of Ruanda in no way object to twins. Only infant mortality hinders the growing of pygmy populations. They have on the average two and a half living children per married woman. Natural conditions in the hard life of these

196

nomad hunters enforce this limitation. The Bambuti re-inforce it by killing the cripples. And one twin is killed, because the mother cannot fulfil her social obligations in addition to child breeding. A too great increase in the mouths to be fed would also strain the scanty subsistence of the group. Their number thus remains stable . . . . Better living conditions, such as settlement, has immediately produced a growth of population in the Bacwa of the Equator. The Bambuti never hunted men for cannibalism, but they devour warriors killed in battle.

M. BARTOUT (1934, p. 89) indicates that the Ba-Binga pygmies of the High-Nyong valley of the Cameroons 'live mainly on meat. When this is lacking, they collect everything edible: reptiles, fish, shrimps, frogs, birds, insects, their larvae and caterpillars, etc.'

Y. GANDON in a novel 'Le Dernier Blanc' (about 1949, p. 112) describes white grubs and termites among the many delicacies which the pygmies receive by air from the Ituri forests.

In a lecture given in May, 1950, Doctor R. HARTWEG at the Geographical Society in Paris described the life of the Ba-Binga Negrillos of the Central Congo. They subsist mainly on game and fish, yet the women and children collect berries and caterpillars. In his films he gave some beautiful illustrations of the collection of caterpillars in baskets, how they are roasted on a fire and, when removed, are eaten in large quantities with great delight. HARTWEG tasted them himself and found them good with a nut-like flavour. He described these caterpillars as the regular dessert of their meals.

The women also have to collect the small animals, such as snakes, most other reptiles, shrimps and crabs from the rivers, etc. In the forest eyes must always be kept open: this patch on the trunk of a tree is really a colony of caterpillars, which will be speedily collected into the provision baskets. Roasted they are a delicacy and crushed among the vegetables they form a savoury seasoning (SCHEBESTA 1940, p. 31 f.).

Termites are not found everywhere, but every termite hill has its devotee. When the time for this 'crop' approaches, the camp is in full movement, the women busily plaiting baskets for the event. Every termite hill has its owner, just as has a caterpillar- or a honey-tree. The discoverer takes possession by stamping and breaking all the bushes around. The real moment is that of the swarming. The termite-hill is often visited, in order not to miss the event. The thin earthen layer of the hill is scraped away, and the galleries full of movement are inspected. Small ferrules are put into the openings of the hills; by taking them out, they learn when the swarming approaches. The flight occurs in the evening. The family is ready for the event, waiting nearby. A small screen of leaves is stretched above the termite-hill, at its base a burrow is made, where resinous wood is piled. The swarming begins with a grating noise. With spread wings the insects rise, strike against the roof of leaves and fall down to the ground. The fire is lit in the burrow: the termites are attracted by the light and fall into the trap. The squatting women collect them by the handful and throw them into their baskets, which they cover with leaves. The crop is divided, packed and brought for the festival dinner into the camp. The termites are fat and this is one reason why they are so prized. There are many ways of preparing them: one may tear off their wings and eat them alive. It is considered preferable to roast them with their wings, and to stuff them into one's mouth, until one's chin is dripping with their good fat. In order to understand the real pleasure of such feasts, one needs to hear the pygmies chewing termites. A further method is to tear off the wings and grind them to powder with salt in a small mortar.

Both sexes delight in honey. The women collect it from bees nesting in the ground, digging it out with fingers and knives. The men collect the tree-nests of bees. The nests are discovered when a swarm leaves or by the guidance of the honey-bird. The lucky discoverer marks his property by stamping the bush, and speedily calls his family. The boys climb up the branches with special hatchets in quest of the nest. The men bring fire in improvised baskets, which are hooked up on lianas, and which soon produce a heavy smoke. With its help a hole may now be cut into the nest in which one's arm may be inserted. The first comb is thrown to the ground with a solemn gesture, as an offering to the goddess. The other combs are wrapped between leaves, and later divided among the members of the family.

7. MADAGASCAR

A. and G. GRANDIDIER, who have collected all early works on Madagascar, gave a striking illustration of the importance of locusts as food by mentioning that a battle between two tribes was once interrupted by the sudden appearance of locust swarms. The fighting ceased immediately and both sides were intensively occupied in collecting this 'manna'.

Father COMBOUÉ (1886) made the following observations. The locusts generally pass in spring over the high plateau of the Central Provinces of Madagascar. The *valala* are simultaneously a pest and a benefit, as they provide valuable food for animals and men. When the natives observe a cloud of locusts, they hurry to places overgrown with high plants where they will pass. As soon as the locusts arrive, the plants are set on fire. The combined shock of the fire and of the smoke causes many locusts to drop, whereupon men, women and children collect them eagerly for provisions. The locusts are thrown into big pots, where they are well stewed and then spread on mats to dry in the sun. Wings and legs are removed, and the insects are pounded or stored as they are for the needs of the household or to be taken to market, where they can always be found. Thus dried, the locusts keep for a long time. The natives eat the *valala* either seasoned with pimento and salt, or, better, roasted in fat, or boiled with rice and meat. The last way is preferred. They also make a bouillon of it, which they season with rice. They are to be found even at the royal table at Tananariva. The late queen Ranavalona II kept, in addition to her hunters and fishermen, some women who merely scoured the fields to collect locusts. Other grasshoppers are also consumed. Father COMBOUÉ reiterates these statements in 1888, when he says that after the removal of head, wings and legs, the *valala* are soaked in very salty water, before being fried in their own fat.

Locusts, especially the migratory locust *Locusta migratoria capito* Sauss., the *valala* of the natives, still form, together with other grasshoppers, an important Madagascan food (R. DECARY 1937). When the huge flights arrive and rest on the ground, the women collect them from their clothes and put them in baskets. The preparation varies. First, there is simple desiccation and preservation in big baskets. Then they are either pulverized until they look like tobacco powder, or are eaten after cooking as a seasoning of rice. Or, the wings and legs are removed, the body is soaked for half an hour in salt water and then roasted in fat; in this form they are served even at the table of the nobles. The French railway administration taxed the transport of dried locusts 1.30 fr. per ton per kilometer, i.e. the same tax as is levied on game and fish.

The same author (R. DECARY 1950, p. 173) later reiterates the importance of locusts in various preparations as food on Madagascar. Their taste is by no means disagreable and is reminiscent of hazel nuts. The railway tax as well as the common occurrence of locusts on the inland markets sufficiently indicate the extent to which they are eaten.

Many other insects are collected and eaten on Madagascar (DECARY 1937, 1950). To our astonishment termites are not among them. They are reserved as food for the domestic fowl (DECARY 1950, p. 147). Larvae of dragonflies (*Odonata*), wich abound in the swamps, are freely eaten, but not highly appreciated. Among beetles the big *Cybister hova*, the grubs of *Proagosternus* sp. and of the Lucanid *Cladognathus serricornis* Latr. are eaten fried, mainly by the rich. Custom forbids the collection of *Cybister* in the ricefields, and they often have to be gathered far away. The custom of eating wasp maggots, mentioned by GRANDIDIER, has now been abondoned.

Other beetles eaten on Madagascar include: the fried larvae of *Scarites sp.* (CARAB.; DECARY, 1937), of palmicolous *Passalidae* (PAULIAN, vide THEODORIDES 1949, p. 128) and *Tricholespis sp.* (DECARY, 1937). The palmworm (*Rhynchophorus, Eugnoristus monachus* Ol., *Rhina sp.*, all weevils) is consumed raw or fried. PAULIAN (vide THEODORIDES 1949, p. 135) correlates the high culinary interest in the palmicolous beetle grubs to their abundance in a well defined habitat close to the settlements.

The pupae of some silk-producing Bombycids, mainly of *Borocera madagascarensis* Boisd., the *landibé*, of *Bombyx radama* Coq. and of *Libethra cajani* Vins., are eaten on the high plateau and in the West of the island. The silk is used for spinning, while the pupae are taken out, boiled in water or roasted in ashes. The *landibe* was even served in 1894 on the table of the French residence at Tananariva. In the North the pupa of *Coenostegia diegos* Mab., the caterpillars of which live in big white silken webs, are much sought after. The caterpillars and pupae of a common pest of peach-trees and mimosas, the Psychid *Debarrea malagassa* Hy., the *fangalabola*, is eaten, boiled or roasted, near Tananariva. The Sakalava of the West consume the pupae of large hawk moths, roasted, and the moths themselves. During an interview of the French Ambassador with the ill-fated King Radami I of Madagascar, his son, a boy of ten years, had his pockets filled with the roasted chrysalids of the wild silkworm *Libethra cajani*, with which he regaled himself during the interview (vide SIMMONDS 1885, p. 355).

The Fulgorid *Pyrops tenebrosa* F., the *sakondry*, is often eaten in the North. The waterbug *Nepa* and other bugs are eaten on the high plateau, but are not held in much esteem. The larva of another Cicadid, *Phremia rubra* Sign., which lives in west and south Madagascar on *Combretaceae*, freely secretes a white, slightly sugary substance, a kind of dulcine or mannite. This appears in droplets which fuse on branches or on the soil into masses which may be almost fist-sized. The Sakalava, Bara and Mahafaly are fond of this sugar, which they call cicad-honey, *tantey sakondry*.

Gathering of honey and wax is one of the main occupations of the forest tribes of Madagascar, such as the Tanala, the Metsimisaraka and other southern tribes (R. DECARY 1950, p. 151). The honey-hunters, the *mpantantely*, look for the flying bees of *Apis unicolor*, the mother of honey or *renitantely*. When the bees fly on a zig-zag course they are on their trip from the hive, but when they fly in a straight line, they return to it. The *mpantantely* then go in that direction and by a number of successive observations of various returning bees are directed towards the nest. In the forests most nests are found in hollow trees, and sometimes also in crevices of rocks.

Also beginnings of a primitive apiculture are observed on the island. Certain of these hives, like those around Ambatalaona, resemble primitive European hives. They are simple boxes covered by a small roof. In the forest villages the hives are made out of hollowed tree trunks. The wood is pierced at both ends and then closed by removable discs, each with an entrance hole. These logs are laid on the ground near the hut of the owner, or hung at a height of 3 to 4 meters in the fork of tree branches, usually in November, when the hot season begins. The bees then begin to build and the honey crop is in April. The bee works throughout the year, but the honey is slightly bitter and always of an inferior quality in the cold season, when flowers are rare.

To obtain the honey crop the logs are opened at one end, the bees fumigated and the combs, except those in the centre are removed. The same operation is repeated from the other end. When the log is on a tree, green wood is set on fire; the rising smoke drives the bees away, and they return after the end of the operation. The honey, as far as it is not consumed on the spot, is extracted in a primitive press, where it is also separated from the wax.

A kind of mead, the *toaka tantely*, is prepared from a watery solution of the honey cakes which ferments for about a week. Boiled and fermented for one day only and then filled into hermetically closed bottles, an agreable lemonade is prepared from honey.

The tiny stingless bee of southern Madagascar, the *Trigona madecassa*, the *sihy* inhabits crevices of branches of big trees. It produces only little honey, which has however an excellent taste.

The inhabitants of Madagascar are ill fed for half the year; they are very fond of fried grasshoppers and silkworms, esteeming the latter a great delicacy (SIMMONDS 1885, p. 368).

Captain GREEN (vide KIRBY and SPENCE 1823, I, p. 364) tells about a bee honey (of *Apis unicolor*) which is green, sweet and aromatic, of the consistency of oil. This green honey was exported to India, where it was bought at a high price.

A longicorn larva is prepared at Mauritius under the name of *moutac*, which white inhabitants as well as the Negroes ate greedily (ST. PIERRE Voyage, p. 72; vide COWAN 1865, p. 74).

8. LOCUSTS AND OTHER INSECTS IN NORTH AFRICA

There are many reports about locusts in North Africa. THOMAS SHAW in his 'Travels in Barbary' (1738) devotes two pages to the ravages of the locusts and continues, that when sprinkled with salt and fried, they taste like crayfish. And BENNET declares that locust soup reminded him of nothing so much as crayfish bisque which is so highly esteemed by connaisseurs. He would gladly see it on his table every day, if he could get the insects. He boiled the locusts on a brisk fire, having previously seasoned them with salt, pepper and grated nutmeg, the whole being occasionally stirred. When cooked they are pounded in a mortar with bread fried brown or with a purée of rice. They are then replaced in the saucepan and thickened to a broth by placing them on a warm part of the stove, but not allowed to boil. For use, the broth is passed through a strainer and a few croutons are added (vide P. L. SIMMONDS 1885, p. 330 f.). And SIMMONDS (1885, p. 359) was informed by T. BLACKMORE of London that locusts invade Morocco every year and are used extensively as food whenever they abound, thus reducing demands on the ordinary food supply. They are habitually roasted for eating, and brought into Tangier and other towns by the country people, and sold in the market and in the streets. The Jews collect the females only for this purpose, having an idea that the males are unclean, but that beneath the body of the females there are some Hebrew characters which make them lawful food.

E. DAUMAS and A. DE CHANCEL (1848, pp. 301–313) encountered locusts at the foot of the Hoggar Mts. A great cloud of them passed and the Negroes began eagerly to collect them, when Chegueneun said to them: 'Leave them alone! They come from the water and there we will collect them in the evening!' The locusts are good for men and camels, fresh or preserved, with legs, wings and head removed, grilled or boiled and prepared with cous-cous. Dried in the sun, they are made into a powder which is mixed with milk, or it is kneaded with flour and boiled with fat or butter and salt. The camels like them very much, dried or boiled in a large groove between two layers of charcoal. The Negroes also eat them in a similar way and our camp was instantly full of smoke from their improvised fires. The locusts are lawful food, provided they are taken alive and killed by Moslems. Yet when they were killed by frost or by infidels, their meat is impure (Imam El Malak and El Hanebali). An illustration of this is found in Verse 19 of the Koran. Many hadiths prove that God has given the locusts to man for food. EL ASMAI quotes a fellah as saying:

'The locusts settle down on my wheat field, and I tell them:
'Do not eat my property, and don 't destroy it.'
One of their sages answered me from an ear of wheat:
'We are your guests, and you must satisfy our appetite."

J. B. LABAT (1728 II, p. 177) states that the Moors take revenge on the locusts by eating them. They collect them carefully, put them into leather bags, pound them and boil them in milk. This they regard as an excellent dish. They are also fried on stoves and then make a delicate and valuable food.

CHANIER (1788) says that smoked locusts are brought in prodigious quantities into the markets in Morocco, but that they have 'an oily and rancid taste which habit only can render agreable'. The Moors use them to the present day in the manner described by JACKSON (vide SIMMONDS 1885, p. 360) by first boiling and then frying them. But the Jews of Morocco, more provident than the Moors, salt them and keep them for use with the dish called *Dafina*, which forms the Saturday's dinner for the Jewish population. The dish is made by placing meat, fish, eggs, tomatoes, in fact almost anything edible, in a jar which is put in the oven on Friday night, and taken out hot on the Sabbath, so that the people get a hot meal without the sin of lighting a fire on that day. Abbé GODARD (1860) mentions that in Morocco locusts are placed in bags, salted and either baked or boiled. They are then dried on the terraced roofs of the houses. Fried in oil they are not bad.

The Arabs of Morocco consider locusts a great delicacy (J. G. JACKSON, 1811, pp. 104 ff.). During the summer of 1799 and the spring of 1800, after the plague had almost depopulated Barbary, dishes of them were served up at the principle repasts. They were usually boiled in water half an hour, then sprinkled with salt and pepper and fried, with a little vinegar added. The body only is eaten and tastes like prawns. For their stimulating qualities the Moors prefer them to pigeons. A person may eat a plateful of them containing two or three hundred without any ill effects. JACKSON says, however, that in another place the poor people, when obliged to live entirely on this kind of food, become emaciated and indolent. STEEDMAN (1835 I, p. 137) even states that in Morocco the price of provisions falls when locusts enter the neighbourhood.

J. RILEY (1850, p. 237) records that locusts are considered very good food by Moors, Arabs and Jews in Barbary. They catch large numbers of them in their season, and throw them, while alive and jumping, into a pan of boiling argan oil, where they are allowed to remain, hissing and frying, till their wings are burned off and their bodies sufficiently cooked; they are then poured out and eaten. RILEY compares them in consistency and flavour to the yolk of hard-boiled eggs.

Capt. BEECHY (vide COWAN 1865, p. 123) tells us he saw many asses, heavily laden with locusts for food, driven into Misurata, a town in Tripoli.

A. ROBBINS (1851, p. 172) saw the Arabs of the Sahara welcome the approach of locusts, which often save them from becoming famished with hunger. They prepare them for food by digging a deep hole in the ground, building a fire at the bottom and filling it with wood. When the soil is heated as much as possible and the coals and embers have been removed, they prepare to fill the cavity with the live locusts, which have been kept in a bag holding about five bushels. Several people hold the bag perpendicularly over the hole with the mouth near the surface of the ground, while others stand around with sticks. The bag is then opened, and the locusts shaken out with great force into the hot pit, while the surrounding Arabs immediately throw sand on them to prevent their flying off. The mouth of the hole is now completely covered with sand, and another fire is built on top of it. When the locusts are thoroughly roasted and have cooled, they are picked out by hand, thrown on tentcloth or blankets and placed in the sun to dry. During the process, which requires two or three days, they must be watched with the utmost care, to prevent live locusts from devouring them, if by chance a swarm should pass over. When perfectly dry, they are slightly pounded, pressed into bags or skins and are ready for transportation. To prepare them for eating, they are pulverized in mortars and mixed with sufficient water to make a kind of dry pudding. They are, however, sometimes eaten singly without pulverizing, after breaking off the head, wings and legs. ROBBINS tasted them and considered them nourishing food. At Wadinoon (ROBBINS 1851, p. 228) locusts are sometimes boiled as food for men and beasts.

J. KÜNCKEL D'HERCULAIS (1891) wrote a special paper on the locust-eating populations of the far South Algeria. 'The natives are well disposed to carry out orders for the destruction of locusts, since they use them for food. Around Tougourt every tent and house has prepared its store of locusts, on the average about 200 kilo to each

tent. Sixty camel loads (9000 kilo) are the quantities of locusts accumulated daily in the Ksours of the Oued-Souf. They are a valuable
resource for the poor population. To preserve them, they are first
cooked in salt water, then dried in the sun. The natives collect and
prepare such considerable stocks that apart from their own needs,
they have some for trading on the markets of Tougourt, Temacin,
etc. I have in my hands now two boxes of freshly prepared locusts
and I convinced myself that they are quite an acceptable food.
The taste of shrimps is very pronounced; with time they lose their
quality'.

This is not the place to give an account of bee-keeping in North
Africa. WESTERMARCK, however, has made a collection of many
ancient beliefs concerning bees, some of which have very old traditions. The inhabitants of Morocco, for instance, believe that there
is much *baraka*, holy blessing, in bees and in honey (1926, I, p. 104;
II, pp. 47 ff.). Certain Moroccan tribes perform mid-summer ceremonies to increase the supply of honey. People will burn cow-dung,
the smoke of which will make the honey plentiful and protect the
bees from thunder storms. Another ceremony is to take earth from
a place where three roads meet and throw it over the bees to keep
them in good condition. In another tribe a honeycomb is cut into
two pieces on mid-summer day and eaten if there is honey in it
(WESTERMARCK 1905, pp. 28 ff.). WESTERMARCK was told that if
these ceremonies were not performed, the bees would have no honey.

We quote the following from among the relatively few reports on
other insects consumed in North Africa:

M. V. MOTSCHOULSKY (1856, p. 77, vide Prof. LICHTENSTEIN
of Berlin) states that the eggs of the water-bug *Corixa esculenta*, having
the appearance of manna, serve as food in Egypt. Unfortunately,
I have not been able to find any other reference for this observation.

At Fezzan (N. Africa) cakes, tasting like caviar, are produced
from the eggs of insects collected in certain desert pools (vide
SOUBEIRAN 1870, p. 722). This unauthentic note may refer to eggs
of water-bugs. HUMBOLDT already knew the *Ahuatle* of Mexico, and
asks whether the *Loul* collected by the Arabs from pools in the
deserts of Fezzan are not the same (vide BARGAGLI 1877, p. 6).

Dr. CLARKE reports that *Ateuches sacer* L. in Egypt is used against
sterility. LETOURNEUX (Pet. Nouv. Ent. 1875) mentions that *Blaps sp.*

is used by women in Tunis if they wish to put on weight or beautify themselves. NIEBUHR mentions the use of *Tenebrio* and *Pimelia* in Turkey for the same purpose (see also BARGAGLI 1877, p. 4).

The larvae of the longicorn beetle *Dorysthenus forficatus* F., the *korta hlima*, are eaten in Morocco, fried in oil. They are collected with the help of a short-handled rasp (GHESQUIERE 1947).

V. ASIA

1. GENERAL INTRODUCTION

From a zoogeographical and ecological point of view, Asia is not one but a number of continents. From the Arctic Circle of Siberia and from the tundras, taigas, steppes and deserts of Central Asia we have almost no information on insect-eating habits, apart from a few notes on curiosities. Possibly insect food was actually of little or no importance. Hunters in regions richly stocked with game have little need of additional food and this holds also for highly specialized nomads. The remainder of Asia may be divided roughly into three major sections. The Middle East, a zoogeographical mixture, offers as a very old centre of highly specialized agriculture and animal husbandry neither in its historical development nor in its present condition much inducement to entomophagy. The occasional locust swarms are eagerly acclaimed by the nomads, but certainly not by the fellahs. Wild honey is utilized wherever it may be available, and bee-keeping itself has been long established. The various mannas are more a curiosity than a source of food.

In tropical Asia the Indian sub-continent is very distinct from the primeval forest worlds of Malaya and Indonesia. Yet in all districts many insects are regularly gathered and consumed as a welcome addition to a one-sided diet in the monsoon areas. Termites, locusts and crickets, ants, beetles, many insect larvae, bugs and cicadas, etc. are eaten. In the forest districts honey-hunting is widespread, and for the Veddas it is one of the main sources of livelihood and barter.

In China and Japan also the eating of insects, from silkworm pupae to water beetles, is widely accepted. With the development of a highly specialized agriculture this is no longer of basic importance, but certain insects are everywhere sold in special shops in the towns and gathered by the peasants as prized delicacies.

Asia as a whole yields little information on insect-eating by primitive man, but shows that even with very intensive agriculture a one-sided and unbalanced diet may induce conspicuous entomophagy through the instinctive craving to make good any inadequacies of diet. Before discussing the information available on insect-eating in specific regions, we give some general facts about certain beetles and stinkbugs in the diet of Asiatic peoples.

Coleoptera. Cybister and other water beetles are not only consumed in China, but GHESQUIÈRE (1947) also mentions that fried *Cybister* are sold in special shops in the Chinese quarters of California, being imported from China.

The larvae of *Passalus interruptus* L., a pest of potatoes , is consumed in Bengal (BRYGOO 1946, p. 63). In the Shan states of Burma the population searches from March to May for the pupae of *Heliocopris*, the *shwe-po;* these are a delicacy which are even widely exported (GHOSH, 1924). *Megasoma actaeon* L. is consumed in Malay (NETOLITZKY, 1920). The beetles of *Xylotrupes gideon* L., which are prized in Siam as fighting insects, and also their larvae are another Siamese delicacy.

Among longicorn beetles *Batocera albofasciata* Deg. and *B. rubus* L. are eaten in Indonesia and Ceylon (NETOLITZKY, 1920).

The common palmworm of Indonesia and the Sunda islands is *Rhynchophorus ferrugineus* Oliv., and in Siam *R. schah* F.. GHESQUIÈRE (1947) says that in the Dutch East Indies the palmworms are recognized as a most fortifying, easily digestible food which is given with preference to weak and consumptive persons. Malayans in Europe often order them to be sent from their home.

Many Melolonthid beetles are eaten in Indonesia and Malaya. The habit of Turkish and North-African women of eating adult Tenebrionid beetles (*Blaps, Tenebrio, Pimelia*) and also dung-beetles (*Scarabaeus*) when they wish to put on weight, must be considered more as a curiosity. Fat women were the beauty ideal of the Levant. NIEBUHR, SHAW, and many other travellers, particularly of the 18th century, report on such habits.

Rhynchota. W. L. DISTANT reports in the Fauna of British India: Rhynchota I (1902, p. 283) that, according to Capt. GORMAN, *Aspongopus nepalensis* Westw. is found under stones in the dry riverbeds of Assam. The bugs are much sought after by the natives, who use them for food, pounded up and mixed with rice.

C. STRICKLAND (1932, pp. 873 ff.) obtained from Dr. CONNOR of Upper Assam a related bug, *Aspongopus chinensis* Dall., with the following letter: 'I send herewith a beetle which was found in one of the Assam Frontier Tracts. It lives under large stones in the Lohit river and appears to be edible. The Mishmis eat it and before doing so, they remove two little red bags which they say contain poison; these bags appear to lie between the thorax and abdomen. Another gentleman informed me that he knew of this matter and that if perchance the natives forget to extract the poison glands they get paralysis of the neck from wich they inevitably die'. A number of officials confirmed the known edibility of these bugs. Mr. FURZE even sent *A. chinensis*, *A nepalensis* and a new bug, which STRICKLAND described as *Cyclopelta subhimalayensis*. These are all eaten by the Miris, Mishmis, Abors and Nagas tribes. In the case of the Mishmis, who scrape a bare existence in very precipitous country, this is easily understandable. FURZE also heard about a well-authenticated report where its eating caused paralysis. STRICKLAND fed complete bugs to monkeys, which did not become paralysed nor showed any measurable physiological change.

HOFFMANN (1947, p. 236 f.) reports the use of *Coridius chinensis* Dall. as a widely used aphrodisiac all over China, where it is called *Chu shan chung* or *Hai tao chung*, and is mentioned as early as 1590 by *Li shi chen*. *C. nepalensis* Westw. is widely used in Assam, pounded up and mixed with rice dishes to improve their taste. Another stink-bug, *Erthesina fullo* Thnbg., is eaten by the Nagas, an Assamese tribe. This widespread Oriental Pentatomid is of variable colour and is found on the trunks of many trees.

2. THE MIDDLE EAST
(*Locusts, Mannas, Honey and Honey-hunting*)

a. *Locusts*. Many of the travellers in the Levant and in N. Africa during the 18th century commented on the eating of locusts by the Arabs. One of the best and earliest of these reports is that by the Swedish traveller FREDERIC HASSELQUIST in 1750 (1766, pp. 230 ff.):

,During my stay in Egypt I used every means to learn whether locusts are eaten to this day either in this or the neighbouring countries. I was the more solicitous to be informed of this, as I thought the answer would determine what St. John lived on in the desert . . . . If it can be demonstrated that locusts are to this day eaten in the neighbourhood of the place where St. John dwelt, the impossibility and unnaturalness of this diet can no longer be asserted . . . . A traveller is the only person who can learn whether locusts are still eaten in the East . . . . Reliable information and suggestions, therefore, produced by persons who have visited and seen the customs of the country, are the only available means for obtaining the truth; and these I have earnestly endeavoured to obtain during my travels in Asia and Egypt. I have asked Franks who have long lived in the East whether they ever heard that locusts were eaten there. They all answered in the affirmative, and those of greatest veracity I got in Egypt (CHASSIN) and Aleppo (BONARD), being the places where such information may be easiest obtained. I have asked native Christians, Armenians, Greek, Copts and Syrians who were born here and travelled in Syria, near the Red Sea and Egypt, if they knew whether the Arabs eat locusts. All answered that they have both seen them eat them and heard that they were a common food among the Arabs. But the information I had from Greeks who have travelled to Mt. Sinai is that which I can most depend on, for the Greek Church has a noted convent there . . . . I at length met with a person who gave me better information and stronger assurances than all the rest. This was a Sheikh with whom I was acquainted in Cairo, a most learned and ingenious man who had been six years at Mecca. In the presence of M. LE GRAND and M. FOURMONT he said: 'At Mecca which is furnished with corn from Egypt, there frequently rages a famine, when there is a scarcity in Egypt. The people here are then obliged, as in all other places of the world, to support life with unusual food. Locusts obtain a place then among their victuals. They grind them to flour in their hand mills or powder them in stone mortars. They mix this flour with water forming a dough and make thin cakes of it, which they bake like other bread on a heated griddle; and, for want of something better, this serves instead of bread to support life'. On my question, whether the Arabs do not eat locusts without being driven by necessity, he answered that it is not uncommon to see them eat locusts when there is no famine; but then they boil them a good while in water, afterwards stew with butter and make a sort of fricassée, which, he says, according to his own experience, does not taste bad. The locusts of Arabia do not differ from those of Egypt. Small and large locusts are eaten without distinction. At certain seasons these insects are as common in Arabia, as they are rare in Egypt at this time'.

LE CHEVALIER CHARDIN (1711 IX, p. 227) observed in late March near Bender Abbas in S. Iran that the sky appeared to be obscured by clouds owing to the locust swarms flying 60 to 70 feet high, and wherever these swarms passed an enormous quantity of locusts fell to the ground: big, red insects, so heavy that they could not rise again. The peasants catch them as they drop. At that season similar clouds passed almost every evening, and the locusts are caught, dried and salted, but also eaten raw. They are sold on the markets of S. Iran as a common food.

In the Oriental Memoirs of J. FORBES (1813 I, p. 40) we read about locusts being sold in the markets and eaten with rice and dates, and sometimes flavoured with salt and spices. His *Acridites lineola*, which is the species commonly sold for food in the markets of Baghdad, is, of course, *Schistocerca gregaria*.

ANNA BLUNT in her Pelerinages in the Nedjd (1881) writes: 'Locusts have become our daily lent-fare, and they are excellent to eat. After having tried them in different forms of preparation, we concluded they be best when boiled. The legs are pulled out, they are taken by the wings, dipped into salt and eaten. They have a vegetable taste, like green wheat. They replaced us all the missing vegetables. The red locust is better than the green one . . . . WILFRID thought they would not be bad among the hors d'oeuvres of a restaurant at Paris', whilst Lady BLUNT was not so sure thereabout.

S. HEDIN (1918, p. 270) reports that along the shores of the Euphrates and the Tigris, the Arabs tear off the wings and legs of the locusts and roast their bodies over the fire, just as in the days of St. John.

For more recent times we quote only a few of the great classics of Arabic lore and travel. CHARLES M. DOUGHTY (1926, I, pp. 203, 336, 472; II, pp. 245, 323, 332, 436) has described the inter-relations between Bedouin and locusts in a masterly fashion: 'The locusts *(jarad)* devour the Beduw, and the Beduw devours the locust'.

The Nedjd husbandman every year suffers from locusts, which may be locally bred or merely passers by. 'This year were hitherto few and weak flights, but sometimes with the smooth wind locusts drove in upon us; then the lads with palm branches of a spear's length ran hooting in the orchard and brushed them out of the trees and the clover. The good lads took up the bodies of the slain crying: 'They are good and fat', and ran to the arbour to toast them. If I were there, they invited me to the feast'. On the market of Boreyda in the Middle Nedjd Doughty observed in many shops baskets of dried locusts.

But the true importance of locust food is with the Bedouin (I, 472): 'I suffered their summer-famine with the nomads. They who are brought low by hunger in so serene and cherishing an atmosphere are not soon carried into wasting diseases . . . . It is, they believe, of that little camel-milk they have to drink, that their bodies are made nimble and light, and hardened to a long patience of fatigue and hunger . . . . It is seldom in their lives that they must make a shift to endure with a squalid diet of locusts; which they say may hardly hold life in them until better times'. One evening the wind brought a few locusts into Doughty's camp. These locusts were toasted at all watchfires and eaten. The women on the morrow had gathered great heaps, and were busy singeing them in shallow pits, with a weak fire of herbs; they give up a sickly odour of fried fish oil. Thus cured and a little salt cast in, the locust meat is stored in leather sacks, and will keep a good long while. They mingle this, brayed small, with their often only liquid diet of sour buttermilk. Locust powder is not a food to set before guests; and Doughty saw poor nomads a little out of countenance when confessing they appeased their hunger by eating this wretchedness.. The best is the fat spring locust. The later broods are dry and unwholesome. The early locust, toasted, is reckoned a sweetmeat in town and desert.... The children bring in gathered locusts, broached upon a twig, and the nomads toast them on the coals; then plucking the scorched limbs, they break away the head, and the insect body which remains is good meat; but not of these later swarms, born in time of the dried-up herbage . . . The poor lad drew out parched locusts from

213

his scrip and fell to eat again: locusts clouds had passed over the mountain, he said, two months before . . . In the Nedjd DOUGHTY's guide turned his attention to some trenches which were dug by locust hunters.

P. W. HARRISON (1924, p. 289) explains that roasted locusts taste better than they look. 'We do not have roasted locusts every year in Arabia, but when we have them, we have lots of them. In such years the locusts may come over the country in great clouds which obscure the sun for two or three days'.

The famous Swiss explorer of Arabia, J. L. BURCKHARDT (1831) says that all the Bedouins of Arabia as well as the town people of Nedjd and Hejaz are accustomed to eat locusts. At Medina and Tayf he saw locust-shops, where these insects were sold by measure. In Egypt and Nubia they are eaten by the poorest beggars only. The Arabs, in preparing them for food, throw them alive into boiling salty water. After a few minutes they are taken out and dried in the sun. The hard legs and wings are then torn off, the bodies are cleaned of the salt and completely dried, after which process whole sacks are filled with them by the Bedouins. They are sometimes consumed broiled in butter, and they often form part of the breakfast, when they are spread over unleavened bread mixed with butter. Of all the Bedouins encountered by BURCKHARDT, those of the Sinai alone did not use locusts as food.

GUARMANI (1938) who travelled in N. W. Arabia about the middle of the last century, writes: 'At Tueie the inhabitants were engaged in gathering up locusts which they had roasted in deep holes in the sand. I bought four sacks full. They are a valued source of supply to the inhabitants of the Nedjd. Their flights across the sky are watched by many anxious eyes and they are followed wherever they settle. Holes are dug in the ground, wherein they are roasted with all speed'. GUARMANI claims that experience had taught him that locusts are not to be recommended as food for man, in spite of the enthusiastic remarks to the contrary made by the greedy orientals. When roasted they are tasteless, and when boiled they become watery, although for horses they are as good as oats. They fill their stomachs and increase their muscle without making them fat . . . . The traveller should have them gathered and after the legs and wings are taken off should buy them at the price of barley. The peasant reaps an immense harvest from this plague.

214

And T. CANAAN (1928) writes of a visit to the Azazima Bedouins of the Negeb: 'Locust swarms passed from the South with a heavy grinding noise . . . . Many Bedouins roasted the adults over a small fire. After a light roasting the insect is gently rubbed between both hands, to break off the wings and legs. Roasted locusts are a highly esteemed dish with the Azazime'.

H. B. TRISTRAM (1873) found locusts very good to eat, when stewed after Arab fashion with butter. They tasted somewhat like shrimps, but with less flavour.

The delight shown by Arabs towards locust food is doubtless essentially a reaction towards the attitude of the visitor. Let us hear what W. G. PALGRAVE (1865, II, p. 138 f.) has to say. When the locusts arrived in the Hofhof, 'I really thought they would have gone mad for joy. Locusts are here an article of food, nay, a dainty, and a good swarm of them is begged of Heaven in Arabia no less fervently than it would be deprecated in India or Syria. This difference of sentiment is grounded on several reasons; a main one lies in the diversity of the insects themselves. The locust of Inner Arabia is very unlike whatever of the same genus I have seen elsewhere . . . . The locust of Arabia is reddish-brown, twice or three times the size of the green northern grasshoppers; they resemble a large prawn in appearance, are as long as a man's little finger . . . . This locust when boiled or fried is said to be delicious, and boiled and fried they are accordingly to an incredible extent. However, I could never persuade myself to taste them, whatever invitations the natives, smacking their lips over large dishes full of these 'delicatesses', could make me to join them. BARAKAT ventured on one for a trial; he pronounced it oily and disgusting. The swarm now before us was a thorough godsend for our Arabs, on no account to be neglected. Thirst, weariness, all was forgotten, and down the riders leapt from their camels; this one spread out a cloak, that one a saddle-bag, a third his shirt, over the unlucky creatures destined for the morrow's meal . . . . Thus we left our associates hard at work'.

CARL R. RASWAN (1934, p. 59), who spent many years living with the Bedouins, writes that while camping near the oasis al-Jauf, reddish locusts began to 'rain' upon the tents. Immediately, men, women and children hurried to collect them from the ground which they covered with a red carpet. Locusts were roasted on all fires.

Children, women and men sat around the fires, and ate them, holding them by their wings, pulling off the legs, and dipping them into salt. Boiled, RASWAN did not like them, as they tasted like insipid cabbage. Yet roasted they are crisp, the interior having the taste of spinach. They are clean animals, not at all repugnant, but one soon tires of them, when they are served every day. Women and children continued to collect them, and on the next morning mountains of locusts were spread for drying in the sun. When we left the camp a few days later, we had no empty sacks or bags. The camels were fully loaded with dried locusts. Men, dogs and camels delighted in them, but for a few days only, when they became repugnant. The remainder of the dried locusts was kept for later days of famine, as huge locust swarms always foretell drought and famine . . . . Faris told him that tens of thousands of Bedouins have to subsist often for weeks only upon locusts, and camels and horses have at times to be fed with them. Four days after breaking camp every pasture was destroyed by the locusts.

J. J. HESS (1938, p. 110) gives the following recipe in use with the Arabs of Central Arabia for the use of locusts as a dainty dish. The locusts are collected early in the morning, while still drosy with cold, in bags which are sown up. Then they are thrown into boiling salty water, which later is poured away. The insects are then spread upon old tent cloth to dry in the sun. For eating they are pounded in a mortar and the flour is mixed with salt and fat, and perhaps also with dates. Sometimes the locusts are roasted in a special cooking pit, the *zibweh*.

This attitude of the Bedouin of Kuwait and Saudi Arabia has scarcely changed in our days. H. R. P. DICKSON (1949, pp. 448, 453 f.) writes: 'The Bedawin proper do not mind locusts much, in fact they rejoice when they see them, as they take them as an omen for a good year, since locusts come only in prosperous years. These camel-nomads are not worried if large tracts of the country are denuded of fodder; they simply move off to unvisited country. They know that the damage is more or less local, and they will migrate northwards. They also enjoy the wonderful opportunity to satisfy for once in a while their ever prevailing hunger. Also their camels and dogs will get their fill, as well as foxes, bustard, monitors, kites, shrikes and many other denizens of the desert . . . .' DICKSON

216

remembers the arrival of a swarm of red locusts in a camp (*Schisto-cerca gregaria*, not *Nomadacris*, as DICKSON assumes on p. 448). The whole camp turned out to collect the insects for food and from the numbers collected, it was evident that they had a most successful day. The roofs of all the black tents were strewn from end to end with dead locusts spread out to dry; every available mat, rug and blanket was also laid out for the excess catch. Camels, dogs, mares and men were hard at work eating the insects, while cooking pots, bags, paniers and water-skins were crammed to overflowing with dried ones. DICKSON did not fancy locusts much, whilst Mrs. and Miss DICKSON revelled in them. He was told that even the Bedouin enjoy them for a few days only, after which their gorge seems to rise against the food, and they give the remnants of their catch to the animals. DICKSON states that the red locust is the stage most eaten in Arabia. The females only are eaten, and when properly prepared, fried in butter with salt, or boiled, he found them quite good! They taste much like roast chestnuts. Yet we doubt the motive given by his Bedouin friends, that in eating them they feel they are taking revenge for all the damage they have caused. Such may be quite properly the reaction of the Fellah, but not of the Bedu.

b. *The Mannas of the Middle East.* Sweets are the delight and the culinary dream of the nomads of the desert. No wonder that the manna of the Sinai provoked excitement and joy among the Children of Israel while migrating in the Sinai mountains (Exodus 16, Numbers 11).

The most widely accepted view in textbooks is that the lichen *Lecanora esculenta* is identical with the manna of the Bible. This lichen grows on rocks and produces fructifications in the form of pea-sized globules which are light enough to be blown about by the wind. The natives of the Irano-Turanian steppes, from Central Asia to the high Atlas plateau, collect these globules for their sweetness (*'halva'*). The idea of a manna rain based on this phenomenon is impressive, even though the collected bits remain as delicacies and do not appear in quantities large enough to replace bread. However, these rains of *Lecanora* are rare and unique occurrences and usually happen during the day. The Bible reports that the manna appeared at night, and also strongly indicates a regular recurrence of the event.

Another fact which refutes the idea of *Lecanora* as the manna of Sinai is its apparent absence in the region of Sinai during the last 150 years when many collections of rock lichen were made. During this same period not one traveller has ever reported a manna rain or heard about it from the local monks or nomads.

The record of the oldest local traditions of Sinai comes from FLAVIUS JOSEPHUS and the early monks of the St. Catherine monastery followed by an almost unbroken line of reports by travellers (BODENHEIMER 1929, 1950). These reports link the manna with the tamarisk thickets in the wadis of the Central Sinai mountains. Here year after year in June appears a granular type of sweet manna from pinhead to pea size. It appears on the tender twigs of tamarisk bushes for a period of three to six weeks. The quantity of this manna fluctuates according to the winter rainfall. The crop may fail entirely in one wadi and at the same time be plenteous in others. Certain wadis such as Wadi Nasib and the Wadi esh-Sheikh are especially famous for their manna production. Usually the annual crop does not exceed several kilograms, but one steady man may collect over a kilogram a day at the peak of the season. This certainly does not allow for the 'bread' or daily food of the wandering Israelites. However, we must note that *lechem* does not have an original meaning of bread, but of food in general. Otherwise, it could not have come to mean 'meat' in Arabic. All in all, the nutritive value of these few kilograms of manna could not have been important enough to deserve a recording in Israel's history. There must have been a special quality to justify its inclusion in the chronicle. This special quality was its sweetness. A chemical analysis reveals that this type of manna is mainly a mixture of three sugars (glucose, fructose, saccharose).

This manna was regarded until recently as a secretion of the tamarisk. EHRENBERG, however, about 120 years ago connected it with an insect which he described as *Coccus manniparus*. His description is so mixed in character that we must suppose he described the larva of a lady beetle within the ovisac of a scale insect together with the insect itself. EHRENBERG assumed that the insect sucked on a plant, and from these sucking holes or punctures oozed the manna.

In June of 1927 the writer visited the valleys of Central Sinai in order to study this manna. It soon became obvious that two closely related species of scale insects produce the manna by excretion. These are *Trabutina mannipara* Ehrenberg, the short, brown, sticky ovisac of which was described by EHRENBERG, and the *Najacoccus serpentinus* Green, which is easily recognized by the very long, snow white, narrow and cylindrical ovisac. Of these two, the former is the manna producer in the mountains, the latter in the lowlands. This manna is none other than the well known honeydew excretion of so many plant lice and scale insects. The honeydew drops are excreted mainly by the growing larva and immature females. Rapid evaporation due to the dry air of the mountain desert quickly changes the drops into sticky solids. These manna pieces later turn a whitish, yellowish or brownish color. The reason that insects excrete and waste high-valued foodstuffs in great quantities, such as pure sugars, can be explained on a physiological basis. These insects suck great quantities of plant saps which are rich in carbohydrates but extremely poor in nitrogen. In order to acquire the minimum amount of nitrogen for the balancing of their metabolism, they must take in carbohydrates in great excess. This excess passes from them as the honeydew excretion.

Now, we shall compare briefly the Biblical report with the actual manna phenomenon in Sinai. But first, we must make clear one thing about the Biblical report. All the statements about manna in the early formations of Scripture, the Elohim or Yahweh codes, agree closely with our observations. However, many commentaries of the Priestly code, which were added hundreds of years later and which are based on conjectures or on misinterpretation of the oral tradition, show definite divergences (F. S. BODENHEIMER 1929, pp. 79 ff.).

We begin with the criteria of space and time. The location of the manna excretion which is given in the older codes as beginning at Elim (near Wadi Gharandel) and ending at Rephidim (the oasis Feiran), agrees well with the northern limits of the manna excretion in our day. Manna first was discovered on the 15th day of the second month after the Exodus from Egypt. This would be the middle or end of Siwan, which is late May or early June. This date agrees with the natural season of manna production. The description of manna in the Bible, which likens it to small, light brown cummin seeds and to the stickiness of bdellion resin, is a remarkably suitable description of the tamarisk manna. In the Bible its taste is described as like that of *tsappihith bidvash*, wich easily may refer to the crystallized grains so often found on the surface of honey. Exodus 16 : 14 and Numbers 11 : 9 state that the manna fell from heaven during the night. Actually, most of the dropping of manna, or at least its accumulation on the soil, occurs at night when the ants are not collecting it. And in many countries the honeydew of aphids is called 'dew of heaven', because the causal connection between the insects and the honeydew is not recognized.

An easy explanation can be offered for the worms and stinking decay of the manna which was collected in excess of need (Exodus 16 : 20, 21). The manna grains are eagerly collected by ants, and in a primitive tent there is little protection against them. The worms can be called ants, while the addition of stinking decay is a later misinterpretation and interpolation. The Bible knows a special word for ant (*nemalah*), but in this place it uses the general word for worms and vermin (*tola*). However, a personal experience may explain this. When I asked our nomad guide for the name of the many manna-collecting ants, he called them *dudi*, which corresponds to the Hebrew *tolaath*. When I asked him if they might be called *nimleh* (*nemalah*), he answered that it was possible. Since he knew both designations he had used the more general one. This was almost certainly the procedure of the oral tradition. Because the late editors of the Priestly code did not properly understand the situation they added the misleading interpolation.

Exodus 16 : 21 tells us that the manna was collected early in the morning and that it 'melted when the sun grew hot'. We must regard the melting as a late and mistaken interpolation. The ants begin to collect the manna only when the temperature of the soil surpasses 21° C. (70° F.). In most of the wadis at the time of our visit the rays of the sun usually accomplished this about 8 : 30 a.m. This activity of the ants ceases in the early evening. In the lowlands the ants begin much earlier. All those manna grains which drop from late afternoon to early morning remain until the beginning of ant activity in the morning. Then, however, they are speedily collected and carried away.

These and a number of other agreements between the Biblical report and the tamarisk manna have led people to connect them since olden times. We have seen that all the eyewitness reports of the Bible can be taken as literal descriptions of the tamarisk manna of Sinai.

The tamarisk manna of the Sinai is no isolated event. It is not even the most abundant manna known to us. If it were not for the Biblical record we would know very little at all about it. *Man* is the common Arabic name for plant lice (*Aphidae*), as well as for the honeydew which they excrete. The Kurdish *man-es-simma,* the manna of the skies, will be discussed later. A number of other minor manna secretions have been observed in the Sinai peninsula. A. Kaiser (1924, 1930) discovered them on the shrubs of *Haloxylon* and *Artemisia,* without being able to find the insects responsible for their appearance. We have been able to identify two small cicadas, *Euscelis decoratus* Hpt. and *Opsius jucundus* Leth. (*Jassidae*), as additional manna producers on tamarisk (Bodenheimer 1929, pp. 75 ff.). Yet the quantity of manna produced by all these forms is minimal. The *gez* production (Persian for manna), which Th. Hardwick (1822) described and illustrated from an unknown tree at Hussainabad on the Irano-Baluchi border, excreted by a Psyllid larva must be greater. In Iran a number of other mannas are popularly known and used officinally; these, however, have not yet been properly studied. According to reliable reports, a tamarisk-manna is known from one of the western mountain ranges of Iran which apparently is produced in some quantity by a species of *Trabutina* and by *Najacoccus serpentinus.* The manna produced by the big *Cicada orni* L. on ash trees in Italy was also of officinal use. It remains to be discovered whether this manna is a product of the plant or of the insect. We also discuss briefly another so-called manna of the Middle East, the *Trehala*-manna, which are the pupal cocoons of weevils of the genus *Larinus,* to which the name of manna should, however, not be applied.

Accordingly, we find that manna production is, essentially, a biological phenomenon of dry deserts and mountain steppes. The liquid honeydew excretion of a number of cicadas, plant lice and scale insects speedily solidifies there by rapid evaporation. From remote times the resulting sticky, and often hard or granular, masses have been collected under the name of manna. An interesting field of research is still open, yet apart from the Kurdish *man-es-simma* and the Sinai manna none has ever attracted attention beyond the local populations who use it widely for officinal purposes. This chapter is included in the present discussion more as a curiosity than as a pretence that these mannas have ever formed an essential article of food. The mannas of the Middle East are a complete parallel to the Psyllid mannas of Australia and Africa.

The manna second in importance is the Kurdish manna, which is found from Elazig to Urmia and south at least to Sulaimaniya. This is the *man-es-simma*, the manna from the skies, so often mentioned in the Persian and Arabic pharmacopoeas. It is sold in large lumps of hard consistency, stone-like, and always mixed with small fragments of oak leaves. When I obtained the first specimen in Turkey, I was certain that it was a stone covered superficially with a small amount of hardened honeydew mixed with the fragments of oak leaves. We learned much later only that the entire 'stone' was manna. In October, 1942, GELAL BEY, then Qaimakam of Shuarta, treated us to some manna in the same condition as above and also with a purified morsel, which had been boiled and pressed through cloth. It is collected in the surrounding oak forests and is consumed by the peasants as a sweet for breakfast in the form of sherbet drinks, or as a popular medicament. Mixed with flour the manna is turned into delightful cakes. J. LEIBOWITZ (Nature 152. 1943, p. 414) analyzed some of our samples and described them as a half syrupy, half crystalline mass. One of the two specimens was purified. The sugar component mainly consisted of a rare disaccharide, called trehalose, making up 30 to 45% of the total dry matter, or 70 to 80% of the total carbohydrate content, the remainder being formed by the monosaccharids. Trehalose had previously been known only from the pupal cocoons of a weevil (see later), from yeasts, some fungi and from *Selaginella*. The Kurdish name for this manna is *gezo*. In Persian Kurdistan *Quercus mannifera* Lindl., *Q. persica* J. and S. and *Q. taurica* Kl., in Iraki Kuristan *Q. infectoria* are known as sources of the manna.

There is not the slightest doubt that the producer of this manna is an aphid. We twice visited the Jebel Manimenga near Penjween in the quest of this aphid in 1943. A heavy manna crop was expected by the local authorities after heavy winter rains and a cool spring. The normal season is June, with a second minor peak in September.

Unfortunately no manna production was observed in 1943 at all and we could not detect any aphids in numbers which would help us to identify the species. We hope that future observers will pay attention to this oak aphid.

In the files of the Directorate General of Agriculture we found a report by Mr. JAFAR AL KHAYAT in 1937. JAFAR regards the manna as a plant secretion of oak leaves induced by the feeding of an insect, which he supposes to be a small green aphid. The manna first appears on the side under of the leaves as a gummy liquid, which drops on the upper surface of the lower leaves, on branches and on the soil. It is collected in a great many places of the Liwa Sulaimaniya and also in the Halebje district. It is entirely restricted to the forests of *Quercus infectoria* in the higher altitudes. Its normal season is from the middle of June to the second half of July, and it is collected in the coldest hours of the early morning. The peasants believe that it drops from the sky on the leaves and soil. When rains are heavy in spring and in June, the manna and the insects which produce it are washed off the oaks, and the manna production is small. Cold winds increase, hot winds and warm, cloudy weather decrease its quantity. When the weather has been favourable and much manna has formed on the trees, the collectors begin their work. They cut large numbers of the branches on which manna has been formed in any substantial quantity. The branches are then beaten until the manna has dropped off. It is gathered into skin bags and brought to the market as lumps of crystallized manna mixed with pieces of oak leaves and dirt. Very rarely pure, white manna is found. The confectioners who buy it there, beat it into pieces until it becomes soft. Then it is filled into jars, mixed with water and left for 24 hours. The liquid is poured into bags (*al shal*), which are suspended above vessels. The bags are pressed and the liquid which passes out is collected in the vessels. This liquid is then mixed with eggs, 50 eggs for each 400 gr. of manna, with almonds or nuts and with some essences. The whole is boiled, cooled and cut into pieces, which are covered with fine sugar powder. This is the manna which is sold in the markets and streets of Baghdad. The Iraki authorities estimate that annually about 30,000 kg. of manna are sold on the markets throughout the country, two-thirds of which come from the Iranian side of Kurdistan.

From Turkey the writer was able to obtain much less precise information. It is common in the vilayets of Mardin, Van, Siirt and Elazig, where, suddenly in May or June, clouds are assumed to appear over different parts of the country which shed a certain liquid, which solidifies into a sugary solution, looking just like hoarfrost. It is white. What drops on the soil is lost. As no forecast of the manna dropping can be given, much manna is lost in this way. The manna which falls on the oaks solidifies and the villagers collect it without delay, as it soon dissolves and its collection becomes impossible. Within the same village the manna may fall on one field, yet not on that adjoining. Manna collects also on other plants, such as *Thuja*. The manna from walnut trees or from tobacco leaves (other honeydew excretions of known aphids) has a bitter taste. The green colour of the sample comes from the leaves of the oaks. Manna itself is a sugary syrup, white and transparent. After melting, it becomes reddish. In certain years manna may be found in the vilayet of Mardin, but is absent from Siirt and with very little only at Van. In some years it does not fall at all. To conclude, writes our correspondent, the fall of the manna remains an inexplicable mystery.

We may add some information on the *Trehala* cocoons of certain weevils which occasionally have also been reported as food. However, we have found them only in medicine use. These cocoons are sold in the bazaars of Baghdad and other towns of Irak under the name of *teehan*. These are the pupal cells of weevils collected from thistles, mainly *Echinops*, in the mountains of Kurdistan and in the steppes. They measure about two cm. in length and have a calcareous, roughly granulated surface. The inner wall is smooth. The cocoons have a sweet taste and dissolve in water into a thick slime. GUIBOURT (1858) found them to contain 66.5% amylum, 4.6% gum and 28.9% sugar. The sugar was described by BERTHELOT (1858) as a new disaccharose, which he named trehalose $(C_{24}H_{42}O_{21})$. AVICENNA included it into his pharmacopoea, and the Pharmacopoea Persica of Father ANGE (1681) recommended it against coughs and other diseases of the respiratory organs. It is still applied widely by physicians and in popular medicine all over the Middle East. The cocoons are boiled in water until the dissolve and the solution is drunk. The beetles from the cocoons which we bought on the

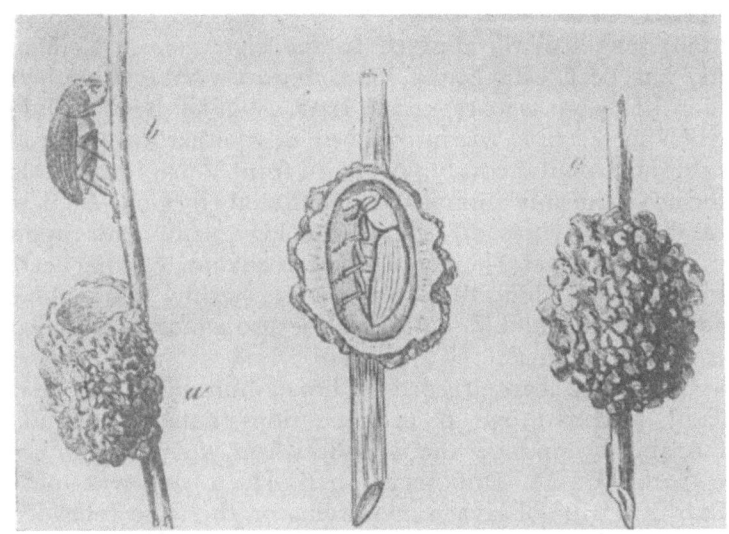

Fig. 31. The *Trehala* the pupal cocoon of *Larinus nidificans.*
From R. BLANCHARD 1890, Fig. 781.

bazaar of Baghdad belonged to *Larinus onopordi* F. This species is
widely distributed, from Turkestan to the Mediterranean shore. In
the Zoological Museum at Berlin there is an old specimen of this
species with a note that the cocoons of this beetle are used medici-
nally in the Levant (SCHUMACHER, D. Ent. Zeitschr. 1921, p. 98).
We have found similar cocoons of *Larinus rudicollis* Petri in Israel
(BODENHEIMER 1935, pp. 247, 249). In Iran *L. onopordi* (= *L.*
maculatus) and *L. mellificus* Jekel (= *L. nidificans* Cap.) are known
to be used in the same manner. *L. syriacus* Gyll. also belongs to the
same group. The materials for the cocoon are massed in the hind-
intestines of the larva before pupation, and are perhaps prepared
in the Malpighian tubules. The cocoons are also known as *Trehale*
manna. At Teheran they are called *tiqal*, i.e. sugar of nests (HOOPER
and FIELD, Field Mus. Nat. Hist. Chicago. Bot. Ser. 9. 1937, p. 195)
or *shakar elma asher*, which has the same meaning.

c. *Honey and Honey-hunting in the Middle East.* In Arabia honey has been prized since olden times. Wild honey was collected in the mountain ranges of the deserts. In the highlands of Yemen bee-keeping has been established since antiquity. We find there the Egyptian bee (*Apis unicolor fasciata* Latr.), but we have no information when bee-keeping was introduced, or whether it was introduced or established spontaneously. We learn from Verse 17 of the Koran that the bee is the only animal accosted directly by God. AL BEIDAWI, one of the authorities of Islamic theology, comments upon this verse: 'The bee secretes honey of different colours, varying according to the flowers on which they feed – white, yellow, red or blackish. Honey is not only good for eating, but is also a most useful medicine against many diseases'.

There are only scarce records of honey-hunting in the deserts of Arabia. J. J. HESS (1938, p. 118) mentions that honey is taken in Inner Arabia by smoking the wild bees from their nests in crevices in the rocks. C. M. DOUGHTY (1926, II, p. 90) was told near Kheybar of a tribe of savage men living on the wide Jebel Rodwa, 'who are very long-lived and of marvellous vigour in their extreme age, as they are nourished of venison and wild honey'.

The bee of S. Arabia is the Egyptian bee *Apis unicolor fasciata* Latr., the most northern race of the Ethiopian bee of Africa, This has been confirmed by the well-known authority on bees, Prof. W. W. AL-PATOV of Moscow (1935, pp. 284 ff.) who obtained a good sample from the high Yemen (coll. N. N. PHILIPOFF). This was extremely close in appearance, measurements and cubital index to the Egyptian bee. Bees from Lahej (Aden) earlier identified as the Indian bee *Apis indica* F. were apparently not this species (BINGHAM 1898, p. 114).

VAN DER MEULEN and VON WISSMANN (1932, p. 67) remark that no dates are exported from the Wadi Do'an of Hadramaut, while up to 500 camel loads of honey are sent every year to the coast. Bee-keeping in the Wadi Do'an is confirmed by H. INGRAMS (1942, p. 172 f.), where the famous luxury honey is the only product for export. 'When we returned to the castle, we were shown a beehive, a pipe fitted into the wall consisting of circular sections about a foot in diameter. Outside there is a small hole through which the bees enter and leave the hives'.

'The bees have a father', said Ahmed. 'Sometimes a new father arises and it leaves the hive and goes away a short distance followed by some of the others. Then you take a mat and make a roll like a hive, and close one end, and sprinkle inside a perfume of the perfumes used by the ladies. Then you go to the place where the new father has gone with the other bees round it, and you pick up the father gently in your finger – it does not bite – and put it into a small cage like this. (He produced a little wooden cage like a tea infuser). Then you put the cage inside the rolled mat and get someone to beat a tin or a copper tray and the bees leave the place where they have swarmed and come to the father. You carry the bees to a hive, put the father in its cage into the hive and in go the bees after it'.

On INGRAM's reply that the father is a queen, Ahmed answered: 'But it is the leader and who ever heard of a woman leading an army like that?' Referring to the males, he answered: 'But they are soldiers, they have the swords to sting with. The bee women (i.e. the drones) are bigger and don't sting'.

C. E. V. CRAUFORD (p. 110) warns against the heavy consumption of honey from hives in the date gardens of the Aden Hinterland; it is luscious, but strongly laxative.

BURY (1915, p. 44) found the honey-bee of the Yemen usually a quiet insect, even when the hives were in the yard close to the house. Yet near Dthen in the Aden Mts. (1915, p. 67) BURY's men piled some logs into the fire, which actually were hives, hollowed out of two-foot sections of trunks of small trees, strongly settled by the *Apis fasciata;* the bees resented this treatment and caused much more perturbation than the sniping from the hills.

He describes the bee-keeping in the Yemen as follows (G. W. BURY, 1915, pp. 311 ff.): 'By far the most important and characteristic product of the Yeshbum Valley is its honey, which is exported all over S. Arabia, and as far as Zanzibar and N. India. There are three honey yields in the year, but that which is gathered after the giant 'elbs' have blossomed is the best and most plentiful. The bee is the *Apis faciata* and as in other parts, it is the *elb*-blossom, that gives the honey its special virtue which is partly of a medicinal character. It is packed in gourds and goat-skins for local and retail traffic, and in empty kerosene tins – two to a case – for the export trade via Shehr and Makalla across the sea.

The bee-hives are taken to flowery valleys high up in the mountains after the *elb* has blossomed. If encountered on a narrow hill
path, it is as well to give the camels bearing the log-like hives, plenty
of sea-room, especially if you are mounted, but the bees are a peace-
ful lot if left alone.

The hives are placed in the loop-holes of Yeshbum towers during
the *elb*-season (the nights are too cold for them in the open). Consequently the busy insects are much to the fore in every room, but
you soon get used to them and they to you, although they are suspi-
cious of strangers. I was informed that they never stung the pure
in heart, and though fairly well broken to insects that sting and bite,
felt quite nervous at the innuendo implied by their attacks. A boy
who was stung, explained to me between his sobs, that he wished
to catch a bee and watch it making honey, that he caught one, but
that it wasn't one of its honey-making days'.

H. SCOTT has been able to complete this description during a
recent expedition into the High Yemen (1942, p. 58), noting that
the dwellers in the solitary house in lower Wadi Leje were bee-
keepers. In their small garden stood a 'battery' of seventeen hives,
raised a few feet from the ground on a wooden platform, in a building consisting of three rough stone walls, the fourth side being open,
and the whole roofed with brushwood. The hives, long hollow
cylinders laid horizontally and filled in at the ends with clay or
plaster, were in four tiers, comprising respectively five, five, four
and three hives, from the lowest to the highest tier . . . . I think they
were made out of hollow tree-trunks (SCOTT's Fig. 31).

In spite of the name of Canaan, the 'Land of Milk and Honey',
so often quoted in the Bible, almost all experts are now united in
their judgement that in Biblical times the collecting of wild honey
was obviously amply practised, but that bee-keeping in tubes of mud
and straw was introduced only in the Roman period. AUHAGEN and
others (cf. BODENHEIMER 1928. I, 1930), two generations ago, gave
reliable descriptions of honey-hunting in the calcareous hills of
northern Palestine which are strongly reminiscent of the Bible
references to 'honey from the rocks' (see, for instance, Deut. 32:137),
which should be translated: 'and he scooped honey from the slabs
for him'. Only L. ARMBRUSTER (1932), in happy ignorance of
Hebrew, has recently aimed to establish the record of a well de-

veloped bee-keeping from the verses Samuel I, 14 : 25 ff., which always had been, and quite correctly, translated as: 'And all they of the land came to a wood; and there was honey upon the ground... Jonathan put forth the end of the rod that was in his hand and dipped it in an honeycomb'. None of ARMBRUSTER's philological speculations, e.g. that the word *yaar* i.e. wood, should read: a 'pile of bee (hives in the form of) pipes', are tenable. To add a few further observations on honey-hunting in Palestine: The Rev. W. M. THOMSEN (1887) found immense swarms of bees nesting on the gigantic cliffs of Wadi Kurn in Galilee. The people of M'alia several years ago let a man down the face of the rocks by ropes. He was protected from the bees by cloth and extracted a large amount of honey. In the densely wooded gorges of Mt. Hermon and the southern Lebanon he found plenty of wild bees, nesting either in old hollow trees or in horizontal clefts of the rocks. B. NEUMANN (1877), one of the early Jewish physicians in Jerusalem, states that bees are more common wild than in hives. G. SCHUMACHER (1924) writes that even in 1924 nests of wild honey-bees are common in the northern Ajlun, where bee-keeping also is widespread.

The present writer has devoted special attention to honey-hunting and to true primitive forest bee-keeping (BODENHEIMER 1942, pp. 44–50) in a monograph on primitive bee-keeping in Turkey. This was based upon two questionnaires sent out while serving with the Turkish Ministry of Agriculture at Ankara, where he left a fairly well arranged museum of local bee-keeping. There were only very few vilayets which reported that no bee-hunting was practised in them. The wide distribution of this custom is also illustrated by a short note from the Turkish daily *Cumhuriyet* of the 25th October, 1940: 'A terrible accident has occurred at Kozan. A boy of twelve, named Mustafa, climbed up the old fortress, where he was told bees made honey. Unfortunately his foot slipped, he fell down into the ditch and was killed. It is a strange coincidence that Mustafa's father was also killed by a fall, when he was climbing after a nest of bees'. BAY CELAL DAVUT, an old and experienced Anatolian beekeeper gave us many details of the widespread distribution of honey-hunting and of the important quantities of honey sometimes gained from hollow trees or from horizontal rock crevices. The bees are usually destroyed when the nest is taken out. From certain vilayets up to 5000 kg. of wild honey is reported as an annual crop (Mardin), but it is obvious that only a very small fraction of the wild honey crop is registered. The contents of one wild bees'- nest were usually estimated at from 10 to 30, in exceptional cases up to 80, kg. of honey. We give below two reports on wild bees from Adana and Kastamonu.

Wild bees are found in Adana mostly in hollow trees or in crevices amongst the rocks, very rarely in subterranean cavities. When the nest is in the trunk of a relatively thin tree, this is cut both above and below the nest. The bees are drowned – except in the rare case when the man intends to keep the nest as a hive –, the log is split and the honey is taken out, as the entrance hole is usually too small to permit extraction by hand. When the tree is thick, the entrance hole is stoppered and the tree felled. Nests in rocks are very often just about midway between the top and the bottom of the rock, so that they cannot be reached by hand or ladder. When the rock is convex, one of two persons who climb to the top of the rock ties a thick, strong rope round his waist and his companion lowers him down. When he reaches the nest, the rope is held fast and the man who has been lowered collects the honey which he sends up by means of a pail attached to another rope. When the rock is concave, one end of the rope is fixed to the top of the rock, whilst the other end is passed over the entrance hole of the nest and let down to the ground where it is again firmly attached. The length of this rope depends upon the height of the rock. The second rope is tied around the waist and feet of one of the party, who is lowered down. Holding on to the first rope and guiding himself by it, he reaches the nest and collects the honey which is hauled up in the same manner as above. Nests containing 30 to 60 kg. of honey have been collected in this way.

In Kastamonu wild bees are met with in the hollow trunks of old trees, as well as in crevices in rocks. While the nests in rocks are usually destroyed during collection of the honey, those from trees are usually cut out and taken home as hives, if the bees are not too savage. 15 to 50 kilos of honey may be obtained from one wild nest. At Antalya also wild bees are common, usually in crevices amongst the rocks, and collecting with help of ropes is common. There the average crop is 10 to 15 kilos per nest. We found that wild honey is an important source of food in the mountains, but rare in the coastal plains or the high plateau.

Primitive forest bee-keeping is described from the Black Sea vilayet of Çoruh, as follows: wild bees are found at Borçka and Hopa, where the peasants do not destroy them, but are glad to keep their bee-hives in the forests, up to four hours from the village. The forests are divided between the various villages. When any of the villagers trims a young tree into a bee-hut, he becomes the owner of the tree. He builds a kind of bee-hut in the tree, 2 to 4 m. above the ground, and fills it in early May with empty hives containing about 100 gr. of comb-wax. He inspects these empty hives in June and takes down those in which bee-swarms have settled. These he puts into special bee-huts, high on a pine tree, which are out of the reach of bears, with ten to fifteen hives on one tree. These are the *canli kovan*, or living hives, in contrast to the *konar kovan*, or settling hives, which are put out for the new swarms. Setting bees into trees or rock cavities which are already in use as hives is prohibited by traditional law, yet the swarms which fly out from recognized property are free and do not belong to the owner of the hive from which they swarm. The peasants never steal honey from each other's hives. The rest of the report describes the various log-hives which are in use in Çoruh.

3. INDIA

Intensive agriculture and one-sided vegetarian diets have greatly limited the eating of insects in India. This is practised mainly in the desert and forest areas and during periods of famine. Yet a number of reports indicate that in earlier times it was widespread; local observations may still add to our present scanty knowledge. Locusts, termites, and honey, also ants and bugs are still consumed. The following story of an early traveller, M. I. Pinto (vide Cowan 1865, p. 295) also hints at earlier conditions: 'In our travels with the ambassador of the King of Bramaa to the Calaminham we saw in a grotto men of a sect of one of their Saints, named Angemacur. These lived in deep caves in the rocks, according to the rule of their wretched order, eating nothing but flies, ants, scorpions and spiders with the juice of a certain herb, much like our sorrel'.

Col. Burke (vide Bargagli 1877, p. 8) states that the Indian sepoys make a famous curry with locusts as ingredients. Also C. Horne (Science Gossip 1863) reports that locusts are eaten in India. One evening he offered to two guests a curry and croquettes of locusts for dinner. The latter passed for Kabul shrimps, which in flavour they much resembled. A locust leg, which the cook inadvertedly left in the dish, revealed the secret, much to the disgust of the one and to the amusement of the other guest. Many more reports should be available, as Das quite recently discussed 'Locusts as food and manure' (1945), and Mustafa and Janjua (1942) published a photograph taken during a locust campaign in Baluchistan, showing four Baluchis sitting around and eating locusts.

König (1775) in a lecture to the Naturforschende Gesellschaft of Berlin states that in some parts of the East Indies live termite queens are given to the old to strengthen their backs, and that the natives catch the winged sexuals, which they call females, before swarming. They dig two holes into the termite nest, one to windward, the other to leeward. At the leeward opening they place the entrance of a pot which has previously been rubbed inside with an aromatic herb, the *bergera*, which is more esteemed than laurel leaves in Europe. On the windward side they make a fire of stinking materials which not only drives the insects into the pots, but frequently also the hooded snakes, on which account the natives are obliged to be cautious when removing their pots. By this method they catch great quantities of termites, and with these, together with flour, they make a variety of pastry, which they can afford to sell very cheaply to the poor classes. He adds that in seasons when this kind of food is very plentiful, termites eaten in too great quantities may bring about an epidemic colic and dysentery which may cause death in two or three hours.

Other reports about termite consumption in India are not nearly so numerous as from tropical Africa. Broughton (1813) writes from the Mahratta country that the prime minister Surgee-Rao during a severe illness was nourished at great cost mainly with termite queens. In Java and in Annam reports on termites as popular diet are more numerous. The queens are generally regarded as a strong

232

aphrodisiac. BUCHANAN (I, p. 7) says that the white ant is a common article of food among the Hindoo tribes, and J. FORBES (1813 I, p. 305) confirms this for the low castes in Mysore and the Carnatic. Captain GREEN (vide KIRBY and SPENCE 1822, I, p. 340) adds that in the ceded districts of India the natives place branches of trees over the nests, and then by means of smoke drive out the insects, which break off their wings by the mere touch of the branches when attempting to fly.

Many reports confirm that the green weaver-ant is eaten throughout almost the entire Orient. C. I. BINGHAM (1903, p. 311) mentions that at Kanara in India a paste of these ants is eaten as a condiment with curry. A. M. LONG (1901, p. 536) vividly describes the use of the red ant as a regular article of food by the Murries of Baster in the South of the Central Provinces. Throughout the year, especially during the dry season, they collect these nests, tear them open, shake out the contents into a cloth, and beat the insects, mature and immature, into a pulpy mass with a stone, and when all are dead, enclose them in a packet about the size of a goose's egg made of *sal*-leaves. In this condition the article is taken to the bazaar and sold, about 16 being sold for 4 cowries each. To prepare the crushed ants for food, they are mixed with salt, tumeric and chillies and ground down between stones, and are then eaten raw with boiled rice. They are sometimes cooked up with rice flour, salt, chillies, etc. into a thick paste and in this condition the food is said to give great powers of resistance against fatigue and the sun's heat. O. BECCARI (1904, p. 161) mentions that the Dayaks of Borneo also eat this ant or rather they mix it with rice as a condiment. It has a pungent, acidic taste and smell, which evidently pleases the Dayaks.

In Siam ant-maggots mixed with pork ragout are much appreciated. The Chinese also regard ant-maggots as a delicacy. The peasants of Tonkin are fond of ant-'eggs'. (E. BRYGOO 1946, p. 51).

A paste of the green tree-ant (*Oecophylla smaragdina*) is added, together with curry, to the rice dish in Kamara and other parts of India, Burma and Siam.

The Sama Nagas are a primitive tribe inhabiting the hills between Assam and Burma. They recognize the wild rock bees as private property, the first finder acquiring the rights to the nest, which is collected every year for the honey as well as for the grubs. If any person who has helped to take the nest dies during the year, it is put down to the bees and the nest is not disturbed again. Chastity must be observed the night before taking a bees'-nest; if not, the bees will sting the taker, and he would also be liable to be killed by his enemies. Before the bee-gatherers leave their houses early in the morning to secure the nest, nothing whatsoever must be removed from the house. Should a domestic animal give birth to young, or a fowl hatch chickens within three days of going to the bees'-nest, the owner may not collect the honey as planned. The Sama Nagas also have a taboo of honey and grubs for the first reaper at the harvest season (HUTTON, 1921, p. 72).

The men of the related Angami Nagas, in order to gain immunity from stings while gathering honey, must abstain from sexual intercourse from the day on which the honey-expedition is fixed. On the morning of the expedition they are not allowed to speak; they leave in complete silence and make the ladders for collecting the combs. Then they are not stung, while any breach of these rules leads to fearful stinging. With the Lhota Nagas also the man who intends to take a bees'-nest must remain chaste the night before (HUTTON, 1921, p. 236).

The Todas are another primitive tribe living on the plateau of the Nilgiri hills. They have a number of legends about honey-gathering (see RIVERS 1906, pp. 191, 198).

In Assam the spirits of men are said to become honey-bees. An annual ceremony is held in honour of all who died during the year, only the spirits of infants going into house-flies instead of into honey-bees (J. SHAKESPEARE 1912, p. 466). This belief of bee-souls is fairly widespread. It is also found in Timor concerning the spirits of princes and of warriors fallen in battle (BASTIAN 1885, p. 4). In Hindu ritual the use of honey always refers to it as a produce of bees, which are the vehicles of the souls of the departed (for bibliography see H. RANSOM 1937, p. 289).

Among the Andaman islanders magic beliefs about bees are common (A. R. BROWN 1922, pp. 104 ff., 183, 357, 152 ff.). Honey abounds during the hot season, from February to May, while it is scarce and of inferior quality in the rainy season. The wax especially is gifted with many magic properties. The 'honey-eating'

234

is a very important ceremony among them. The initiates sit cross-legged and honey is rubbed over their shoulders. When the honey fast is to be broken, a quantity of honey-combs are procured for the appointed day. The elder goes to the novice, who sits in the centre, with a big honey-comb wrapped in leaves; he helps the novice to eat a mouthful by holding the comb to his mouth and lets him consume the remainder in silence. He eats what he can and preserves the rest wrapped in leaves for later consumption. The chief then an-noints the novice by squeezing another comb over his head and rubbing the honey well into his body. A bath follows. Apart from silence and abstention fom the kidney fat of pig, the novice may eat, drink or sleep as he pleases. With women the ceremony cannot take place before they have borne their first child. Women abstain from eating honey during pregnancy.

In India the bee is the totem of the Bhramada clan and of one of the Juang clans (J. G. FRAZER 1910 I, p. 242, 315). Honey is the totem of one of the Manda tribes in India (ibid. II, p. 292).

4. INDONESIA

C. L. VAN DER BURG (1904, pp. 37, 90 f.) in his book on the food of the population of the Dutch East Indies first mentions the taboos, which among Muslims include ants, bees, flies, worms and water animals. Termites *(rajap, aniani;* mainly *Termes sumatranum, T. fatale, T. mordax, T. atrox, T. destructor)* are much sought after by the natives. The wings of the flying sexuals, called *laron, raron, reraron,* are removed and the insects are roasted with flour and baked into a kind of cake. They are very fatty, so that even a kind of butter is obtained from one species. The roasted queen *(ratoe rajap)* is a special delicacy, tasting, as with many other insects, like almonds. The honey of certain stingless bees (the *tawon* or *leba* of *Melipona minuta* and *M. vidua*) is eargerly eaten by the native women, together with the bee-maggots and pupae *(tawon moeda, gono, tawon nom)*. These latter are wrapped with the comb into a piece of leaf and roasted *(pipit)*. The eating of the pupae is regarded as un-healthy and causes pains in the throat.

In Preanger (W. Java) the natives eat grasshoppers and locusts

(*walang, belalang, djankrik, gansir, walang gapoek, belalang gambar*)
– the latter being names for *Acridium aerigonosum* and – molecrickets
(*Gryllotalpa, andjing tena*). Even the stinking species are eaten.

Cockchafers also (*legi*) are roasted and are cooked in the native
shops in portions of ten beetles each. The waterbugs *Leptocorisa acuta*
and *Stenocoris varicornis* (*walang sangit*) which smell like candle-fat,
are prepared as *sambal*. VAN EEK (Bull. Kol. Mus. Haarlem,
p. 149) analysed them chemically. They contain: N 6.1; fat 29.2;
water 23.4; protein 38.1; ashes 3.0%, with a total food value of 276
calories. The following *Orthoptera* are known as poisonous: *Poeci-
locerus punctatus* (*walang peloes*), *Acanthoderus bifoliatus* (*onggas*) and
Bacterina nematodes (*walang kandel*), according to Greshoff (Indisch
Vergift rapp. Nos. 15, 16 and 109). Women eat head-lice. The
cicada *gareng*, the caterpillars of *Hyblea puere* (Lep.), the *entong* or
oengker, and the grubs of the palm-weevils *Rhychophorus palmarum*
and *R. ferruginea*, the *oelar rotan, gendon, sabeta, oelar sagoe* or *olakna
mandjalin*, are also eaten. The latter are very fatty and are eagerly
eaten when roasted with *ketan*. The caterpillars of *Euproctes* ?
mulleri, the *olak samenet*, are also reported by GRESHOFF as poisonous
(Rapp. No. 35). (Cf. also STILBE 1922, p. 599).

Prof. S. LEEFMANS (Amsterdam) has kindly provided certain information based on his personal observations in Indonesia. In Java
the flying sexuals of termites (*larum*, mainly *Macrotermes sp.*) are
caught on small candles attached to pieces of bamboo. The insects
burn their wings on the light to which they are attracted and are
then avidly eaten. Crickets, mainly the large *Brachytrypes portentosus*
Licht. (= *B. achatinus*), are caught by boys in Java and Sumatra by
digging in their galleries in the ground. The boys put a number of
them on a stick cut from the midrib of palm leaves, and so roast
them, eating them with delight. The males of the common beetles.
Leucopholis rorida F., which swarm in the evening, are attracted in
numbers to stones which are dyed red by the fruit of *Capsicum*.
There they are collected, roasted and eaten in Java. The larvae
of big beetles, such as of *Rhynchophorus, Psodocerus*, etc., which
develop in palm stems are prized as food on the Moluccan Islands.
Other beetles commonly eaten in Java are the swarming *Lepidiota
hypoleuca* Wied. on *naron*-trees.

SCHELTMA (p. 379) states that the Batakkers in Tapanoeli

236

(Sumatra) consume as animal food the meat of buffaloes and pigs, chicken, cats, mice, rats, frogs, the larvae of crickets, wasps and an insect pupa. POSTMUS and VAN VEEN (1949) in their dietary survey of Ceram state that the protein value provided by the large-scale consumption of vegetable food, sago and tubers, is very low. B_1 (thiaminase) is also lacking, even in fresh fish or in shellfish, while dry fish has none at all. They do not mention insects. Dr. HILLE RIS LAMBERS informed us that the Moluccan fishes are especially poor in fat. The palm-borer larvae hence are a very essential food, offering animal fat and protein in a diet otherwise poor in these components. And in the Research Institute of the Indies in Amsterdam we were given a photo from Buru, a Moluccan island, showing a wood-boring insect larva which is, for the same reasons, eaten as a delicacy with the sago porridge. During the night preceeding the day when this dish is to be eaten, the larvae are put into a bowl with ground coconut and after being boiled, are eaten with great pleasure.

K. GIESENHAGEN (1902, p. 78) once saw a termite-swarming on Java. Men, women and children rapidly congregated in front of the exit hole of the nest and caught them. After removing the normally deciduous wings, the bodies were gathered into pots and cauldrons. Even the dogs caught them eagerly. Termites in various preparations are considered a delicacy by the Malays.

The Javanese 'moutouke' is the grub of another big beetle (Sketches of Java, p. 310; of unquoted author, vide COWAN 1865, p. 70): 'A thick, white maggot which lives in wood .... It is as big as a silkworm and very white, a mere lump of fat. Thirty are roasted together threaded on a little stick and are delicate eating'. The natives of Bali catch and eat adult dragonflies (W. R. MOORE, Nat. Geogr. Mag. January 1951, p. 18–19).

A. R. WALLACE relates that the natives of Lombok catch dragonflies on twigs smeared with birdlime. The bodies are torn off and fried with onions and preserved shrimps. It sounds a queer blend, but is considered a great delicacy (vide BURR 1939, p. 213 f.).

On Timor the natives prepare wholesome cakes from pounded locusts (BRYGOO 1946, pp. 34, 50). Bee-maggots are highly esteemed by them. BRISTOWE (1932, p. 397) observed Javanese women at Cheribon picking lice from the heads of their friend and eating them. WALLACE mentions that in the Moluccas the palmworm is regularly taken to market on bamboos and sold as food.

The only information which we were able to obtain about the Philippines, where obviously interesting documents concerning the Moros might be expected, deals with locusts. DAMPIER (vide J. PINKERTON vol. XI, p. 49) says that locusts are eaten as a regular food in the Philippine Islands. The natives catch them in small nets, when they come to devour their potato-vines, and dry them over the fire in an earthen pan. When thus prepared the legs and wings fall off, and the heads and backs, which were previously brownish, turn red like boiled shrimps. DAMPIER once partook of this dish and liked it well enough. When their bodies were full they were moist to the palate, but their heads cracked in his teeth.

Honey-hunting in Indonesia. Information on the biology of the true honey-bees of Indonesia is found in G. SCHNEIDER (1908) and W.ROEPKE (1930). The main species concerned are *Apis dorsata* F., *A. indica* F., *A. florea* F. and *A. zonata* Sm. G. SCHNEIDER describes a honey-hunt in the forest of Sumatra. Sixtyfive enormous nests of *A. dorsata* were hanging on one great *tualang*-tree. A few days before the hunt bamboo pegs were driven into the stem of the tree, to enable the honey-hunters to climb up. On the day of the hunt, towards 4 p.m., the hunters arrived with torches, ropes and baskets. Everyone climbed up one side of the tree, holding a burning torch, consisting of a certain bark which produces heavy smoke, beneath the hanging combs of a single hive, which immediately turned from black to white, while the smoke drove away the bees. With a sharp knife the combs are then speedily cut away, put into a basket and the climbers go to the next nest. When the basket was full, it was let down and exchanged for an empty one which was hauled up to receive the combs of the following nests. Honey and wax are the monopoly of the local rajahs. The value of this single large *tualang* tree was estimated at 300 Swiss francs. Little honey is exported from East Sumatra, as most of the crop is consumed locally. Honey in water is a common drink there.

DE MOL (1934) describes a region in Western Borneo which is entirely under water for most of the year, yet in the dry season which lasts for a few months only, the numerous flowers attract an abundance of bees (*A. dorsata*). Of the local Malayan fishing population

238

about 500 collect wax and honey at this time and also continue with their fishing. They place young tree-trunks halved lengthwise and then hollowed out (*tikoeng*) on the crowns of the trees, and in these the bees build their nests. During the dry season the honey-combs are gathered. Wax and honey are pressed out by hand, and collected into earthen vessels. The number of *tikoengs* per family ranges from 40 to 150, but some families possess from 1000 to 2000. Outside the Lake Region the Dajaks cultivate bees in a similar way. One tin of honey (32 pints) is sold locally for one guilder, yet formerly up to five guilders were paid for the same quantity. In the same years the prices of wax ranged from 60 to 70 guilders per picul (62 kg.), while in August, 1932, only 28 guilders were paid. Tools for collecting honey-combs and the process of honey-collection are illustrated by two figures. The *tikoengs* are 160 to 225 cm. long. Swarms are followed, tracked and gathered.

Attempts to encourage experiments to domesticate wild bees in Indonesia have also been made (see R. A. M., 1924).

Another description of about 25 nests of *A. dorsata* on a big *Canarium*-tree on Java, which was cut down by the natives, is given by VAN DER SWAAN (1934).

A honey-hunt on Timor is described by WALLACE (1902, p. 153 f):

'The bees'-wax is a still more important and valuable product, formed by the wild bees (*Apis dorsata*), which build huge honey-combs, suspended in the open air from the underside of the lofty branches of the highest trees. They are of a semi-circular form, and often 90 to 120 cm. in diameter. I once saw the natives take a bees'-nest and a very interesting sight it was. In the valley where I used to collect insects, I one day saw three or four Timorese men and boys under a high tree, and looking up saw on a very lofty horizontal branch three large bee-combs. The tree was straight and smooth-barked, and without a branch, till at about 25 m. from the ground it gave out the limb which the bees had chosen for their home. As the men were evidently looking after the bees, I waited to watch their operations. One of them first produced a long piece of wood, apparently the stem of a small tree or creeper, which he had brought with him, and began splitting it through in several directions, which showed that it was very tought and stringy. He then wrapped it in palm-leaves, which were secured by twisting a slender creeper round them. He fastened his cloth tightly round his loins, and producing another cloth wrapped it round his head, neck and body, and tied it firmly round his neck, leaving his face, arms and legs completely bare. Slung to his girdle he carried a long thin coil of cord; and while he had been making these preparations one of his companions had cut a strong creeper or bush-rope, 8 to 10 m. long, to one end of which the wood-torch was fastened and lit at the bottom, emitting a steady stream of smoke. Just above the torch a chopping-knife was fastened by a short cord.

The bee-hunter now took hold of the bush-rope just above the torch and passed the other end round the trunk of the tree, holding one end in each hand. Jerking it up the tree a little above his head, he set his foot against the trunk, and leaning back began walking up it. It was wonderful to see the skill with which he took adventage of the slightest irregularities of the bark or obliquity of the stem to aid his ascent, jerking the stiff creeper a few feet higher, when he had found a firm hold for his bare foot. It almost made me giddy to look at him as he rapidly got up, 10, 20, 30 m. above the ground. And I kept wondering how he could possibly mount the next few feet of straight, smooth trunk. Still, however, he kept on with as much coolness and apparent certainty as if he were going up a ladder, till he got within 3 to 5 m. of the bees. Then he stopped a moment, and took care to swing the torch (which hung just at his feet) a little towards these dangerous insects, so as to send up the stream of smoke between him and them. Still going on, in a minute more he brought himself under the limb, and in a manner quite unintelligible to me, seeing that both hands were occupied in supporting himself by the creeper, managed to get upon it.

By this time the bees began to be alarmed, and formed a dense buzzing swarm just over him; but he brought the torch up closer to him, and coolly brushed away those that settled on his arms or legs. Then stretching himself along the limb, he crept towards that nearest comb and swung the torch just under it. The moment the the smoke touched it, its colour changed in a most curious manner from black to white, the myriads of bees that had covered it flying off and forming a dense cloud above and around. The man then lay at full length along the limb, and brushed off the remaining bees with his hand, and then drawing his knife cut off the comb at one slice close to the tree, and attaching the thin cord to it, let it down to his companions below. He was all this time enveloped in a crowd of angry bees, and how he bore their stings so coolly, and went on with his work at that giddy height so deliberately, was more than I could understand. The bees were evidently not stupefied by the smoke or driven away far by it, and it was impossible that the small stream from the torch could protect his whole body when at work. There were three other combs on the same tree, and all were successively taken, and furnished the whole party with a luscious feast of honey and young bees, as well as a valuable lot of wax.

After two of the combs had been let down, the bees became rather numerous below, flying about widly and stinging viciously. Several got about me, and I was soon stung, and had to run away, beating them off with my net and capturing them for specimens. Several of them followed me for at least half a mile, getting into my hair and persecuting me most pertinaciously, so that I was more astonished than ever at the immunity of the natives. I am inclined to think that slow and deliberate motion, and no attempt at escape, are perhaps the best safeguards. A bee settling on a passive native probably behaves as it would on a tree or other inanimate substance, which it does not attempt to sting. Still they must often suffer, but they are used to the pain and learn to bear it impassively, as without doing so no man could be a bee-hunter'.

Another description of the honey-hunt in Malayan Java has been given by C. J. van der Zwaan (1934). Bee-trees are usually isolated forest giants, which may also form a small group of trees, amongst which *Komparsia malaccensis*, *Alstonia sp.* and *Dipterocarpus hasseltii* are most common. Zwaan was present at such a honey-hunt in moon-light. The villagers, for whom the honey-trees were common property, had built a pent-house for the night, Of the thirty natives who took part in the expedition, three only would actually climb up the tree. The disturbing undergrowth was chopped away. Long rotan-lianas were hanging from the branches, which had been left over from a previous visit. About 200 bees'-nests were clearly visible in the moon-light. Yet the actual business began only with darkness, in order to avoid the pursuit by the angry insects. The honey-hunters had smeared their bodies with oil, which contained a sting-averting charm. Apart from a loin cloth they were naked.

241

Torches of *Scaphium*-bark, 2 m. long, were hung with a rope around their shoulders. The torches were lit before the ascent began. The *Scaphium*-bark breaks easily into smaller pieces, which glow but never burn and are followed to the ground by the bees, which thus do not molest the hunters. These climbed up on about 20 cm. long blocks which had previously been driven into the trunk of the tree. When they reached the crown they crept upon their bellies along the branches, until a nest was beneath them. Then they began to sing and to swing their torches to and fro around the nest, whereupon the bees followed the rain of sparks down to the ground, a beautiful sight in the dark night. When all the bees had left the combs, these were cut off with a sharp bamboo into empty tins, which were lowered down when full. When all the three hunters were at work, there was a loud buzzing of the bees beneath the tree, but their buzzing was drowned by the loud singing of the hunters. The tins were hastily carried away as they were lowered. In the meantime the reckless hunters attacked a second nest in the same way, always loudly singing. There were twenty to thirty bees'-nests on the big branches, one being collected after the other, always accompanied by the beautiful rain of sparks. About four o'clock the hunters climbed down. The bees remained buzzing above the ground and returned only in the morning light to what had remained of their nests. New white honey looks good, tastes good and is merely a little sharp for the throat. The natives eat it together with the larvae. Meanwhile they removed up to a hundred stings from the bodies of the three honey-hunters. The honey was cooked for preservation.

In Malaya many tales are current about bees. We select one of these, the story of Rakian (I. H. N. Evans, Folk stories of the Tempas-uk and Tuaran Districts of British Noth Borneo. Journ. R. Anthrop. Instit. 43. 1913 and P. Hambruch, Malaiische Maerchen. Jena. 1922, pp. 116–121): There was once a mango-tree in the country with many large nests of wild bees. And when the bees had collected sufficient honey, a man, called Rakian, came to the tree and knocked into its trunk a number of bamboo-sticks, in order to climb up to the nests. Once, the sun was setting, when he climbed up. Among the many bees'-nests Rakian observed one, built upon the highest branch, which was white. He desired to collect this nest, because he never had seen white bees. He drew his sword to cut the branch. The bees did not swarm out of the nest, but when he began to cut the branch, he heard the bees calling out 'Oh, that hurts'. Rakian was astonished and returned the sword into its sheath, when he heard: 'If you want the nest, take it carefully'. This he did and brought the nest home in his satchel.

The next morning Rakian worked from dawn to evening on his field. When he returned, his fish and rice were already cooked on the stove. The rice was his but the fish did not belong to him. He ate and was contended. The rest of the cooked food remained for his early breakfast. This repeated itself for many days. One day he resolved to return early, in order to see who was the cook. He returned in the morning and hid himself. After some time the door of the house was opened and a very beautiful woman came out, to take water from the river. During her absence he looked into the bees'-nest and found it empty. He hid the nest and himself within the house. The woman returned and began to cry, not finding the nest: 'Who has stolen my wardrobe? I fear that Rakian may find me, when he returns'. Rakian left his hiding-place towards evening, and asked the woman: 'What are you doing here. Perhaps you will steal my bees?' 'I know nothing of your bees'. After some discussion, the woman agreed to marry Rakian on condition that he would never call her bee-woman because of the shame of it. They married and a child was born. Some time later there was a feast at one of their neighbours. When Rakian was drunk, he answered an inquiry about the home of his wife: 'She was really a bee, before I married her'. His wife received him with reproaches, and told him she would return home, but the child she would leave with him.

Fig. 32. Bamboo cylinder of Malaya, which is hung on the hut when the owner is in need of wax (Seyffert, Bees in Africa 1940, p. 45, fig. 5.)

After some days Rakian saw a white bee flying around the house, which was the father of his wife. His wife changed into a bee and flew away with him. Rakian took his child and followed the bees for seven days. Then he was tired and hungry and slept on the shore of a river. A woman awakened him and asked him why he did not sleep in the house of his wife. The woman showed him the way to a long house with eleven rooms, and pointed to the middle room. Rakian climbed into the house, which was full of bees. The child began to cry, and his wife entered the room. All the bees dropped from the roof and turned into men. Rakian and the child remained in that village and never returned home.

While most tribes of Borneo are agriculturists and fishermen, the Punans are nomad hunters living exclusively on jungle produce, both animal and vegetable. Apart from hunting mammals, from the wild pig to small rodents and jungle birds, they gather all edible fruits and collect gutta-percha, rubber, camphor, various rattans, bees'-wax and honey, vegetable tallow, wild sago, gums from various trees and the famous edible birds'-nests. Small parties of men and boys go up-river for some days before striking into the jungle of the drier upland forests. The party may stay several weeks or months away from home.

The gathering of honey and bees'-wax is described by C. Hose (1926 pp. 109, 113). The combs are usually found on the high branches of the large *tapang*-trees, sometimes 50 to 60 combs on one tree. To reach the nest, which usually is attempted after nightfall, the men climb the trunk by an improvised ladder. A large number of sharpened pegs of ironwood having been driven into the softer bark and sapwood of the stem in a vertical row about 60 cm. apart, long bamboos are lashed to them, and also to the stems of the lower branches. The ladder is thus built up until at 18 to 21 m. from the ground it reaches a branch bearing two or three combs. A man now ascends the ladder, carrying in one hand a burning torch of bark, which gives off a pungent smoke, and on his back a large hollow cone of bark. Straddling out along the bough, he hangs his cone of bark beneath the honey-comb, smokes out the bees, and cuts it away from the bough with his knife, so that it falls into the cone of the bark. Then, choosing a piece of comb containing grubs, he munches them with gusto, indicating in pantomime to his envious friends their delicious quality. After thus gathering two or three nests he lets down the cone with a cord to his expectant partners, who then feast upon the remaining grubs and squeeze out the honey into jars. The tree having been clared of nests in this way, the wax is melted by boiling in an iron pot and moulded into balls. The honey is eaten in the houses; the wax is sold to Chinese traders.

'The larger branches of a tall *tapang*-tree were literally covered with clusters of wild bees clinging to the underside, at least a hundred families. When I reached the foot of the tree, to my astonishment a half-grown honeybear, about the size of a tall bulldog, came sliding down the trunk, tail foremost, growling fiercely, and using his powerful claws as a brake' (HOSE 1929, p. 22 f.).

5. HONEY-HUNTING BY THE VEDDAS AND OTHER ASIATIC PYGMIES

Honey-hunting is one of the main occupations of the pygmies of the interior of the Asiatic primeval forests, as with those of South America and Africa. The Negritos, the pygmies of S. E. Asia, are advanced in hunting and collecting (SCHEBESTA 1940, pp. 108–118). The forest is a good country for lovers of molluscs and of honey ... The Semang of Malacca are not passionate hunters. The small animals which live in the trees and the monkeys satisfy them ... They do not like insects and molluscs, as the Bambuti do, and do not delight in termites or in caterpillars, They are mainly vegetarians, but they all like honey.

The Aeta of the Philippines have a special preference for honey. The children eat this delicacy with the comb ... The Andaman Islanders do not smoke out the bees, when they collect honey. They are stunned by the sap of a kind of *Alpipia*, which is obtained from the leaves of this plant. The men smear this sap over their face and they also sprinkle it over swarming bees'-nests.

R. PERCIVAL (1803 II, p. 62) says of the Veddas that they live almost only upon game, being skilled hunters. Yet honey which abounds in the forests of Ceylon is also part of their food and replaces for them the salt which they cannot procure. Honey is also used by them for preserving meat. They spend much time in collecting honey, which they sell in considerable quantities to Kandy. The people of Kandy are said to embalm their dead in honey. The wild honey is sold in the combs.

R. KNOX (1817, p. 48) says that the natives of Ceylon hold burning torches under any bee swarm which they discover on a tree, to make them drop. They catch them and carry them home, to cook and eat them. And when they fell trees containing bees'-nests, they not only collect the honey but also the bees, which when cooked are an esteemed dish. *Xylocopa* and other large bees are also eaten. We have, however, no later reports of this kind.

The following description of honey-hunting by the Veddas of Ceylon comes from R. L. SPITTEL (1924, pp. 204–213). 'There are five varieties of honeybees in Ceylon. The large *bambara* (*Apis dorsata*) which builds huge combs of very pure wax on bare hillsides and on the large branches of lofty trees; the *mee massa* (*Apis indica*), the sweetest honey-bee of all, whose combs are to be found in hollow trees, rock-clefts and within ant-hills; the *dandual massa* (*Apis florea*), the smallest honey-bee in the world, which hives on rocks and bougs; the *kenava massa* (*rosin* or *Dammar*-bee), a docile, stingless, friendly little creature whose long black combs within tree crevices yield a delicious, if heady, honey; the *kotha massa*, an even smaller species, whose puny comb is hardly worth the taking. *Kenava* and *kotha* combs are available all the year round. They are the *pathinchikarayas*, the stayers. When the Vedda is hungry, he does not disdain even their honey. The *bambara* and *mee massa*, the most abundant honey makers are seasonal in their habits. June and July are their great honey months. As the fruit season falls at the same time, these months mark the festive time of the jungle. Then the Veddas leave their homes for caves and stream beds, take their fill and have a surplus for barter.

On the flavours of honey the Veddas are epicures. They can tell by tasting a comb the flower or medley of flowers that contributed to its making. The honey in which the nectar of the *mora*, *veera* and *galseem bela* flowers prevail is the most relished one.

Honey gathering is *the* master art of the Vedda. Supreme as he is in other jungle crafts, he is superlative in this. As he travels through the forest he is alive to the faintest hum, and no hive, however concealed, escapes him. His adroitness in tracking a bee to its lair is well described in a delightfully euphonious song they have (beginning: *Thomba, thomba, gomba me-na-na*) which tells about a Vedda of the name of Thutha who goes on the search for honey with his brothers Gomba and Naga. Thutha leads on the trail of a particular bee; and as he goes he sings to his brothers indicating the various trees in which bees are busy; but not decoyed by these,

246

Fig. 33. Honey-hunting Veddas. Left: the honey-hunter with the leaves for smoking out the bees. Right: the honey-hunter at work. This picture shows a remarkable similarity with the Palaeolithic picture from Spain in Fig. 2. (From SPITTEL).

and with his eyes steadily on the bee he first sighted, he follows it, till eventually it guides him to its hive within the skull of an elephant, a *kolanathe gatche*, a tree without leaves, as he quaintly describes it.

The taking of any hive, other than that of the *bambara*, is a very simple thing for the Vedda. All he does is to climb the tree, steady himself at the hole where the bees are, and blow into it to drive the bees away. Then with his axe he enlarges the opening súfficiently to enable him to introduce a hand and to secure the comb, stopping every now and then to blow inside or to slap an offending bee. He eats the comb at his perch, grubs and all, throwing down fragments to his friends below.

The *bambara*-comb is a very different thing. These great bees build high on the precipitous sides of vast rocks, their security lying not in their seclusion, but in their inaccessibility. These hives are the prize of the bold and the skillful only. A dangling cane at 'Westminster Abbey' gave us some idea of the hazards of the task. To reach the combs ladders of cane are lowered from the summit of the rock and down these the collector descends. As this does not bring him within arm's length of the hives, he usually had to rely on the use of a long stake. Sometimes the hives are unapproachable even with this. The wax may be obtained by being dug out with long bamboos and collected at the foot of the hill. Such combs are called *iti bambara* (wax hives). There are others, so safe under overhanging cliffs, that they are absolutely inaccessible; these are called *vakini baliya* (she-demon's portion).

When the rock is not too high, the ladder is made long enough to reach to the ground, where it can be held slantwise close up against the face of the rock, bringing the collector within easier reach of the hive. But where the hill is too high for this, as usually happens, the ladder hangs free and the collector may even have to swing to and fro to enable him to get at the combs. Nor is this all; the *bambaras*, several of which may kill a man, have to be reckoned with. They can only be dislodged by intensive smoking; and what is worse, when several hives are clustered together, the work has to be done on a pitch dark night.

Imagine then the picture of the man smothered in smoke, assailed by angry bees, swinging in the darkness 70 m. or more aboven the stony earth, with neck and knee hitched to the ladder so as to leave the arms free to manipulate torch and prong. Is it a wonder that not every Vedda is a *kapunkaraya*, a cutter, but only the boldest of them? And that when climbing down to his work, he does so with the though in his heart that he now has no use for life or father or mother?

We will now describe the scene as it actually takes place: At Panihela were five large combs which the Veddas intended to collect one night. In the morning they assembled at the rock. Their first business was to collect the canes and vines that go to the making of the ladder. Two canes, 7 to 10 m. long and about 2 to 3 cm. thick, were cut, dragged to the top of the hill and lined out. Knotting the end of one to a tree, they split it in half. One man, standing at the end, held these apart, while two others stood in between, constructing strong rungs at intervals of 60 cm. with *kirriwel*-creepers, which they first looped round the two halves of the cane and then bound together with a running twist. A ladder of any length may thus be constructed; for, should a single cane not be long enough, its ends are doubled up and tied into loops, in which similar loops of another split can engage. The ladder is finished off by overlapping the free ends in the form of an U and binding them together. This is the type of ladder employed in the perilous descents. Where the drop is easier, they do not go to all this trouble, but merely split the end of the cane for about one meter and tie a stick across. Standing on this the man is let down.

The ladder having been completed, the *hulas* (torches) and the *matha* (forked stake) are still to be made. The torches, four of which are required, are designed to give a maximum of smoke, and are therefore constructed by enclosing dry *mana*-grass, leaves and bark within a sheet of green leafy twigs. The *matha* is a stake, 180 cm. long, made of the light *velang* or *walumala*. It is pointed at one end and forked at the other by being split cross-wise into four pointed prongs which are kept apart by two pieces of stick wedged in cross-wise and tied. Both *matha* and *hulas* have long loose loops attached to them by which they are slung to the arms when not in use. Lastly, there is the *hangotuwa*, the leather receptacle, a vessel of deer-skin to receive the honey, also called *yakka-katte*, the devil's mouth, when in use, lest the spirits be angered and send the cutter to his doom.

A heap of firewood is now collected on top of the rock to protect the guardian of the ladder and another heap below to help in smoking out the bees. All being ready, they await the protecting darkness. When this comes, the commotion begins. The great fire at the foot of the rock is lit and fed abundantly with green leaves. Large volumes of smoke lick uncannily up the side of the hill towards the hives. Now there is a stir among the bees and, as those below hear it, their merriment waxes great and they sing: 'The bees, the bees, the bees! Pile the fire, pile the fire, pile the fire! Ho, the bees run!' '*Giddi, giddi, giddi, rung, rung, rung*' they jeer in imitation of the flight. Meanwhile the fire above has been lit and two smoking *hulas* have been lowered towards the hives. The air is heavy with the ominous hum of angry bees.

Now is the time for the descent. The ladder, firmly secured to a tree or stone and guarded by the brother-in-law of the cutter (the only person not eligible to marry his wife in the event of an accident), is let down with a propitiatory incantation to the demons and ancestral spirits of the rocks. This being uttered, the hand grip is said to be strong and the feet to cling like jak milk to the stone. With smoking *hulas* looped to the forearms and *matha* on shoulder, and hair dishevelled over his face, so that the bees may not get at his eyes and nose, the cutter, gripping the cane with a single hand, disappears over the edge of the cliff into the darkness and the smoke. The ladder creaks as he feels his way down, for he sees nothing. He has reached the level of the hives. At a signal the deer-skin receptacle is lowered to him and this he secures to the ladder. Thrusting a leg between two rungs so that his foot is on the lower one and his bent knee against the upper one, and his head between two others, so that the back of his neck is braced against a rung, he is free to use his hands. First he smokes the hives, covering them with the *hulas* (usually mounted on the *matha*), and drives away the lingering bees. Then, slinging the *hulas* on his forearms, he takes up the *matha* and, with its pointed end, cuts horizontally through the lower part of the comb, separating off the useless *yotha* containing grubs, which the bees stoutly defend. He next cuts the comb in vertical sections, and loosening each from its attachment, splits it with the pronged end of the stake, removes it, and shakes it into the receptacle at his knees. Thus he deals with all the hives.

As he works, he sings the cutter's song and many a vulgar song besides. For this is the epic event of the Vedda's life, and never is he so happy, as when engaged in the perilous work. Wild is the jungle night and the rocks around re-echo to his song. When the 'devil's mouth' is full, he releases it from the ladder and shouts '*dapo*' (lift) to those above. They pull on the attached rope. With a series of short jerks, the cutter climbing and helping from below, the honey is lifted. As each pull brings the vessel to the level of the cutter's head, he calls out '*ho-ho*' and the hauling is stopped. Then, climbing a rung or two, he gives a helping hand and cries '*adapo*' again. And so, with *adapo* and *ho-ho*, following each other in quick succession, the precious freight is safely hauled over the edge of the cliff. It may be that the cutter has to descend to his work again. With the last load he climbs over the top, a figure to behold – sweating, exhausted, honey-bedaubed, bereft of all the stamine that has sustained him so long. He throws himself down. Nor have all his arts availed him against the bees, for he is covered with stings. Solicitously his wife and perhaps a friend or two group round him and with pitying words pick off the stings. But, impatient of this, he besmears himself with sand so that the ants may not worry him, and is soon asleep. Such is the story behind the potful of *Bambara*-honey which the Vedda offers for a trifle of grain or a fragment of cloth'.

The Veddas live for half the year wandering from one rock shelter to another and depending largely on honey for their food, which they collect from rock crevices. Each family is the recognised possessor of one or more rocky hills, while the whole community joins to collect honey from each hill, and the honey is shared equally. In this connection they perform a most important honey-gathering ceremony. Honey-gathering is a risky occupation, as the rock bees are very fierce, and hence spirits are invoked beforehand for protection, which are fully described by the SELIGMANNS (1911, pp. 42, 163, 252-9, 327):

The *maha yakino* are the spirits of old Vedda women, one of which gives special luck in honey-gathering and causes bees to build many combs. All *maha yakino* are connected with honey, as they haunt the rocky crests of hills, and so offerings of honey are made to them.

The spirits of dead Veddas, especially that of *dola yaka*, are invoked to safeguard the men when they are collecting honey and to prevent them from falling. The ceremony takes place in the early afternoon, when the bees are actively foraging. A place is cleared for a dance and two arrows are placed in the centre. A betel leaf is put on the top of each and a small bead necklace looped over the leaf. These leaves symbolise the large bundles of leaves with which the Veddas use to smoke the bees from the comb, and the necklace the creeper by which the twigs would be tied together and by which they would be lowered over the cliff. Small leafy twigs are placed on the ground round the arrows, and on these a number of betel leaves and areca nuts are placed as an offering. The twigs are to keep the offering from touching the ground. The adult men take part in the dance, and only those possessed by *dola yaka* would obtain favour and help from him in gathering honey. The men walk round the arrows singing an invocation, bending their bodies towards the arrows as if listening for the hum of bees. Then they dance widly round and cry: 'We hear many bees, there will be plenty of honey!' They beat their bodies with their hands, driving away imaginary bees, even feigning to pick some off their bodies. Other spirits, such as the *rahu yaku*, are called upon to give good luck for collecting honey from trees, whilst *dola yaka* specialised in rock-honey.

The Veddas also make ladders of cane to descend the precipices for the cutting of the honey-combs adhering to their sides. They use, too, smoke when doing this, but this dangerous task is undertaken only by the most athletic. While doing it they sing special invocations to appease the spirit of the rock. A song is also chanted and a little honey sprinkled for the spirits before the combs are cut from the rock'.

'The people in one village on the borders of the Vedda country lived upon cakes from the nuts of the wild areca-palm, pumpkins, wild fruits and berries ,river fish and wild honey. The latter is very plentiful throughout Ceylon, and the natives are very expert in finding ou the nests, by watching the bees in their flight and following them up. A bee-hunter must be a most keen-sighted fellow, although there is not so much difficulty in the pursuit as may at first appear. No one can mistake the flight of a bee en route home, if he has once observed him. It is no longer wandering from flower to flower in an uncertain course, but it rushes through the air in a straight line for the nest. If the bee-hunter sees one bee thus speeding homeward, he watches the vacant spot in the air, until assured of the direction by the successive appearance of these insects, one following the other nearly every second in their hurried race to the comb. Keeping his eye upon the passing bees, he follows them until he reaches the tree in which the nest is found' (S. W. BAKER 1881, pp. 244 ff.).

BAKER describes four species of honey-bees. The largest and most extensive honeymaker is the *'bambara'*, nearly as big as a hornet. It forms its nest upon the bough of a tree, from which it hangs like a Cheshire cheese, being about the same thickness, but 5 or 6 inches larger in diameter. The honey is not much esteemed, as the flavour partakes too strongly of a particular flower which the bee has frequented. The wax of the comb is the purest and whitest of any kind produced in Ceylon. The next honey-maker is similar in size and appearance to the European honey-bee. It forms its nest in hollow trees and in holes in rocks. Another, similar bee, but only half its size, suspends a most delicate comb to the twigs of a tree, of the size of an orange. The honey of the two latter species is of the finest quality, and quite equal in flavour to the famed 'miel vert' of the Isle de Bourbon, although it has not its green tinge. The last of the Ceylon bees is the most tiny, although an equally industrious workman. He is a little smaller than our common house-fly, and he builds his diminutive nest in the hollow of a tree, where the entrance to his mansion is a hole not larger than would be made by a lady's stiletto. Its honey is a thick, black and rather pungent, but highly aromatic molasses. The natives, having naturally coarse tastes and strong stomachs, admire this honey beyond any other.

The wax export from Ceylon is trifling. The Cingalese waste or consume all the bees'-wax. They cram the comb with the honey. Some few natives in the poorest villages save a small quantity, to exchange it for cotton, cloths, etc. BAKER once permitted a native to collect honey in his forests in return for the delivery of the wax. This bee-hunter delivered him after a few days 72 pounds' weight of well cleaned and perfectly white wax, made up in balls of 18 pounds each.

6. SIAM

One of the best studies on insects for human consumption is a paper by W. S. Bristowe (1932, pp. 387–404) on his observations in Siam. On his arrival in that country Bristowe was told that the Laos ate insects. He collected much pertinent information and paid special attention to discovering whether the Laos really like to eat insects or eat them from economic necessity. He came to the conclusion that the Laos without doubt like the insects they eat. 'Some fetch high prices and the capture of others is fraught with considerable risk. What is more, so would we like them if they were suitably disguised and if we gave ourselves the chance of acquiring the taste. By ourselves eating spiders, dungbeetles, waterbugs, crickets, grass-hoppers, termites and cicadas, we found none distasteful, a few quite palatable, notably the giant waterbug. For the most part they were insipid, with a faint vegetable flavour, but would not anyone tasting bread, for instance, for the first time, wonder why we eat such a flavourless food? A toasted dungbeetle or soft-bodied spider has a nice crisp exterior and soft interior of soufflé consistency which is by no means unpleasant. Salt is usually added, sometimes chilli or the leaves of scented herbs, and sometimes they are eaten with rice or added to sauces or curry. Flavour is exceptionally hard to define, but lettuce would, I think, best describe the taste of termites, cicadas and crickets; lettuce and raw potato that of the giant *Nephila*-spider, and concentrated Gorgonzola cheese that of the giant waterbug (*Lethocerus indicus*). I suffered no ill effects from the eating of these insects'.

That the Lao is driven by force of necessity to eating insects seems unlikely, living, like the Siamese, as poor peasants and padi-growers; the Siamese do not eat insects. Prof. C. Zimmerman (Siam Rural Econ. Survey 1930/31, Siam. Min. Commerce and Communications) showed that among almost 10.000 personal diets studied in various districts there were 30 clear cases of diet deficiency, 25 of them being ber-beri in districts where milled rice had recently been introduced. This would indicate that enough suitable food is normally available without resorting to insects. Yet up-country the Laos are more energetic farmers than the Siamese, and employers testify to their greater stamina and their harder working. Zimmer-

MAN's analysis shows as average food of one adult Siamese peasant per day:

Rice   693 gr. (glutinous) or 553 gr. (non-glutinous)
Fresh fish   6 to 25 times per month
Fermented fish   27 gr.
Fish- or shrimp-sauce 10 gr.
Shrimp paste   14 gr.
Green beans   6 to 25 times per month
Salt   10 gr.
Eggs, chicken, pork, beef, shrimps and shellfish
                    1 to 5 times per month.

Lao food is similar, with the addition of insects, which are eaten *in considerable numbers* at the particular season in which each occurs. The basis of diet is rice (75% carbohydrates, 13% moisture, 8% protein, 1% fat). Fish represents the main protein constituent of their diet and in the seasons when fresh fish is not available, fermented fish, shrimp paste, or dried saltfish are consumed. The protein content of fermented fish and shrimp are 18 to 23%, fat 6 to 3%. Insects have a high protein as well as fat content.

Some slightly roasted spiders (*Melopoeus albostriatus* Sim.), which are widely eaten by the Laos, contained (analysis of the Bangkok Govt. Lab.) 63.4% protein and 9.8% fat. A Lao and I hunting together collected six such spiders, together ½ lb. in weight, in one hour. This was in the dry season when the giant waterbug and some other insects are unobtainable or very scarce. It would appear that some kind of insect is always in season. The part of man (Laos) in controlling pests is probably greater than might be imagined. Young padi is much damaged by freshwater crabs (*Potamon spp.*), which nip off the young stems. Fortunately they are liked and on sale in some of the up-country villages .... The larvae of timber- and stem-boring longicorns and weevils are sought by the Laos, such as:

Coleoptera	*Rhynchophorus schah* F.	Grub is a severe pest of coconut-stems.
	Xylotrupes gideon L.	Dto. of coconut and sugar palms.
Orthoptera	*Patanga succincta*	Pests on
	Locusta migraroria L.	padi-rice.
Lepidoptera	*Xyleutes leuconotus*	Caterpillar boring in *Casuarina*-trees.
	Zeuzera coffeae	Coffee-borer, larvae occasionally on Bangkok market.
Hymenoptera	*Xylocopa confusa*	Damages wood.
	Xylocopa latipes	Recently did considerable damage to cables in Bangkok.

The Laos are called 'dirty feeders' by the Siamese on account of their insect-eating habits, and in particular, of their liking for dung-beetles and their grubs, so that they are somewhat sensitive of being questioned . . . . Actually the Siamese will themselves eat some kinds of insects, though they do not make such a general practice of it. Giant waterbugs are added to royal sauces, larvae of *Zeuzera coffeae* were much prized by the late king, and certain ants and ant-grubs are pickled for the consumption of the good Siamese families at Bangkok. Wasp-grubs and honey, locusts, crickets and even big mygalomorph spiders are occasionally eaten by some Siamese peasants. The Laos are concentrated in N. and E. Siam, yet live scattered everywhere. I questioned Laos from all parts of Siam. Their choice of insects seems to be remarkably consistent in every area and also the order of preference, the most popular being the giant waterbug, dung-beetles and their larvae, mygalomorph spiders, grasshoppers, and the larvae of *Hymenoptera*, beetles and moths. Up to 4d. may be paid for a single *Heliocopris* or a giant waterbug. Exceptions include scorpions, which do not appear to be eaten in Chieng Kai or Battambang; dragonflies eaten only in the Ubon area, though their nymphs are collected at Hua Hin, and cockroaches which in most areas are said 'stink', but are eaten at Korat and Hua Hin. The eggs of cockroaches are fried and eaten by children in all districts. Of course, some insects, notably a cicada (*Dundubia intemerata*), have a restricted range and are therefore not known in other districts. In spite of individual variation, the average Lao has a remarkable knowledge of the structure and life-habits of the insects. In some cases (*Cicada, Vespa sp.*) this knowledge is used to devise ingenious methods of capturing them'.

All scorpions are eaten in Siam (*Tityus spp., Heterometrus spp.*). BRISTOWE tells of the experience of Miss E. FRISCH, that certain families of 'scorpion-catchers' allure these animals by whistling. Any spider of reasonable size will be eaten. Most popular are the big *Nephila maculata* F. and *Melopoeus albostriatus* Sim. Apart from the freshwater crabs (*Potamon spp.*) mentioned previously, the eggs of the king-crab, *Tachypleus gigas*, are also highly prized. The observations on insect food will be quoted verbatim. 'Coleopterea: *Cerambycidae*, etc. Many Cerambycid and other grubs are eaten by the Laos, the beetles of which I was often unable to determine. A very shy tribe of Laos on the borders of Siam and Burma at Suphan shoots the longtailed *Presbytis*-monkey and extracts its intestines. The body cavity is then stuffed with leaves of *Citrus hystrix* and other herbs and sewn up. A paste from the interior portions of a termite mound is completely smeared over the dead monkey's body which is now hung up on the branch of a tree. A dish is placed beneath to collect the juice that drops from it, the monkey sauce (*Nam phla Kharng*), a great delicatessen of Siam. After one or two weeks blow-fly (?) maggots begin to fall into the dish and the monkey was left hanging until these maggots ceased to appear. Then the 'shell' was cut open and two or three special big maggots of a different kind, with 'legs and a dark mouth with jaws' are found inside, being the larvae of some beetle. As many coconuts as there are larvae are collected and heated in their shells, a hole is bored in the top and, when the liquid is cool, one larva is introduced into each. The hole is now closed with termite paste, the coconut is swathed with cloth and this in turn is enclosed in termite paste. For about three weeks this 'mummy' is stored away, then split open and a white grub, 'the size of a tangerine', is found practically filling the interior of the coconut. The cost of one of these grubs is about 12 shillings. They are eaten at ceremonies and on special occasions. Can the larvae feed on such different things as decaying monkeys and coconut, or does it die within the coconut and get puffed up in these three weeks to the size of a tangerine?

Rutelidae. The beetles of *Adoretus compressus* Web., *A. convexus* Burm. (Lao: *Mang ee noon*) are very popular throughout Ubon, but do not appear on the market. They are caught at night, when swarming around the light; they are roasted and the elytra pulled off.

Buprestidae. The beetles of *Sternocera equisignata* Saund. (Lao: *Mang Khup*; Siamese: *Mang thup*), 3 cm. long, are sought by the Laos for food, by the Chinese for decorative purposes. The Chinese breed them and sell them at Bangkok for $1/_4$ d. per beetle. The Laos pick them off shrubs, roast them over a fire and squeeze out the faeces before eating them.

Dytiscidae. The black, shining waterbeetles *Cybister limbatus* F. (Lao: *Mang mee-eng*), slightly longer than 2.5 cm., are caught in nets, probably as a by-product of fishing, and roasted.

Dynastidae. The larvae, pupae and beetles of *Oryctes rhinoceros* L.

257

(Lao: *Mang bongh*, the larva; *Mang kwang*, the beetle) are searched for amongst cow and buffalo dung, especially in sheds where the animals have been. The dark brown, 3.5 cm. long, beetle has a horn in the male sex. They are usually roasted but sometime fried. Sometimes they are used in curries and sometimes eaten alone, after breaking off the elytra and hard parts. The grubs and pupae are soaked in coconut milk for a quarter of an hour and then roasted. Their flavour is not strong and quite pleasant, but difficult to define; it is vegetable, not animal. Beetles and grubs are highly prized.

Beetles and grubs of *Xylotrupes gideon* L. (Lao: *Mang kwung*) are eaten, but with less enthusiasm. The males are pitted in battle against one another, but so much money changed hands on such occasions, that a law prohibiting beetle fights was passed in Siam recently. The males fight, and once one had managed to outflank the other the sound of rending chitin proclaims the end.

The very large black *Heliocopris sp.*, 5 cm. long, is used in N. Siam as medicine and food. One beetles fetches 4d. at Lampang. It is roasted, pounded up and added to curry.

Onitis virens Lansb. (Lao: *Mang chew chee*) abounds amongst cattle dung at Hua Hin. It is roasted and eaten with salt, as BRISTOWE reports also of *Ateuches sacer* L. by the Bedouin of Egypt. Miss E. FINCH reports its use in the circumcision ceremony. A circle of men squatting on the ground surrounds the boy and a sheikh. The men are shoulder to shoulder, each touching his neighbour on either side. They chant in rhythm the twenty-two names of Allah and bring themselves into a kind of trance. The sheikh remains relatively calm, then reads from the Koran about eternal life and commands the boy to eat from a bowl which contains scarabs, and the boy is then recognised as a man of the tribe or village.

Melolonthidae. Annandale reports that the Laos near Pateling eat four species of Melolonthid beetles, such as *Lepidiota stigma* F. (Lao: *Mang ee noon*). Laos from other parts of Siam also told me that they eat it, as well as *Leucopholis sp. (Curculionidae)*. The big black weevil *Rhynchophorus schah* F. and its grubs are also eaten. The grub is extracted from coconut palms and roasted.

Longicornia. The larvae of various species of wood-boring Cerambycids are extracted in order to be roasted and eaten.

Rhynchota: Waterbugs are very popular, especially the 5 cm. long

Lethocerus indicus L. and S. (Lao: *Mang daar nah*), which is a great delicacy for Laos and Siamese alike, being served also on the royal tables in Bangkok. They are caught in waternets. Their usual preparation is: 1. Steam thoroughly and then soak them in shrimp-sauce. They are served picked into pieces, each piece yielding a little meat from its inner side. The flavour is strong, like Gorganzola. 2. After cooking pound up and use for flavouring sauces or curries. A popular sauce, the *Namphla*, is made by mixing shrimps, lime juice, garlic and pepper, and then adding *Mang daar* to finish up with. Vegetables are dipped into this sauce. The price is one penny to fourpence per piece in Bangkok, where it is unobtainable from December to February. In Indo-China its essence is extracted, and this is sold in small bottles in the towns.

Two small, common greenish waterbugs (*Sphaerodema rustica* F., *S. molestum*, Lao: *Mang kharn*) are knocked down with sticks from bushes, where they rest by day. When several have been collected, they are roasted in a dish and eaten with fingers like shrimps. Another waterbug (*Laccotrephes grisea* Guer.; Lao: *Mang dah*) is eaten toasted on a bamboo skewer'.

The only highly prized cicada in Southern Siam is *Dundubia intemerata* Walk. (Lao: *Tua chuck-a-chun*). ANNANDALE (Proc. Zool. Soc. 1900 : 859) and H. M. PENDLEBURY (Journ. M. S. Mus. 1923: 11) describe its capture, as follows: On darkness falling a fire is lit and the cicada-seekers arrange themselves on the ground round it. They clap their hands together in unison and the female cicadas come in swarms. Mr. SMYTHE mentions that he has seen and heard Laos doing this night after night in Pataling; only in his experience pieces of bamboo have always been used for clapping together instead of the hands. In Laos along the Mekong River H. D'ORLÉANS observed young women, half-naked, hunting cicadas which they catch with bird-lime to sell in the market or to fry for their own food. This is *Dundubia intemerata* Walk. (Vide DAGUIN 1900, p. 19).

Orthoptera: Acrididae. Grasshoppers like *Patanga succincta, Locusta migratoria* L., *Aeolopus tamulus* F., and others (Lao: *Took-ah-tanne*) are eaten everywhere in Siam – roasted, toasted or like shrimps. The Siamese eat grasshoppers too, as do the Philippinos, usually fried in oil with salt. They are pounded up, when appearing in swarms, into a mass, which is buried in the sand and left there for

259

a considerable time. The not unpleasant flavour is said to vary between that of shrimps and anchovies.

Gryllidae. Robust crickets, such as *Gryllus testaceus* Walk. (Lao: *Ching-reep-sigh*), *Brachytrypes portentosus* Licht. (*Ching-reep-ong*) or *Liogryllus bimaculatus* de Geer, are greatly prized, especially the two former. But BRISTOWE found them insipid and lacking any definite flavour. At Hua Hin he dug up several *Brachytrypes*-nymphs from their 23 cm. deep burrows. The head is removed and the insect roasted on sticks over a fire. The Siamese also eat them. The males of *Liogryllus* are used for fighting by the Siamese, those of *Brachy-trypes* by the Malays, much money being wagered on the results.

Gryllotalpidae. The mole-cricket *Gryllotalpa africana* Beav. (Lao: *Kin-ni*; Siam: *Mang-ka-chan*) is dug up from its deep burrows by the Laos all over Siam.

Mantidae. The eggs and the adult green praying mantids *Hierodula sp.* (Lao: *Mang-naap*) are eaten by the Laos at Hua Hin and probably elsewhere.

Blattidae. Among cockroaches *Blatta orientalis* L. and the apterous *Stylopyga rhombifolia* St. were collected in BRISTOWE's presence at Hua Hin. The Laos there and in Korat will eat cockroaches, but elsewhere they are said to 'stink'. Yet everywhere the eggs (sic!) are collected by children for frying. BRYGOO (1946, p. 29) tells about a French colonel who ate cockroaches raw with evident pleasure, whilst the Annamites eat them only after they have been held by a needle over a fire.

Odonata (Lao: *Mang por*): Only the dragonfly *Anax guttatus* Burm. is eaten in the Ubon district, where it is roasted in a dish; but the nymphs of a fat-bodied species are also eaten at Hua Hin. They are boiled and taste like crayfish. In other districts dragonflies are apparently not eaten. In the Ubon district a lighted candle is placed in the middle of a big bowl of water and the *Anax*-dragonflies singe their wings and fall into the water. It seems doubtful whether the Laos would get enough for a meal in this way, were it not for the fact that a host of other insects will be captured at the same time, such as termites, beetles, etc.

Isoptera. Flying termites, perhaps also flying ants (Lao: *Mang mauw*), are caught by the Laos in traps, consisting of candles which will singe their wings, surrounded by water, and in this way they catch large numbers at the swarming periods. The 'catch' is roasted with salt and eaten. This is by no means a bad dish. What little flavour they have is vegetable in nature and the salt brings this out. Eaten raw they are insipid. The termite queen is a delicacy.

Lepidoptera: A number of large caterpillars are eaten but butterflies and moths are, for some reason, left severely alone. Major LADELL mentions that the Laos extract the larvae of *Xyleutes leuconotus* from the stems of *Casuarinia*. A. KERR (1931, p. 217) finds some satisfaction in the fact that in Siam the caterpillars of some pests like the coffee-borer (*Zeuzera coffeae*) are turned to useful purposes. This larva tunnels in the branches of various trees and shrubs, such as *Sesbania roxburghii*. Though not cultivated in the fields, it yields two edible products: the flowers which are eaten and the larvae of *Zeuzera coffeae* which live within the stem and branches. When fully grown these '*Duang sano*' are collected for eating. Aynthia is the province where they are chiefly obtained, the host plant growing plentifully along the river. There is some trade in the larvae, which are sent down to Bangkok alive during September/October. They are prepared for the table by frying. Prince Sithiporn told BRISTOWE that his cousin, the late king of Siam, was very fond of the caterpillars of *Z. coffeae* (Siam: *Duang*), which are extracted from *Sesbania aculeata*, the larvae being roasted and eaten with salt and rice.

Hymenoptera: Formicidae. The weaving ant, *Oecophylla smaragdina* F. (Lao: *Mottdaang*), builds its common nests of tennis- to football-size on shrubs. Man can be numbered amongst its enemies in Siam, for the Siamese will eat its eggs (read: pupae) and the Laos the adults as well. They are said to taste sour. A jar of water is placed immediately beneath the nest and then pushed upwards, so that the nest is submerged. About twenty nests will make a meal for a family.

Another ant which is much eaten by both Siamese and Laos has not been identified. When these brown, medium-sized ants are seen walking on the ground, they are followed to the entrance of their subterranean nests, which is basket-sized. The ants, their larvae and pupae, are pickled in saltwater, tamarind juice, ginger, onion, a little sugar and the leaf of *Citrus hystrix*.

Around Hua Hin the nests of *Crematogaster sp.* (Lao: *Mott dam*) are collected for their grubs which are eaten in curry.

NICOLI CONTI, who travelled in India in the early part of the

15th century, says the Siamese eat a species of red ant, of the size of a small crab which, seasoned with pepper, they consider a great delicacy (HAKLUYT Society II, p. 13, vide COWAN 1865, p. 159). At the present day the pupae of ants are a costly luxury with this people. They are not much larger than grains of sand and are sent to table curried, or rolled in green leaves, mingled with shreds or very fine slices of a fat pork (The Mirror 31. p. 342, vide COWAN 1865, p. 159).

Vespidae. The grubs of *Eumenes petiolata* F. (Lao: *Mang taan)* are fried for food. The hornet *Vespa cincta* F. (Lao: *Tua thor)*, conspicuous by its broad abdominal orange stripe, and its grubs, which are common all over Siam, are eaten everywhere. In N. Siam it is believed that their sting makes the hair turn white. A few stings are liable to cause death. The nests are in hollow trees and the wasps are destroyed by fire and smoke. When all is quiet, the nest is pulled out and grubs and wasps are both fried with a little salt, after the legs and head have been removed. Both Siamese and Laos eat the grubs.

Apidae. No domesticated bees exist but a lot of wild honey is put on the market together with wax.

The big *Xylocopa confusa* Pers. (Lao: *Mang poo)*, which nest in planks, and *X. latipes* Dr., which damage cable encasings around Bangkok, are eaten. The Laos knock them down and then spike or crush the heads of the bees, pull off both wings and heads, and eat the underside of the abdomen raw. Or the abdomen is bitten into from the underside and the raw goodness sipped from it. The honey and wax of *Trichona sp.* (Lao: *Channa-roong)*, a small brown bee, together with its grubs is extracted from the nests and eaten, and is said to taste sweet and sharp. The same holds for the slightly larger *Nomia sp.* (Lao: *Mang mim)*.

The Indian bee, *Apis indica* (Lao: *Mang peung koh)*, slightly smaller than the European bee, builds mud-nests one foot in diameter. The Laos near Hua Hin take the nests, after having smoked out the bees with burning coconut fibre. Honey and grubs are eaten.

Apis dorsata? (Lao: *Mang peung)*. The Laos say that three or four of its stings give fever. They build very large, crescent-shaped nests, five to six feet in diameter, which are fixed to the branches of tall *Ficus*-trees in the forests of N. Siam. The Laos describe their collection as follows: Certain men are not stung by the bees. This immunity is passed down from father to son, and thus all the nests of one district are taken by one or a few families. On the day appointed for taking the nests prayers, ceremonies and incantations follow one after the other until the man is worked up to a pitch where he can feel nothing. Sacred waters are thrown over him. Night has come; he is ready to start. On arrival at a tree with several nests on it, one final assurance that the spirits are willing must be obtained. His followers have brought a number of bamboo-sticks, one end of which has been sharpened and hardened with fire. One by one these are hammered into the trunk, and if any one of them needs more than three blows to leave it fixed there, a sign has been given that the fates are not propitious. Further attempts on the bees must be postponed until another night. If, however, the test has been successful, the man is hauled up to the branch from which the nests can be reached, by means of a rope worked like a pulley. It is time now for his friends to retire to a safe distance. He climbs along the branch towards the nests, lights a big wad of cotton wool or similar material with his flint and steel and waves it wildly round the nests. As the bees come rushing out in a cloud, he drops the flaming wad and they follow it to the ground 80 or 100 feet below. Now is the time to cut down the nests and this he does as quickly as possible, not being content with an evening's work until 200 to 300 have been taken. They feast on the honey and the maggots; the wax, of which there is a goodly quantity, is much in demand for candles for temples and for cremation ceremonies. In the forest it fatches a price of about one shilling and sixpence per cattie, but by the time it has passed through the hands of several Chinese middlemen, and as often as not has been adulterated with pumpkin juice and paraffin wax, it costs at Bangkok five or six times as much.

The grubs of *Apis sp.* (possibly *A. indica;* Lao: *Mangnon when;* Siam: *Mang non won*) are eaten all over N. Siam. The football-sized nests, made of mud or of cow-dung, hang from a branch of a tree. A sentry-bee is always on guard at the lower entrance to the nest. To take a nest, a man advances with a long pole and a bundle of dried grass or old cloth attached to the end, which is set alight. The bees fly out and the nest is knocked down. The heavy impact of the fall usually breaks the mud covering and reveals five to twelve combs of cells. The grubs are all picked out and fried or eaten in curry. A favourite dish is as follows: in coconut milk put onion, pepper, *Cymbogon citratus* and leaves of *Citrus hystrix*. Wrap them in linen, steam them and then add this grub-sauce to the rice.

7. INDO-CHINA

BRÉBION (1913), professor of the college at Baria (Cochin China), states that the Annamese and the many forest tribes of the Mois take revenge for the damage caused by numerous insects by cooking them without any formal procedure. Thus the palmworm, the *con-duong ch-la*, is taken from the roots of a certain palm. In Cochin China and in Annam this grub is introduced, after its capture, into the internodes of sugar-cane. When sufficiently fattened it is inserted into the *nuoc-mam*, the national sauce of Annam, fried in pig fat and wrapped in paste. Roasted in butter or rolled in flour this larva is rather succulent and smells like hazel-nut. Europeans like this dish, while in Annam it is reserved for the royal table only. This palmworm is found only in the maritime districts of Cochin China and its price is always high.

The *con-duong dat* is a beetle larva collected in May among the roots of a green plant at Travinh (Cochin China). In April the Annamese collect the *con-ray*, a cockchafer, which they leave overnight in the *nuoc-mam* sauce, after having first removed the intestines, elytra, wings, antennae and legs, and fry it the next day. The Annamese annals report that a king of Hué once sent with his triennial tribute some *con-ray* as a personal gift to the Emperor at Peking, which pleased the latter so much, that he asked for a second consignment.

The giant waterbug, the *con-bo cap-unoc*, is roasted and consumed in the *nuoc-mam* sauce. At Saigon a pair is sold for 2 fr. 50. Mole-crickets or *con-de-com* are freed of their legs and wings, cleansed, covered with an *Arachis*-nut and cooked in lard. This famous Annamese dish is served mainly in the wet season, from May to October, when these insects abound.

Termite-queens, the *con duong-cha-la*, are also appreciated. No Annamite will destroy a termite hill close to his house. He will cover its top with a piece of red rag, and he regards it as the dwelling of an ancestor in retirement who has approached his house to answer his prayers. At the foot of the hill he will frequently burn incense sticks. Yet in Cambodia many termite hills are used as lime-kilns.

The pupae of bees in the combs are also a popular Annamese dish. In the forest villages scorpions, many big beetles, etc., are also eaten. This entomophagy is well pronounced in the forest-dwelling Mois in Khas and Pnons (Cambodia) where laziness and improvidence make famines an annual event. At such times every larva, every living thing is eaten by them. The lower classes of Cochin China and the neighbouring districts are no less catholic in the choice of their food than these despised natives. They delight in the multitude of lice which all sexes and ages collect from the hair of their heads. This occupation is often observed and the captured prey is immediately cracked with delight between their beautiful teeth. Great use is made of insects in folk medicine.

A. PAVIE (1901, p. 118) mentions a common dish from the same region prepared by his coolies from big spiders, lizards and black scorpions (see the review of DE JOANNIS 1929). NGUYEN-CONG-TIEU (1928) has made a study of the entomophagy of the Tonkinese. The lower classes are not particular in the selection of their food. The sea is poor in fish, molluscs and shrimps. The soil is entirely occupied by rice cultivation and leaves no space for pastures of buffaloes or cattle. Goats also are rare. Pigs and chickens are not uncommon but are reserved only for the table of the rich. Hence insects are much eaten by the poorer classes.

Orthoptera. The grasshoppers *Oxya velox* F., the *chau-chau* or *cao-cao*, is the only one of this group which is eaten in any quantity. From May to December the children knock them down with a triangular slat of plaited bamboo, 50 cm. long. And they are caught in masses by drawing a basket, 120 cm. long, 50 cm. broad and 60 cm. deep, over the soil among plants or the young rice fields. The trapped animals are held back in this basket by the repeated sweeping movements of the hunter. These hunts take place usually by day. At home the basket is put for some hours into cold water, or for a few minutes into boiling water. About 200 gr. are sold, as a rule, for three sous. Females full of eggs fetch a higher price, one

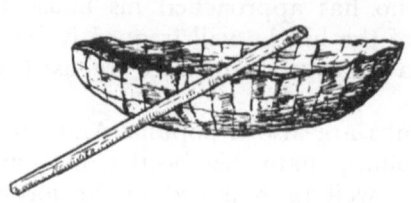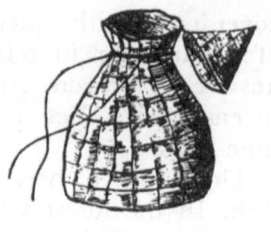

Fig. 34. Net for catching *Oxya*-grasshoppers in Indo-China and basket where the caught insects are kept (From NGUYEN-CONG-TIEN, 1928).

kilo being sold for 15 cents. Freshly killed grasshoppers have the following chemical composition: Moisture 68.9%, protein 8.3%, ashes 1.2%, fat 1.0%, P_2O_5 0.7%, CaO 0.005%, carbohydrates: traces. The grasshoppers are rich in phosphates. The wings, and often also the head with the intestines and the first two pairs of legs, are discarded. Then the insects are cooked in saltwater or fried in pig fat. Spices and lemon leaves are added. They are boiled until most of the water has evaporated and the insects are covered by a thin layer of salt. Served with rice, this is a common, but not a favourite dish of agricultural labourers,

Another grasshopper, *Euconocephalus sp.*, which is too rare for systematic collection, is caught by hand and boiled with cabbage leaves into an excellent soup. The large cricket *Brachytrypes portentosus* Licht. and the mole-cricket *Gryllotalpa africana* Beauv., which do not taste pleasant, are eaten by the poor.

Rhynchota. The large waterbug *Belostoma indica* Vit., the *ca-cuong*, is collected for its meat, as well as for a strongly smelling liquid secreted by stink bags, two long coiled tubes in the thorax. They are 7.5 cm. long and have a weight of 8 to 11 gr. These stink bags are extracted in a very elaborate way, which is described by NGUYEN-CONG-TIEN in detail. Sometimes these sacs are only removed after the wings have been discarded and the chop is salted for storage. The odorous liquid is used for seasoning many dishes, and for some of them this seasoning is held to be indispensible. The bugs are not very fleshy, but the thorax contains a mealy matter. The wings, legs and the cauda are removed, and then they are fried over charcoal or they are stewed in a special pot. In both cases only the soft parts

266

oft he thorax are eaten. Yet when finely chopped and fried in fat, everything is consumed. The waterbugs are caught with special nets and are also taken at light.

Cicadas, the *ve-sau*, are mainly eaten in the mountains. The big larvae and the pupae are rather delicate. Grilled in fat they are highly appreciated.

Ephemeroidea. Mayflies, the *con-vo* or *phu-du*, abound around Tonkin. They form a regular addition to the usual food of the fishermen, after being fried with salt in fat.

Coleoptera. The grubs of the big Cerambycid *Apriona guermari* Hope are mainly used as medicine, to protect children against wind pox. Hydrophilid beetles are rare and are eaten fried. Palmworms are rare and have not the same reputation as in Cochin China.

Lepidoptera. The pupae of the silkworm *Bombyx mori* L., the *con-tam*, contain as a percentage of the fresh matter: moisture 78.8, fat 2.8, proteins 13.0, ashes 1.1, P_2O_5 0.4, CaO 0.05%. Their composition is thus similar to that of shrimps and shellfish, being richer in proteins than chickens' eggs and most shellfish. The pupae, when removed from the reeling basins, are cooked and ready for consumption. Yet, as a rule, they are fried in fat and seasoned with lemon leaves or salted. The pupae must be boiled to such a degree, that they crack between the teeth. When well pounded and cooked with leaves of cabbage, Phyllanthus or bindweed they give a splendid soup. Sometimes they are dried in the sun after cooking and then may be preserved for a considerable time. They are sold in all the markets of Tonkin, one kilo or 2500 pupae fetching 25 cents.

The caterpillar of *Brihaspa atrostigmella* Moore (*Pyralidae*) lives in the terminal leaf of a grass (*Thysanolaena maxima*). The larvae and pupae are rarely consumed as food, but are an appreciated medicine The eating of adult honey-bees by the peasants of Tonkin and their making into an omelette by the Mois has also been reported (Brygoo 1946, p. 50).

Nguyen-Cong-Tien concludes: 'In almost every group of insects edible species are found. Certain of them, such as grasshoppers, giant waterbugs and silkworms, are the object of a regular trade in the great cities as well as in the villages. These insects are thus important for the alimentation of the native population of Tonkin'.

8. BURMA AND MALAYA

C. C. GHOSH (1924, p. 403 ff.) tells us that the Burmeses, Karens, Chins, Kachins, Shans, Talaings, and others are not at all fastidious in their choice of food and that many insects have found a place in their diet. The Buddhists have no objection there to eating animals killed by others. Many Burmeses kill insects for themselves, as well as for sale. GHOSH gives the following incomplete list of insects eaten, in addition to the Dytiscid *Eretes sticticus* and a grub, already mentioned by MAXWELL and LEFROY.

The nests of the weaver-ant *Oecophylla smaragdina* are collected and the ants and the grubs contained in them are suffocated in an air-tight vessel with smoke. The ants, grubs, etc. are then picked off and made into a paste which soon gets sour and is called *Khagyin*. It is eaten especially by the women and believed to be good for the monthly period.

The big brown cricket *Brachytrypes portentosus*, called *payit*, is widely eaten and sold, fried, on the market at Mandalay. They are taken by professional collectors in the villages, where ten big crickets are sold for two annas. In Mandalay a hundred are sold for one rupee and four annas. Baskets full of fried crickets are occasionally eaten during festivals of rich people.

The grubs of *Oryctes rhinocerus* are highly esteemed by the Karens. A Burman turning over a dung-heap for the grubs told GHOSH that he and many others ate them fried. The dark hind-parts are rejected. The grubs of *Xylotrupes gideon* are also eaten.

Those of *Rhynchophorus ferrugineus*, the *on-po* or coconut-insect, are are liked by everyone and, as they are not easily procured, are considered a dainty. They are boiled: the skin separates and is rejected. Their taste is like boiled coconut milk. Several Burmese gentlemen said that they occur in *Phoenix acaulis*. They are fattened by being put inside ripe coconuts, from which the water is first taken out. Such a fattened grub costs eight annas.

The big dungbeetle *Heliocopris bucephalus*, the *shwe-po*, is in great demand amongst the Shans, where each pupa fetches one to one and a half annas. From March to May it is common in the Shan hills, when men, women and children dig over large areas in search of the pupae, which are found inside round balls of agglutinated earth (Pl. 36) one to two feet deep in the soil. They seem to know as if by instinct where to dig for these balls by finding the opening hole of the gallery. The *shwe-po* is dug out at the season when the cuckoo begins to sing. The various species of grubs found in the droppings of cattle in the rainy season are collected and eaten by many. Sometimes various beetles coming to light are collected with lanterns in the fields and eaten or sold as food. Cerambycid grubs extracted from logs of wood are dried and preserved in oil and eaten with Burmese tea.

Winged termites are eaten in many places, boiled or fried. Also the S. Indian coolies indulge in them.

The grubs, pupae and eggs of honey-bees are boiled with parts of their combs and made into a much relished soup. The Shans smoke the nests of *Vespa auraria* on the bushes by night, eating their grubs and pupae. Wasps nesting underground are similarly caught and eaten.

Silkworms are eaten fried and may be stored for future use, being boiled when required to be eaten. Known as *Po-gaung-gyaw*, they sell for 1½ rupees per 3½ lbs. Not a single silkworm pupa is wasted. It is ready to be eaten as soon as it comes out of the reeling pan in the boiled condition. It was delightful to see little children come begging for such pupae from the Indian reelers who were engaged in the Prome district among the Yabeins. The boiled pupae relieved the monotony of the girls being trained, who had a ready dish to be carried home after the day's task was done. In the Karen hills a dish of pupae was offered to and readily partaken of by the Karen town-officer. Presents of such pupae were made him very frequently K. WARD (In Farthest Burma. vide BRISTOWE 1932, p. 397) saw his Nung porters searching in the shingle of a river bed for a species of bug, which when captured was decapitated between the finger nails and dropped into a bamboo tube. These bugs are fried in oil and eaten as a delicacy despite their horrible odour. The Rev. F. MASON (in W. L. DISTANT 1892, p. 22) notes that one of the first objects attracting the attention of an observer in some parts of the Karen jungles, is a clay tube several inches high, raised over a shaft sunk 60 to 90 cm. in the ground, over which may often be seen a Karen bending and inserting the extremities of a long branch of a thorny *ratten*, which after a few twists, is withdrawn, bringing with it a grub that is deemed a great luxury. This is the larva of the cicada *Platypleura insignis* Dist.

A number of reports refer to the Malayan Peninsula. The Reverend FAVRE (1865, p. 58 ff.) describes the food of the Jakuns, who are different from the Malays: 'They have no regular diet, living mainly on game, and when none is available they go hungry to sleep. They live upon the flesh of every kind of animal, including snakes and monkeys .... Yams, plantations containing wild fruits, the leaves of trees and certain roots furnish the principal part of their ordinary food .... For six weeks or two months they eat nothing but durians. When the season is over, the place is abandoned until the next year. One of their most prized dishes is a honey-comb. The time when the honey is in the comb is not considered the proper moment to take the hive. They wait until the small bees are well formed in the cells, and a few days before they are ready to fly away the honey-comb is taken with great care and wrapped in a plantain leaf, is put upon the fire for a few minutes, and then wax and insects are devoured together and considered as an uncommon treat'.

MARTIN (1905, p. 720 ff.), in his monograph on the inland tribes of the Malayan Peninsula, states that form of civilisation and mode of nutrition are well connected in primitive peoples. Unfortunately our knowledge is rather incomplete. Only the southern groups of the Senoi and the Semang apparently devour everything eatable. Vegetarian food prevails. He quotes HALE (J. Anthrop, 1886): 'In spite of their deadly weapon, the *Sumpitan*, they never hunt, except when no other food is available'. Collecting prevails over hunting. The primary vegetable food is composed of various roots, tubers, fruits, leaves, etc. The secondary animal food comprises all mammals, birds and their eggs, almost all reptiles, many fish, shellfish, etc. No taboos or interdictions of food exist. Yet poisonous snakes and elephants, locally also tigers and some insects are not eaten. The Senoi are very fond of the honey of wild bees, which they collect from nests 20 m. high without caring for the stings of the swarming bees.

9. CHINA AND JAPAN

We have a recent report by W. E. Hoffmann (1947 p. 233 ff.) on insects consumed, mainly in S. E. China.

Dytiscid and Hydrophilid beetles are very commonly consumed in the Kwangtung province and in other places, where Cantonese dwell. Although usually kept in separate containers customers very frequently buy some of each family. They care less for the *Hydrophilidae*, which consequently are cheaper than the *Dytiscidae*, and both are cheaper than the giant waterbugs. The common people believe these insects belong to the same species, the *Hydrophilidae* being regarded as the males. Both are eaten as medicine and as confection, being considered as an antidiuretic. They are dropped into a hot brine and appear very greasy as offered for sale, the cooking apparently having hastened the process of oil coming to the surface. The odour of some of these beetles is even more offensive when cooked than when fresh. In eating, the elytra, legs and certain other chitinous parts are discarded. One or two medium-sized species and a number of large ones are used as food, such as *Hydrous pallidipalpis* McLeay of N. China and Tibet, *H. bilineatus* McLeay of S. China and Indo-China, *H. cavisternum* Bdl. of Hainan island, *H. hastatus* Hbst. of Kwangtung and Indo-China, *Cybister bengalensis* Aubé, *C. guerini* Aubé, *C. japonicus* Sharp, *C. limbatus* Falz., *C. sigillatus* Er. and *C. tripunctus* Ol.

Silkworm pupae are eaten extensively in the silk districts of Kwangtung province in S. China. In reeling the cocoons are dropped into very hot water and the reeling girls have a plentiful supply of freshly cooked food before them all day long. They seem to eat off and on all day long since they work rapidly for long hours at a stretch, and the cooked morsels are constantly before them. One gets a pleasant odour of food being cooked, when passing through a reeling factory. The pupae are also roasted and are sold on the foodstalls. The pupae are offered for sale throughout the silk district in the south, and to some extent in other areas. Other ways of preparation are employed in the silk districts of Central China. The pupae, along with waste material from the reeling factories, are used as food for fishes in the fish ponds.

Caterpillars of Hepialidae and others are commonly found infected with *Cordyceps*-fungi. Szechuan province is famous for this material and from there the caterpillars are sent to various other Chinese provinces and abroad. About a dozen of the infected caterpillars, each with a long strand of fungal growth, are tied into neat bundles of uniform size. The shrivelled caterpillar with a fungal filament longer than its own body is somewhat reminiscent of a rat-tailed maggot. These caterpillars are considered a tonic food and are made into a broth, both the caterpillars and the broth being consumed. They are expensive, with the result that only the middle class and the rich buy them, as a delicacy or as tonic food. HOFFMANN sampled this material himself and found it quite tasty, but since he felt fine before and after its consumption, he cannot testify to its tonic effect. The same or a related species of *Cordyceps* attacks also other insects. Some Cantonese peasants had a large number of fresh cicada nymphs infected with *Cordyceps*, which they wished to sell as medicine; but unable to sell them they decided to feast on their surplus. They spent the next day in hospital as very sick men . . . .

In the Canton area some people make a living of rearing fly-maggots for medicine and food. One of HOFFMANN's assistants bought several catties of the dried maggots and some living ones, from which mainly *Chrysomyia megacephala* F. hatched. The eggs of this green-bottle fly are laid on pieces of fish and meat. Perhaps this was only a wartime industry during the Japanese occupation.*) In Canton the medicine shops were great buyers . . . .

The giant waterbug (*Lethocerus indicus*), the *Kwai fa shim* or Henna flower cicada, is widely eaten in Eastern areas. Its odour is not unpleasant and is not unlike that of *Lawsonia*-flowers. These bugs are prepared for eating in Canton by dropping into boiling water with a little salt. In Singapore a specially flavoured salt is sold from these bugs, the *Kwai fa shim im*, which is fragrant and probably contains also henna flowers. Before the war these bugs sold for three or four coppers each and were considered expensive. They are relished as a delicacy, no medical use being intended. These bugs, like the Dytiscid and Hydrophilid beetles, are displayed by the gallon in large cylindrical glass jars in numerous shops and food-stalls in Canton, Hongkong, Shanghai, etc. In Shanghai they are

---

*) Dr. FENG's statement shows that they are eaten as well in normal times (cp. 278).

kept for the great number of Cantonese living there, who are the main consumers. In Peking the people are fond of grasshoppers cooked in sesame oil. The above mentioned bugs and beetles are gathered by professional collectors. HOFFMANN often came across them in the country, carrying their catch and their paraphernalia, but only once observed a collector at work, who promptly discontinued his activity. The professional grasshopper collectors have apparently nothing to do with the collectors of aquatic insects. Not the least interesting fact is that the large-scale consumption of the giant waterbug saves annually several hundred thousand dollars to the pond fish-breeders of Kwangtung, as these bugs are highly predatory on small fish.

At least six species of stinkbugs are eaten in Africa, India and Mexico. *Coridius (Aspongopus) chinensis* Dall. is eaten by one or more Assamese tribes, in China being a pest of beans, Rhicinus and cucumbers, and very commonly used as an aphrodisiac all over China, being called *Chu shan* or *Hai tao chung*.

Mr. A. LUDIN, one of our students who was born in Manchuria and later lived at Peking, informed us that locusts and grasshoppers are widely used there as food. The wings and the legs are pulled off and the remainder boiled in a special oil. Then the oil is drained off and the crisp insect, seasoned by the flavour of the oil, is eaten. Some restaurants prepared fried pupae of bees and roast beetles, but these are dainties which require complicated preparation. Honey is also much used in Manchuria. Many other insects, usually pounded or boiled in water, are in use as popular medicines.

The Chinese regard the locust, when deprived of the abdomen and properly cooked, as passable eating, but do not appear to hold the dish in much estimation (Chinese Repository, vide COWAN 1865, p. 126). L. OLIPHANT (1860, o. 273) saw in Tientsin bushels of fried locust hawked about in baskets by boys in the street. Locust hunting is a favourite occupation of the children. He compares their taste to that of periwinkles. W. WILLIAMS (1851 II, p. 50) makes the following interesting statement: 'The insect food (of the Chinese) is confined to locusts and grasshoppers, groundgrubs and silkworms; the latter are fried crisp when cooked'. The Chinese have eaten locusts since time immemorial and the people were advised to hunt for them on a large scale to diminish their ravages (in: NUN CAN

ZINAN SCIU, the Complete Treatise of Agriculture, transl. by A. ANDREOZZI). **Prof.** GIGLIONI observed the sale of locusts on the markets of Tientsin and Peking (BARGAGLI 1877, p. 7). E. DARWIN (1800, p. 364) mentions that in China items for the table which are considered delicacies are provided by silkworm pupae after the silk is wound off, and by the caterpillar of a Sphingid moth. The missionary FAVAND and many others report upon the use of the *Tsensciong*, silkworm pupae, as a refreshing and strengthening food in China. They are dried over a fire and become a yellowish mass, which is cooked in butter or oil for a short time, pounded and mixed. The mandarins and the rich add *tuorli*-oil, the poor salt, pepper and vinegar (BARGAGLI 1877, p. 9). The discarded chrysalids, after the silk of their cocoons has been reeled off, are hawked about the street of China and sold to the lower classes for about fivepence per pound. At Chinkiang they are sold at thirty-four dollars per 133 lbs. The Chinese also raise the maggots of blue-bottle flies in heaps of putrid fish near the sea-coast (SIMMONDS 1885, p. 349). The Chinese eat silkworm pupae, fried in butter with the yellow of eggs (VERRILL 1938, p. 162). Dr. TOUMANOV (Inst. Pasteur, Paris) has seen in Shanghai, Peking, and almost every other town in China special shops for silkworms, candied cockroaches, Dytiscid beetles and other insects.

Professor A. SEVERINI (BARGAGLI 1877, p. 10) notes that in China honey is added to many medicines and that the maggots of bees and wasps are eaten. Certain larvae from bamboo-stems, as thick as a finger and dark, are eagerly eaten by the Chinese boys, who find them sweet. The mountaineers of China and Japan often dig up the nests of certain ants to collect the pupae for food. SIMMONDS (1885, p. 369) says that ant maggots are a great dainty in China.

DONOVAN (1842, p. 6) in his 'Insects of China', mentions the larva of a big beetle, abounding in an unctuous moisture which is much esteemed as food. A large white grub is found under the roots of canes which, after being fried in oil, is eaten as a dainty by the Chinese. This may be an *Anomala*-grub. ESAKI (1942) reports that *Cybister*-beetles and giant waterbugs are on sale on the markets of Hanoi (Indo-China) and of Shanghai. People eat them in oil, after having discarded the head and the legs. N. C. E. MILLER observed the same at Honkong. BARGAGLI (1877, p. 11) also mentions an Ephemerid which is collected in China, when the mayflies are swarming, and which is pounded and mixed with honey to be made into an acid preserve.

Dr. FEN, Peking was kind enough to send me the following informations concerning the use of insects as food by Chinese people. 'Some of the informations are based on my own observations while others are obtained from friends specialized in the field of entomology. Chinese people eat several kinds of insects; the eating of some of the insects is common throughout the country, while the use of others is limited to certain localities. In most cases the insects are taken as accessory food and used as a dish which sometimes is considered as a delicacy.

1. *Pupa of silkworms:* Silkworm cultivation is an important industry in many provinces, e.g. Shantung, Kiangsu and Chekiang and some parts of Szechuan. In these places country people raise silkworms and make silk themselves. Spring is a busy season in the field. The cocoons of silkworms cropped in the spring are preserved by baking or by pickling them with common salt. Silk is made at leasure during the summer rainy season. Consequently 'large amounts of silkworm pupae are produced during the season. The pupae either from the baked cocoons or from the salted cocoons are then dried in the sun and preserved as food for the rest of the year. Pupae from the baked cocoons are more delicious and are liked most. For eating the pupae are first softened in water and then fried either with chicken eggs in the form of omelette or simply fried with onion and sauce. It is used as a dish in the ordinary meal or on occasions when guests are invited. In all three provinces silkworm pupae are eaten in a similar way. The commonest species is *Bombyx mori* L.

Pupae of other silkworms: In Shantung, people of certain hilly districts cultivate *Antherea pernyi* on oak trees. On account of the large size and thick cuticle, the pupae are generally prepared by frying with onion and sauce, and not with eggs as is done for ordinary silkworm pupae. Since the pupae of this species are rather rare, but of large size, they are especially valued. Farmers who have these pupae may give them to their friends or relatives as a special gift.

2. *Grasshoppers:* The eating of grasshopper is common in various parts of China. Many species are eaten and the species concerned depends on what is available. In Shantung at least 4 species are

275

eaten. The female of the large form is about 3 inches long. The time these insects are collected to be used as food is the late autumn. At that time the female insect contains a large number of eggs and on account of the chilly mornings they are more easily caught. During this time children as well as adults usually collect them while working in the fields. The catch forms one of the dishes of the evening meal. The wings are taken off, the heads together with the intestine are pulled out and the whole insect is fried with the addition of salt and sauce.

In Tientsin and Peking even city people eat grasshoppers. The farmer in the autumn collects and brings them to the market to be sold alive. The grashoppers of a market value are only the locust, *Locusta migratoria*, as this is the only species that can be collected in large numbers in some years. Also regarding this species, those collected in the late autumn containing eggs are especially appreciated. This species is so commonly eaten that during autumn and winter months it can be obtained from any groceries in both Peking and Tientsin. Some are already fried ready to be eaten, while dried ones (they are killed by boiling in water or by steam and dried) can be purchased and fried at home.

While grasshoppers are ordinarily used as accessory food, they, especially the locusts, when the crop has been destroyed by them and the farmers can collect them in large numbers, are used as ordinary food. Families are known who passed the famines due to destruction of crop by locusts by eating the locusts collected in the field with the limited amount of cereal they had on hand. In such cases the collected locusts are either killed by boiling them in water or by steam, then dried in the sun and fried for eating when needed.

3. *Water beetles:* Certain species of aquatic beetles known locally in Canton as '*Lung Shih*' literally meaning 'dragon lice', are used for food by the Cantonese. Two species, namely, *Cybister jáponicus* Sharp *(Dytiscidae)* and *Hydrous hastatus* Herbst *(Hydrophilidae)*, are commonly consumed. These beetles are boiled with salt water and sold in the market. The above mentioned two species can be purchased in any grocery in Canton, they are eaten just as watermelon seeds and peanuts are eaten by the local people. They can be purchased also in Cantonese food shops in other large cities like Peking,

276

Fig. 35. Insects used in China as food. At right: The Giant Waterbug *Lethocerus indicus* Lep. et Serv. At left above: *Cybister japonicus* Sharp *(Dytiscidae)*: *Hydrous hastatus* Herbst *(Hydrophilidae)*. Curtoisy of Dr. FEN, Peking.

Shanghai and Tientsin. They may also be eaten as one of the dishes on the table. Sometimes they are fried.

4. *Other insects:*

    a. *Kwei-hua-ch'an,* a member of Hemiptera, *Lethocerus indicus* Lepeletier and Serville *(Belostomatidae),* is a large aquatic insect also used by Cantonese as food in a similar manner as the water beetles. They can also be purchased in food shops in Canton.

    b. *Cicada.* Any species available, adults as well as nymphs, are eaten, especially the nymphal forms. The latter are either collected on the tree during evening time or dug out from the ground. They are usually eaten after frying.

Many other forms of adult or larval insects are also eaten in certain localities. These include the cockroaches (*Periplaneta americana* and *P. australasiae*), the larvae of beetles (*Melanaster chinensis* Forster, *Psacothea hilaris* Pasacoe), the larvae of *Vespa sp.* the larvae

277

of dragon flies and May flies etc. and even maggots from meat (*Calliphoridae* and *Sarcophagidae* which are called locally 'meat sprouts' in analogy to bean sprouts grown from beans). It may be mentioned that in Japan larvae of *Vespa japonica* are even prepared as canned goods with an annual sale of 200,000,000 tins. (sic!)'.

Miss N. G. SPROSTON (Institute of Hydrobiology, Academia Sinica, Shanghai) wrote to Dr. J. THEODORIDES concerning insect consumption in China: 'Beetles are a very common article of diet in some provinces. They are for sale in Shanghai, but are rather expensive because of the special preparation they require. They are fried very crisp and are eaten with other rich foods along with wine at the beginning of the feast. . . . The rice does not appear till it (all the best dishes except the soups and pork) is nearly over. The beetles are *Dytiscidae: Dytiscus marginalis* is used extensively here, and in Japan and China the equally big *Cybister japonicus* is also eaten. On the whole, the Cantonese are more entomophagous than the other Chinese; next come the Szechuanese from Western China around Chungking. There it is thought, that other water-beetles are eaten. My assistent remembers eating them at his father's table when quite young, but they were small species.

Boys are always lashing at the street trees with long bamboos to bring down the cicadas, etc. Sometimes children eat various insects and pupae they catch. The nearly emergent cicada nymph is eaten raw as a great delicacy, particularly in Shantung . . . . or it was a man from this place who tried to make me eat one!'

In China honey-bees are not cultivated to anything like the extent they are in other countries (A. C. SOWERBY 1925, p. 265). Honey is looked upon more as a medicine than a sweetmeat, and the apothecaries invariably adulterate it heavily with malt or even flour. Certain fruit conserves, notably jujubes or Chinese dates, are treated with honey which makes them extremely palatable. In the wilds of Shansi, Shensi, Kansu and Szechuan many of the villagers keep bees somewhat extensively, while in the Manchurian forest areas the settlers have learnt their value, and set out specially hollowed tree-trunks, stood on end, to attract the wild bees when they swarm. But nowhere is apiculture reduced to an art or looked upon as much more than a profitable hobby.

Little is known about entomophagy in Japan. T. ESAKI (1942)

has written a special paper on it, which unfortunately is written in Japanese. He reports that in some localities of Japan people are 'said to eat certain insects'. The so-called bee-larvae sold on the market are actually the maggots of *Vespa spp.* Silkworm pupae and adults, grasshoppers and the nymphs of *Trichoptera* and *Ephemeroidea* are also eaten, boiled. On Formosa Dytiscid beetles are said to be eaten.

The rarity of bee-keeping in China is also confirmed for Japan by DOFLEIN (1906, p. 90), who writes: 'Honey-bees I saw only very rarely in Japan, where they are little kept. I have never seen them in the coastal districts'.

REMINGTON (1946) reports that Dytiscid beetles are caught by nets in Japan. After having removed the elytra, the beetles are fried and together with sugar made into a sauce. The larvae and pupae of Elaterid beetles are also consumed there. Mrs. TOGA KAUFMAN, one of our students who was born in the Nagano prefecture in Japan, informs us that the larvae of *Vespa japonica* Sauss., the *ji-bachi*, are eaten in this province. The maggots are fried in oil and then soaked in soya-bean sauce, or simply boiled in that sauce. Previously, these larvae were canned and sold on the markets, where at present they have become rare. Adult grasshoppers (*Oxya velox* F., *O. vicina* Br.) are roasted with soya-bean sauce and eaten.

Dr. J. THEODORIDES was kind enough to extract the following notes from REMINGTON (1946) about insect consumption in Japan REMINGTON had first hand information from Prof. TETSUO INUKAI, zoologist of the Hokkaido Imperial University at Sapporo. Nagano, about which we heave heard before, is one of the few wholly inland provinces, segregated by mountains from ready contact with the sea. The people of Nagano are unable to get a sufficient supply of fish and meat to fill their protein requirements and they consume extensively insects. The favourite insect food there is a wasp (*Vespula sp.*) whose maggots and pupae are highly prized. Pupae of all wasps are eaten when found, but *Polistes* seems to rank next to *Vespula* in frequency in Japanese menus. In Sapporo REMINGTON found pupae of *Vespula* preserved in cans and sold in grocery stores. He gives a description of the various methods to catch the maggots from the nest without being stung. Other very popular insects are all species of *Cicadidae* (adults), *Oxya velox*, the *inago*, and practically all other grasshoppers, all crickets (*Gryllidae*) and all praying mantids. These insects are killed in a hot pan and, like the wasp maggots and pupae, they are preferably cooked by frying.

Prof. INUKAI related that all pupae (other than wasps) taken in the soil are carefully avoided. Apparently some are very poisonous. On the other hand, all insects found in fresh water are edible and delicious. Larvae and aquatic adults of *Coleoptera, Hemiptera, Ephemerida, Plecoptera, Trichoptera, Odonata*, etc. are eaten, unsorted. He mentioned particularly beetles of *Dytiscus* and giant waterbugs as aquatic insects commonly eaten.

Larvae and pupae of *Cerambycidae, Elateridae* and certain other wood-boring beetles, as well as the woodboring caterpillars of *Cossidae* are cut out of dead wood and esteemed as food. Lepidopterous pupae, except those from the soil, are said to be very tasty. One of the commonest and easiest gathered is that of the hag-moth (*Eucleidae*), a pest whose caterpillars have stingy hairs. Silkworm pupae are eaten not only in the Nagano prefecture, but throughout Japan. They are fried in fat and salted. REMINGTON was served these pupae, when a dinner guest of zoologists of the Hokkaido University. He found them very delicious, which determined him to try the taste of several big Saturniid pupae of North America.

VI. THE AMERICAS

1. GENERAL INTRODUCTION

America is another continent which by its geographical position, stretching from the Arctic to the Antarctic seas, embraces all the diversity of conditions which terrestrial ecology can offer. We have little information about the northern Arctic areas. Their natives are most eager for fatty food (RILEY 1893). Insects are rare. NANSEN (vide BRYGOO 1946, p. 48) mentions that the Eskimoes did not know fleas before the arrival of white men. They call them Europeans' lice and regard them as true delicacies. They have even invented special traps for them, which are put between the clothes and the skin. Little is also known of insects as food in the areas of South America south of Amazonia. The little we know however points to the conclusion that there, as in North America, insect feeding habits were more common in the arid areas to the west of the Andes than in the pampas of the east.

Within the American continent the Indians lived on whatever the various regions offered them. Among the various important articles of food, we may mention seals and pelicans, shellfish and snails, salmon and turtles, locusts and *Melipona*-honey, reindeer and llamas, dogs and buffaloes, maple sugar and acorns, sunflower seeds and maize, potatoes, various roots, barks, nuts and berries, etc., etc. A. J. FYNN (1907, pp. 87 ff.) has given a survey of some of the food habits of various Indian biotopes. The tribes east of the Mississippi River were primitive agriculturists rather than hunters. Along the banks of the St. Lawrence River with its short summers, hunting prevailed and game was plentiful for the scattered population. On the shores of the Great Lakes and of the Pacific, fishing was a dominant industry, the local differences of which are well brought out by FYNN. From Mexico to the N. W. areas of South America irrigation had permitted the building up of important local agricultures. In the primeval forest of Amazonia we find the most primitive food-gatherers, such as the honey-civilization of the Guayakis, side by side with hunters and primitive agriculturists. Insect-feeding habits have been more prevalent in the arid areas of the S. W. parts of the U. S. A. and Mexico, while they are more incidental in the rest of the continent.

The introduction of the European honey-bee into North America (TEALE 1940, p. 168) in the 17th century brought an important additional insect food into the sub-continent, where the bees speedily spread through the forests. Today over 500 million pounds of honey are produced annually by this 'white man's fly'. In the agricultural areas and partially also in the more accessible forest areas of Central and South America the introduction of the European honey-bee has reduced the importance of the stingless bees (*Meliponidae*), the domestication of which was well under way before the conquest. All information available on this interesting subject has recently been collected in a masterly way by H. F. SCHWARZ (1948).

2. THE UNITED STATES

When A. HRDLICKA (1908, pp. 25, 264) made his survey of the Indians of the S. W. states of the U.S..A and of Northern Mexico, most of these tribes had already developed an important agriculture. In addition, animals from big game to mice and fishes formed an important addition to their diet, while the collection of various ports of cactus species, of tubers, of seeds of various grasses, of mushrooms, etc. either added to the crops or replaced them in years of famine. This famous anthropologist found no evidence that insects are still a food of any importance, but he gives the following relevant information.

The Tarahumare, one of the most primitive tribes living in Chihuahua (N. Mexico), under the pressure of frequent necessity, have learned to eat animal and vegetable substances of great diversity. They raise some potatoes, chile and sugar cane. The flowers of the squashes are dried and kept and a kind of porridge is made from them. Meat, particularly venison, is much relished by this tribe. They also like the flesh of field mice, which they skin and roast suspended on sticks near the fire. They occasionally eat various small animals, including skunks, lizards, locusts, grasshoppers, frogs, water beetles, and even larvae. HRDLICKA quotes HARTMAN (1894): 'They eat almost anything that lives – polecats, mice, rats, snakes, the iguanas, frogs, fish spawn, grasshoppers, and certain kinds of larvae, even those of dragonflies taken out of the water'. LUMHOLTZ (1902) states the same. The Tarahumare prepare tortillas from corn, and bread from wheat flour. They eat the blood of animals after preparing it over the fire. Animal meat is sometimes eaten almost raw, but usually it is well roasted or cooked. The Tarahumare living near streams dive into pools and lance fish; they also shoot fish with arrows armed at the point with a number of nopal spines, catch them with nets or drag for them with blankets. At times they drain the lagoons and kill the fish with stones; and they also have recourse to poisoning fish with certain plants. Crayfish, too, are caught and eaten. The domestic animals kept by this tribe are chicken, cattle, sheep, and some goats; they have also a few turkeys, but no ducks or geese. Besides the domestic fowl they eat various wild birds, and eggs of both. Wild fruits are abundant in season.

Another of the native foods of the Pima is the honey of the wild bees; it is obtained, however, but seldom. A favourite sweet of the Pima children is the honey which a small solitary bee, determined by ASHMEAD as *Anthophora sp.* or *Melissodes sp.*, desposits in mud cells in the soil. The bee digs a tunnel about 15 cm. long below the surface of the ground and there makes one, two or even three little jars of mud, in which it secretes a thick sweet, yellowish juice. The children dig for these little jars of the bee and eat this 'fly syrup' or *mo-wa-li chuh-nie*.

Some North American Indian tribes were in the habit of consuming large quantities of Rocky Mountain locusts (FLADUNG 1924, p. 6). DE SMET mentions that the Assiniboine idea of luxury was an immense dish of pulverized ants, as well as of locusts and grasshoppers dried in the sun. The Shoshocos liked crickets and grasshoppers and strored bags of roasted ants for future use. They dug a hole in the soil and, by beating in concentric circles, gradually drove the grasshoppers through the *Artemisia*-bush into the hole.

The insects were taken out and made into soup or boiled, or into paste for future consumption. The Californian Maisu also relished locusts and grasshoppers. 'Nearly all of our American Indians, in fact, ate some sort of insects, as would do a primitive people with few agricultural pursuits'. The Dog Rib Indians of Athabascan ate the warble-fly maggots of their caribou which they declare a delicacy. HEARNE (vide FLADUNG) says that the Indians could never persuade him to try these warbles, of which they, and especially their children, were remarkably fond. The maggots were always eaten raw and alive, as they were taken from the skin of the caribou, and were said to be as fine as gooseberries. Caterpillars are still today eaten in great quantities (FLADUNG 1924) by the Pai-Ute Indians. The gathering and preparation of caterpillars for food is an industry of considerable importance along the borders of California and Nevada. FROGGATT reports that in Mexico the caterpillars of *Hesperia sp.* burrowing in the leaves of *Agave americana* are cut out of the leaves and sold as a delicacy. He then mentions from the same region the sale of waterbug-eggs, the famous *ahuahutl*, as cakes on the markets. Many authors have reported on the large-scale consumption of the 17-year cicada, *Tibicen septemdecim*, which is much relished when dried and pounded, fried or roasted, or as a soup.

The white population also was sometimes driven to locust eating. SIMMONDS (1885, p. 366) mentions that in 1855 the locusts devastated Utah to such an extent, that the inhabitants had nothing to eat other than the locusts themselves. When the grasshoppers in California were in the best condition, the Indians selected some favourable localities, dug several pits, like inverted funnels, the aperture being narrower at the surface then at the base, to prevent the insects from jumping out of them. Then an immense circle was formed and the surrounding grass was set on fire. The Indians, men, women and children stationed themselves at proper intervals around the fiery belt, keeping up a continual ring of flame, until the grasshoppers were caught in the pits or roasted at the brink. Mixed with pounded acorns, they constitute one of the national dishes.

The editor of the County Argus (California; Rept. Comm. Agric. Washington 1870, p. 426) gives the following vivid description: 'A piece of ground is sought where the grasshoppers most abound, in the centre of which one excavation is made, large and deep enough to prevent the insects from hopping out when once in. The entire party of Diggers, old and young, men and women, then surround as much of the adjoining ground as they can, and with each a green bough in hand, whipping and thrashing on every side, gradually approach the centre, driving the insects before them in countless multitudes, till at last all, or nearly all, are secured in the pit. In the meantime smaller excavations are made, answering the purpose of ovens, in which fires are kindled and kept up, till the surrounding earth, for a short distance, becomes sufficiently heated, together with a flat stone large enough to cover the oven. The grasshoppers are now taken in coarse bags, and after being thoroughly soaked in salt water for a few moments, are emptied into the ovens and closed in. Ten to fifteen minutes suffice, to roast them, when they are taken out and eaten without further preparation, with much apparent relish, or reduced to powder and made into soup'. The editor adds that they are not bad eating. Grasshoppers are also pounded up with service, hawthorn or other berries. The mixture is made into small cakes, pressed hard and dried in the sun for further use.

In the same Report (1870, p. 426) we find information about the eating of ants by the Digger Indians of California. They catch the ants by spreading a damp skin or fresh peeled bark over the ant hills, which immediately attracts the ants to the surface. When the skin is full of ants, it is carefully removed and the adhering insects shaken into a tight sack, where they are confined until dead, and then they are thoroughly sun-dried and laid away. Bushels are thus gathered annually and are not more offensive than snakes, lizards and crickets which the tribe also eats.

C. L. MARLATT (1923, pp. 104–6), in his splendid monograph on the 17-year cicada of N. America, has collected some early references to these insects as food of the Indians, which are locally termed 'locusts'. The Rev. A. SANDEL of Philadelphia (1715) reports on the cicada as food of the Indians. This statement was supported by ASA FITCH, who was informed by Mr. W. S. ROBERTSON that the Indians use the different species of cicadas as an article of diet, every year gathering quantities of them. They are prepared as food by rosting in a hot oven and are stirred until they are well browned. P. COLLINSON (1764) also notes that *Tibicen septemdecim* is eaten by the Indians of North America. The wings are plucked off and the cicadas roasted. Doctor HIDRETH (1830) remarks that when the cicadas first leave the earth, they are plump and full of oily juices; so much so that they are employed in making soup. Doctor PHARES (1859) and other observers that while most domestic and wild mammals and birds eat the cicadas without harm, overating may occasionally cause bad consequences. MARLATT thinks that cicadas as human diet are only of theoretical interest because of their rarity and of the general aversion towards eating insects. Theoretically, the cicada, collected at the proper time and suitably dressed and served – as shown by the experience of RILEY and HOWARD – should make a rather attractive dish, their larvae living only on pure and wholesome vegetable food.

It is generally known that certain American Indians ate ants freely, especially among tribes with little agriculture, where periods of famine are rather frequent, owing to the absence of permanent vegetable staples (A. SKINNER 1910, J. BEQUAERT 1922). JOHN MUIR (1916, p. 46) tells how the Digger Indians of California are fond of the larvae as well as of the adults of a large, jetblack, wood-boring ant (*Camponotus sp.*), of which they bite off and reject the head, and eat the highly acid body with great relish. And H. C. McCOOK (1882, p. 32) mentions the various uses of the honeypot ants (*Myrmecocystus spp.*) by Mexicans and Indians of the S. W. states. 'That they eat it freely and regard it as a delicate morsel is beyond doubt. Prof. COPE, when in New Mexico, had the ants offered to him upon a dish as a dainty relish. The Mexicans (vide LOEW) press the insects, and use the gathered honey at their meals. They also are said to prepare from it by fermentation an alcoholic liquor. Again they are said (EDWARDS) to apply the honey to bruised and swollen limbs, ascribing to it great healing properties'. In 1852 there was some discussion on these Mexican honey ants in the Academy of Natural Science at Philadelphia. LANGSTROTH and LEIDY gave a report on them (Proc. 1852, p. 71 f.), while C. M. WETHERILL made chemical investigations (1852, p. 111). Six of these average-sized ants had a weight of 2.65 gr., of which the body weight made up 0.29 gr., leaving 2.37 gr. honey (or 0.39 gr. honey per ant). The honey in these honeypots thus weighs over eight times the body-weight! The specific density of the body with honey was 1.28, without it 1.05.

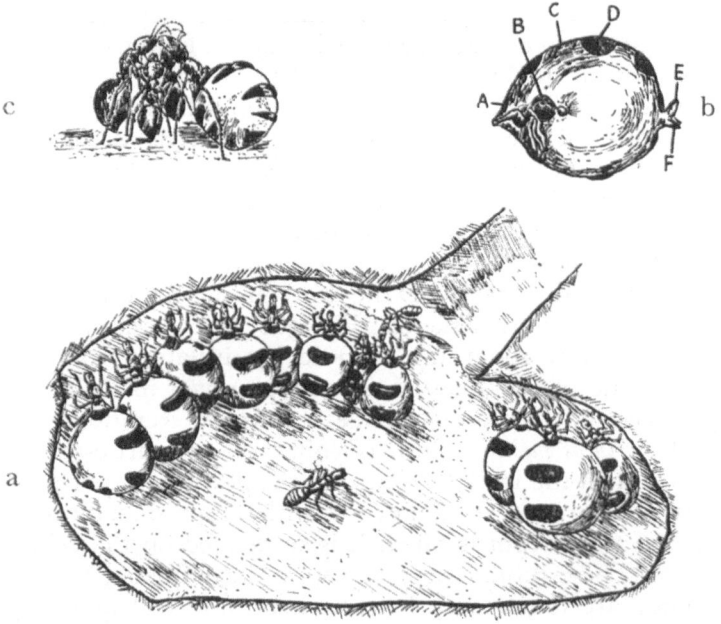

Fig. 36. From the life of the honey ant (*Myrmecocystus hortus-deorum*). Below: View into a storeroom where honeypot individuals are hanging from the ceiling in storage Above left: A honeypot disgorges its honey to three normal workers which ask for it by titillation of the antennae; right: Abdomen of a honeypot individual (For explanation of A to F see text). After McCook.

The syrup extracted from the ants had an agreable sweet taste and an odour like that of syrup of squills, but slightly acid. It is very hygroscopic and shows no signs of cystallization when dried, but changes into a sticky gum. Analysis showed an almost pure solution of fructose (sucre des raisins), which however does not crystallize.

These honeypots, occasionally also called nurses, are large, occasionally also smaller workers of the ant *Myrmecocystus mexicanus var. hortus-deorum* McCook, the crops of which are enormously extended. Fig. 36a shows a large storage chamber of such honeypot-individuals, such as McCook (1882) first described from Manitou in Colorado. Up to fifty and more honeypots may be hung on the ceiling of one of these store chambers like provision bottles. For use they are taken out and after proper titillating by other ants, they are induced in times of scarcity to regurgitate their honey. A. FOREL obtained such specimens from McCook as early as 1877 and convinced himself of the accuracy of Fig. 36a, except that the minute fifth tergite is missing. The intestines of the honeypots are not severed, as some writers assumed, but the enormously expanded crop, filled to capacity with honey, simply presses the mid- and hind-intestines against the posterior wall of the swollen abdomen. In Fig. 36b A denotes the rectum, B the mid-intestine, C the distended crop which fills almost the entire abdominal cavity, D the third tergite, E the petiolus and F the oesophagus. Another sketch (Fig. 36c) shows a honeypot taken out from the store room and induced by titillating workers to regurgitate some of its honey. McCook observed in Colorado that this ant feeds mainly on the sweet nocturnal secretions of a gall caused by *Cynips mellea* Ashm. (Cynip.) on the shrub-oak *Quercus undulatus*. When the workers return to the nest from their nocturnal forage trips, they cram the honeypots with this product by regurgitation. Once a honeypot is filled, it is suspended on the ceiling of a storeroom. These rooms are 7 to 10 cm. in diameter, up to 4 cm. high. They are found only in extremely hard soil at a depth of 20 to 25 cm., and the nests are built only on the summits of dry, stony ridges (WHEELER 1908), which keeps them permanently dry. McCook collected about 600 honeypots in one nest containing some thousands of normal workers. He estimates that about 1000 honeypots are needed to yield half a kilo of honey.

WHEELER (1908) succeeded within four to six weeks of cramming in artificially transforming young individuals of the larger workers of *M.m. hortus-deorum* into completely or half-filled honeypots; this could not be done with older workers.

Other ants were also occasionally eaten. And we have eye-witness reports confirming that even the white lumberjacks of Canada and Maine have not been averse to a meal of the large black carpenter-ant *Camponotus pennsylvanicus* (PROVANCHER 1882, C. V. RILEY 1893, FLADUNG 1924).

Entomophagy was mainly restricted to the tribes west of the Mississippi, especially among those towards the Pacific slopes (A. SKINNER, 1910).

The old missionary DE SMET writes: 'I have seen the Cheyennes,

Snakes, Utes, etc. eat vermin off each other by the fistful. Often great chiefs, while they talked to me, would pull off their shirts in my presence without ceremony, and while they chatted would amuse themselves with carrying on this branch of the chase in the seams. As fast as they dislodged the game, they crunched it with as much relish as more civilized mouths crack almonds and hazelnuts or the claws of crabs and crayfishes . . . . Add to this, by way of an exquisite dessert, an immense dish of crusts, composed of pulverised ants, grasshoppers and locusts (read: cicadas), that had been dried in the sun, and you may be able to form some idea of Assiniboine luxury. The Shoshoco who subsist chiefly on grasshoppers and ants, are miserable, lean, weak and badly clothed . . . . The principal portion of the Shoshoco territory is covered with *Artemisia*, in which the grasshoppers swarm by the myriads, and these parts are consequently most frequented by the tribe. When they are sufficiently numerous, they hunt together. They begin by digging a hole, 10 or 12 feet in diameter, by 4 or 5 feet deep; then, armed with long branches of *Artemisia*, they surround a field of 3 or 4 acres, according to the number of persons who are engaged in it. They stand about 20 feet apart and their whole work is to beat the ground, so as to frighten up the grasshoppers and make them bound forward. They chase them toward the centre by degrees – that is, into the hole prepared for their reception. Their number is so considerable that frequently 3 or 4 acres furnish grasshoppers sufficient to fill the hole. The Shoshocos stay in that place as long as this sort of provision lasts. Some eat the grasshoppers in soup, or boiled; others crush them, and make a kind of paste from them which they dry in the sun or before the fire; others eat them en appalas – that is, they make pointed rods and string the largest one on them; afterwards these rods are fixed in the ground before the fire, and as they become roasted, the poor Shoshocos regale themselves until the whole are devoured'.

RUSSELL (Exploring the Far North 1892/94, p. 228) says of the Dog Ribs: 'A gadfly (apparently *Hypoderma lineata*) deposits its eggs in the back of the Caribou, in some individuals to the number of several hundred, which renders the skin utterly useless for leather. The grubs were well developed in late April when I left the barren ground. The Indians did not remove them from pieces of meat destined for the kettle'.

DIXON (Bull. Am. Mus. Nat. Hist. 17, III, p. 120) writes: 'The Maidu of California hunted locusts similarly, burned the wings off in fire, dried them and kept them for winter food, to be eaten either dry and uncooked or slightly roasted'. He states of the Pima of S. Arizona (ibid. p. 245, vol.. 8): 'The *Makum* are unidentified worms which are plentiful when a rainy season insures a heavy crop of desert plants. They are gathered in large quantities, their heads pulled off, and intestines removed. The women declare that their hands swell and become sore if they come into contact with the skin of the worms. The worms are then put into cooking pots lined with branches of salt-brush and boiled. The skins are braided together while yet soft and dried a day or two in the sun. The dry and brittle sticks are eaten at any time without further preparation'.

F. FREEMAN (History of Cape Cod I, p. 524. Vide SIMMONDS 1885, p. 358) mentions that when Col. N. FREEMIC was appointed in 1777 Commissioner at West Point, his attention was arrested by certain inexplicable movements among the French troops. They had kindled numerous fires in the adjoining fields and were running about in strange disorder. General WASHINGTON and other officers mounted horses and rode to their neighbouring encampment. The French were enjoying rare sport in a campaign against the numerous grasshoppers. These insects, as soon as captured, were impaled upon a sharpened stick on posts and held for a moment over the fire, and then eaten with great gusto.

ALDRICH (1920) discovered in June, 1911, near Mono Lake, California, that certain caterpillars were eaten in quantities by the Pai-Ute Indians. The caterpillars were dried and later cooked. Aldrich found their taste insipid and flavourless. Local information revealed that the caterpillars feed on the needles of various pines. The Indians collect the caterpillars by making a fire under the tree, for which purpose they make a trench fairly close to the base of the tree; this is presumably to guard against the spread of the fire. As the thick smoke rises and envelops the caterpillars, it causes them to lose their grip and drop to the ground, where they are collected by the Indians, killed and dried. The preserved material is called *Papaia*. An inquiry by the Forest Service revealed that similar trenches were widespread and that the collection of this caterpillar for food is an industry of considerable importance in the territory along the Nevada-California border. This caterpillar of *Hemileuca sp.* (*Saturn.*) was unknown at that time and DYAR regarded it as a 'rare' species.

ALDRICH (1912) had previously published a statement of the Indian Chief Ben Lawyer at Oregon about the *Koo-chah-bie*: 'About 40 years ago when the Indians used the *Koo-chah-bie* as food, they would go to Pitt River in Modoc county, Cal., at a point about ten miles from the village of Canby. The time for gathering the flies was in the early summer. The Indians would place logs across the river in somewhat the same manner as a present day log or lumber boom is constructed. Then they would go upstream and shake the flies off the willow bushes growing along the banks of the river. The flies falling on the water would float downstream and lodge against the logs in great quantities. As many as 100 bushels could be gathered in this way in a single day. The Indians used a kind of basket to collect the flies from the water and carry them to the place where they were to be prepared for food. A pit was dug in the ground about 1½ to 2 feet deep and about 2 feet or more square. Then two layers of stones were placed in the bottom of the pit, each layer being about three inches thick. A wood fire was built on these stones and more stones were put around and over the fire. When the fire was burnt out and the stones were hot, all the stones were removed except the bottom layer. The green tules or green coarse grass was spread out on the bottom layer of rocks. The walls of the pit were lined with hot rocks also, and this enclosure lined with tules or grass. The oven-like enclosure was then filled with the flies. These were covered with green coarse grass and the whole covered with more hot stones. Water was then poured on the hot stones of the walls of the pit, the hot stones converting it into steam. As soon as the water was poured on, soil was hurriedly thrown over all to a depth of several inches. The flies were allowed to cook in this manner until the heat was pretty well expended. The soil and grass then removed from the top and the mass allowed to cool. When sufficiently cooled the product was taken from the oven and was ready for use as food. In this state it was called by the Modoc and Pitt River Indians '*Koo-chah-bie*'. When cold, it had about the consistency of head-cheese, having a reddish brown colour and could be cut into slices with a knife'. Similar reports from two other

Indians from the same region enlarge on and confirm this statement. The fly is doubtless a species of *Atherix* (*Leptidae*), the females of which oviposit gregariously.

The Pinos Indians collected the caterpillars of *Macrosila carolina* for a soup and they also fried them crisp and brown, An U.S.A. agricultural officer records having seen this tribe collect bushels of these larvae for immediate consumption, or to be kept for the winter after they had been dried and pounded (SIMMONDS 1885, p. 355). The Californian Indians ate many caterpillars. ENGEL-HARDT (1924) mentions in this respect especially the eating of a common giant silkworm, *Coloradia pandora*, which lives on the needles of pine trees. These and also the pupae are dried and roasted or stewed. BRUES (1942 p. 420, pl. XXII) states that this habit still survives with the Piutes of California and figures a basket used by them for gathering and drying caterpillars for food.

A common brine fly, *Ephydra hians*, breeding in countless numbers in California in the Mono Lake is collected by the Indians as pupae. These are dried and furnish a nutritious, fatty, though hardly appetising food (BRUES 1942, p. 419). In other salt lakes of western U.S.A. similar flies offered an appreciated food. Thus SIMMONDS (1885, p. 356) tells of the *koochabe*, a whitish maggot which at certain seasons is carried by the waves in huge masses on the shores of Queen's Lake, Cal., in layers which sometimes are several inches thick. These were formerly collected and, after being dried in the sun, were rubbed between the hands and roughly winnowed, then crushed in a stone mortar and baked into cakes which were an important article of food. And the same writer mentions a report by PACKARD, that the maggots of *Ephydra gracilis* were similarly used in Utah. Some of these maggots were eaten raw by the Indians of western Nevada, in spite of a rank and oleaginous taste, whilst others were made into soup. Other insects eaten were the maggots and pupae of wasps in South Carolina (DUGUIN 1900, p. 14). FROST (1942, p. 63) reports that in the western states the big, white wood-boring grubs of *Prionus californicus* (*Cerambycidae*) were a favourite food because of their size. He also mentions bees and wasp maggots as a tempting food. More details on the locust drives of the Indians in California for food gathering are described by ESSIG (1934). BRUES (1942, p. 419) remarks that the Indians of the S.W.

areas of the U.S.A. were not ethnologically different from the other Redskins, but they seem to have utilized insects for food on a larger scale than many others.

3. MEXICO

C. H. Curran (1937) has studied ancient Aztec insect-lore as expounded in an old travel book by Bernardino de Sahagun (around 1560) at the time of the conquest. The Indians ate wild honey whenever they could obtain it. *Nequaz-catl* or honey ants are well described. With regard to bees, Curran is of the opinion that bumblebees are referred to in the following paragraph: 'There are some drone bees in this country that make honey and construct nests in the earth where they make it . . . . They sting like bees, pityful, and the sting swells'. Their honey is as highly esteemed as that of the stingless bees and that of the honey-making wasps. The honey of the latter is very yellow and very good to eat. Wild honey was often consumed together with the bee-maggots. To the Indians this made no difference, since many insects and insect larvae formed part of their diet. Thus the Aztecs ate the corn on the cob together with the corn earworm with relish. Among their insect food grasshoppers are notable and all of them were considered edible. They undoubtedly formed an important part of the diet during seasons when they were abundant. If the Aztecs followed the practice of the other Indians they removed the legs, wings and perhaps the head, dried them and later stewed them. They were considered to be rather tart, also not as desirable as some of the caterpillars. Curran underlines that these are only chance observations of the punctilious Sahagun.

To this may be added the notes of A. S. Packard (1885), which he obtained from Mexico through Mr. J. M. Carter. The maggots of an ant are eaten as food by the Otomite Indians. These ants live in oven-like hills. Also a caterpillar, about 5 cm. long, which abounds in July and belongs apparently to a Noctuid moth and which lives in the thick leaves of the *maguey* or century plant is eaten, either raw or cooked by the Mexican Indians.

Rev. T. Smith (1807 vol. XII, p. 198) enumerates the following insects as eaten by the ancient Mexicans: 'The *Atelepitz*, a marsh

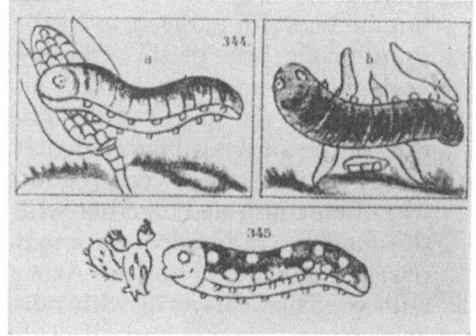

Fig. 37. Drawings of Mexican Insects eaten by the Incas. 336: caterpillars from the century plant; 344: the Corn Ear Worm; 345: a different species of caterpillar; 324: grasshoppers. After SUROGA (15th century) from Nat. Hist. N.Y. 39. 1937, pp. 302, 303.

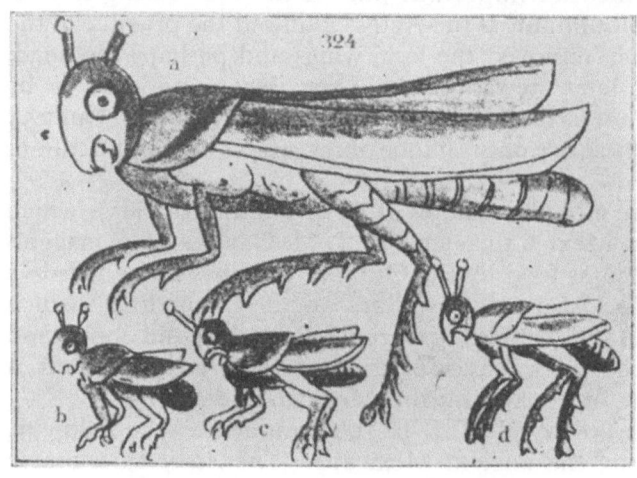

294

beetle, resembling in shape and size the flying beetles, having four (sic!) feet, and covered with a hard shell. The *Atopinan* is a marsh grasshopper of a dark colour and great size, 15 cm. long and 5 cm. broad(!). The *Ahuihuilla* is a worm inhabiting the lakes of Mexico, 10 cm. long, thick as a goose quill, tawny above and white below; it stings with a tail, which is hard and poisonous. The *Ocuiliztac* is a black marsh worm which becomes white on being roasted'.

The Zapotec Indians near Oaxaca (Mexico) eat larvae of the agave caterpillar *Aegiale hesperialis* Walk. as appetizers. Two good photographs of the prepared dish and the hostplant are given by W. H. HODGE (Nat. Hist. N.Y. 1949, p. 385).

Prof. A. BALACHOWSKI (Paris) observed a trade in Mexico of caterpillars boring in cactus, which are a common dish served even in fashionable restaurants.

The early travellers in Mexico had already noted the *ahuatle* or *bledo del agua*, waterbugs which are sold in various towns as food. Thus FRANCESCO HERNANDEZ (1649) states that these are seasonally abundant and collected in some lakes. The insects are rubbed and pressed and thus sold in the markets. TH. GAGE (1625), F. J. CLAVIJERO (1780), L. COINDET (1867), E. CHINA (1931) and others mention them as locally important food. F. E. GUÉRIN-MÉNEVILLE (1857, pp. 578–81) devoted a special paper to 'Three species of bugs, the eggs of which are used to prepare a bread called Huatlé in Mexico'.

'These small insects provide food for man by the prodigious quantity of their eggs, which are collected as if they are a regular crop. A flour is made from them, used for the baking of cakes, which are eaten by the natives and form even now the object of a small trade in the markets of Mexico. Prof. CRAVERI of the Medical School of Mexico sent to his brother a sample of this flour. The bugs and their eggs abound in all freshwater lakes around the city. In the lagoon of Chalco a species of rush, the *toulé*, is sought for, on the stems of which the insects prefer to oviposit. Many bundles of this rush are collected and put in great numbers into the water of the lagoon of Tescuco. The bugs immediately lay their eggs on these rushes. About a month later these bundles are taken out of the lake, dried and then beaten on great sheets of cloth to separate the myriads of these eggs with which they are covered. The eggs are then cleaned and sifted, put into sacks like flour and sold for the

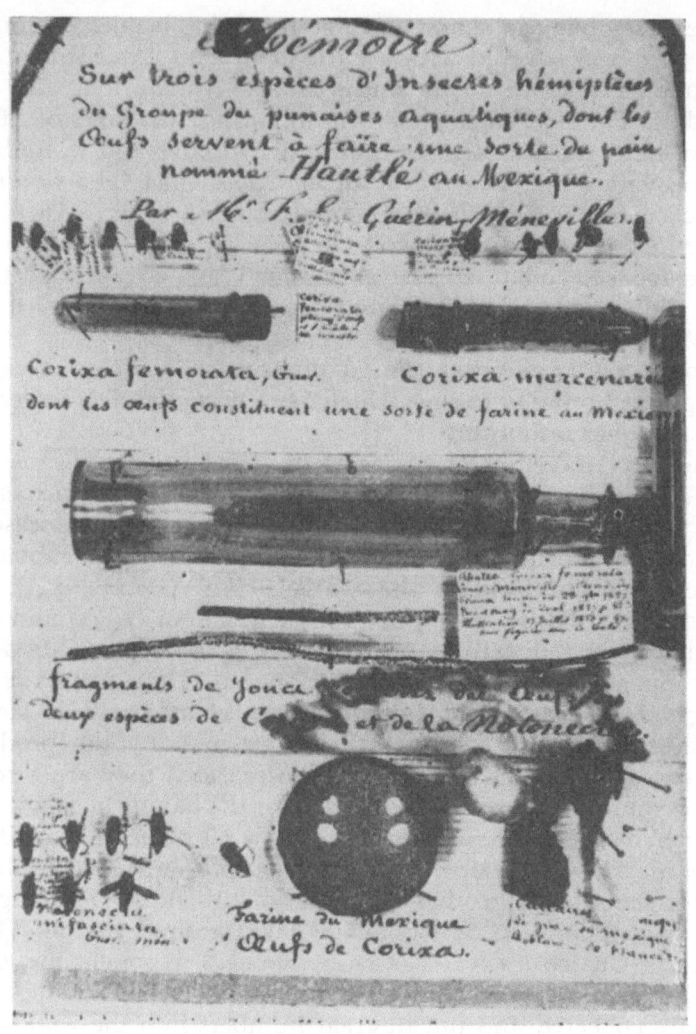

Fig. 38. Collection box in the Musée d'Histoire Naturelle de Paris where GUÉRIN-MÉNEVILLE put together all the materials for his paper on the edible eggs of Mexican waterbugs (From PORTEVIN 1933, p. 337)

making of the *hautlé* cakes, which are fairly good to eat. They have a pronounced fishy taste and are slightly acid. The bundles of rushes are again put into the lagoon to provide another crop, and this continues indefinitely'. Prof. CRAVERI says that the Indians take great quantities of these insects with nets. These bugs are dried and sold in the streets as food for birds by people calling: '*moschitos, moschitos!*' This usage must be very old.

BRANTZ MAYER (1844, p. 218) states: 'On the lake of Tescuco I saw some people occupied collecting the eggs of flies on herbs and from rags which are planted in long rows as traps for these insects. These eggs, called *agayacatl*, were a favoured food of the Indians long before the conquest. When they are cooked in pastry, they do not differ from the eggs of fishes, having the same flavour and appearance. They are rather a delicacy and I find that they also are found on the tables of the rich in the capital'.

DE SAUSSURE, SALLÉ and others confirm these reports, and eggs on the rushes have also been received in Europe. Mr. VIRLET d'AOUST, the geologist, has established that the immense layers of oolite which are found in the same localities are composed of myriads of these eggs deposited centuries ago.

L. ANCONA (1933, pp. 51–69) states that the old lake of Texcoco is at 2237 m. altitude, and A. PENAFIELD (1884) estimates that in spite of the many natural enemies of the *ahuatle* they form a mass of 3.650.000.000.000 cu.m. on the bed of the lake. These bugs are chiefly *Corixidae*. Their main species are:

Krizousacorixa azteca Jacz.	in winter 31%, in March 72% of the total catch.	
Notonecta unifasciata Guér.	33%	18%.
Krizousacorixa femorata Guér.	12%	7%
Corisella texcocana Jacz.	7%	2%
Corisella mercenaria Say	17%	1%

TH. GAGE (1721 I, p. 144) saw on the markets of Mexico a peculiar soil for sale, which is obtained as follows: At certain seasons the inhabitants collect with nets a lemon powder which accumulates on the waters of the Lake of Mexico, and which resembles meerschaum. This they collect in great pots and prepare brick-shaped pastries from it. These bricks are not only sold in Mexico, but in many distant towns as well. They eat them with as much appetite as if they were the best European cheese.

VIRLET d'AOUST (vide COWAN, 1865, p. 276) states that in October the lakes of Chalco and Texcuco near Mexico City are haunted by millions of 'small flies', which after dancing in the air plunge down into the water, to the depth of several feet, and desposit their eggs at the bottom. An anonymous writer in the Journal de Pharmacie (vide COWAN 1865, p. 276) says that these Corixid eggs are attached in innumerable quantities to the triangular leaves of the carex (!), forming the bundles which are deposited in the water. They are oval with a protuberance at one end and a pedicle at the other extremity, by means of which they are fixed to a small round disc, which the mother cements to the leaf. Other larger, cylindrical eggs, mixed with them, belong to *Notonecta*.

D'ANCONA (1933, pp. 103 ff.) states that the *jumiles*, a Pentatomid bug *Euschistus zopilotensis* Dist., together with many other species of *Heteroptera*, are extensively used as food in Chautla in the state of Morelos (Mexico). Mr. L. N. H. KRAUSS in a letter to Dr. H. J. SAILER stated in 1945 that he saw them sold alive in baskets at Cuernavaca (Morelos) by the Indians, who say that they scrape the bugs off trees in the nearby mountains. He further states that they are usually sold in paper cones in handful lots, the price being about two pesos (0.42 U.S. dollars) per kilo. They are supposed to cure complaints of the kidney, the liver and the stomach, and are eaten alive or dropped into a stew just before serving. Only a few are used since they have a strong taste. D'ANCONA calls this species the *jumiles de Cuautla* and mentions'that they are sold on the markets to flavour other dishes, and are also regarded as an excellent remedy against rheumatism. In a further paper D'ANCONA (1932, pp. 149 ff.) states that another Pentatomid bug (*Atizies taxcoensis nov. sp.*), the *jumiles de Taxco*, is sold on the markets of Taxco and other small towns of Mexico and is eaten fried.

KÜNCKEL D'HERCULAIS (1882 II, pp. 130 ff.) discusses the honeypot ants and on p. 148 the leaf-cutting *Atta*-ants. The abdomens full of eggs of the females are one of the greatest dainties of the Indians who eat them with a little salt. After a rich *Atta*-crop, they are fried and powdered with salt, and in this form the Europeans also appreciate them.

The honey ant of Mexico, *Myrmecocystus melliger* (Llave) Luc., was first discovered by Llave in 1832, as *basileiras* or *mochileras*, yet

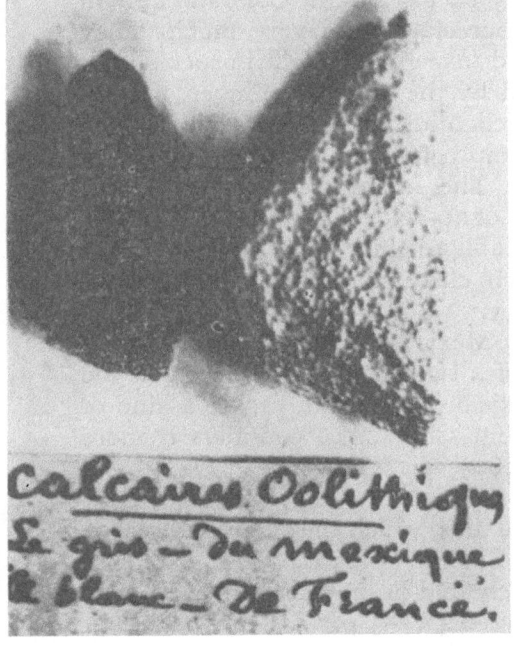

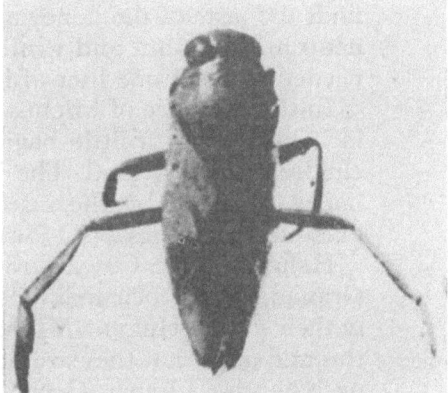

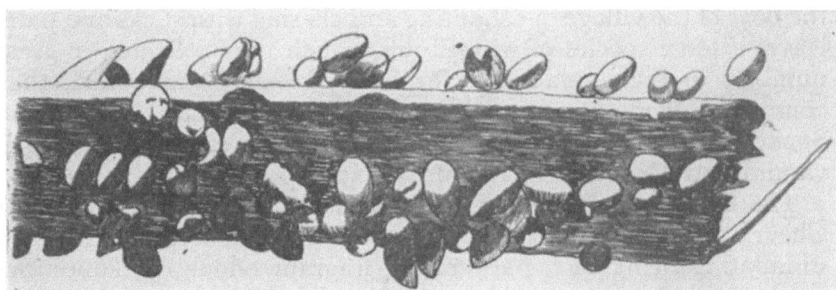

Fig. 39. a. Part of a rush with eggs of waterbugs. b. Limestone formed by agglo-meration of egg masses of waterbugs. c. *Notonecta unifasciata* Guér., one of the Mexican waterbugs producing these egg masses (From Portevin, after the material used by Guérin-Méneville).

their character was recognized by R. P. COOK in Colorado only in 1882. Indians and Mexicans appreciate them very much. They search for them in their nests and take great delight in sucking the abdomen, which is mainly filled by the crop. A number of early travellers also regarded them as delightful. On the market in Mexico City they are sold by dozens, simply fastened to square pieces of paper (SIMMONDS 1885, p. 369). The Indians also fermented the honey into an alcoholic drink. LOERO (vide BRYGOO 1946, p. 53) finds the taste of the honey agreeable, slightly acid in summer, but neutral in autumn and winter. He claims that 1000 honeypots are needed to yield one liter of honey.

In the province of Michuacan, Mexico, the honeypot ant carries in its abdomen 'a little bagful of a sweet substance, of which the children are very fond. The Mexicans suppose this to be a kind of honey collected by the insect, but CLAVIGERO considers it more likely to be its eggs'. (T. SMITH 1807 vol. XII, p. 197).

HERERA (vide COWAN 1865, p. 160) says the natives of New Granada made their main food of ants, which they kept and reared in their yards (HERERA VI, 5,6). SLOANE (1725 II, p. 221) confirms this and adds that they are publicly sold in the markets. ABBEVILLE DE NOROMBA (Journ. Geogr. Soc. 1841, X, p. 175) mentions that these large ants are fricasseed. He also quotes SCHOMBURGK, who on his journey to the sources of the Essequibo one evening saw all the boys of the village out shouting and chasing with sticks and palm leaves a large species of winged ant, which they collected in great numbers in their calabashes for food. Roasted and boiled, they considered them a great delicacy. A. v. HUMBOLDT (KIRBY and SPENCE 1822 I, p. 241) informs us that ants are eaten by the Marivatano and Marguitares Indians, mixed with resin for sauce.

The Cicindelid beetles *Cicindela curvata* Chevr. and *C. roseiventris* Chevr. are macerated in water or alcohol and fermented into a stimulating drink of a particularly fragrant odour (THEODORIDES 1949, p. 2). SIMMONDS (1885, p. 353) informs us that the larvae of another beetle, *Trichoderes pini* Chevr., form part of the food of the Indians in Mexico.

Lice and 'long worms' were sold for food in Mexico (PURCHAS vol. III, p. 113). When the Indians of the Province of CUENA are infested with lice, 'they dress and cleanse one another; and they

300

that exercise this are for the most part women who eate all that they take' (PURCHAS vol. III, p. 975). The natives of the Isthmus of America have lice in their heads, which they feel out with their fingers and eat as they catch them (WAFER, vide DAMPIER 1729, III, p. 331). And DOBRIZHOFFER (II, p. 396, vide COWAN 1865, p. 318) also mentions lice as being eaten by the Indian women in S. America.

The main food of the Rama-Indians of Nicaragua is now by primitive agriculture, especially tubers of manioc, some fishery and the hunt of tortoises and mammals. Wild honey (*nalali*) is a delicacy and a remedy; it is taken from free nests of: *Melipona*-species or those hidden in hollow trunks of trees. In earlier times the fatty females of an *Atta*-ant were collected by the women and consumed.

4. THE WEST INDIES

We have learned from the note of PETER MARTYR (see p. 25) that before the European occupation of the islands the eating of insects was widespread, and that stores of locusts, etc. served for barter trade between the various districts. PETER MARTYR (1612, p. 274) also adds that the inhabitants of the Caribbean islands 'willingly eat the young bees raw, roasted and sometimes soaked'. J. REMY (vide J. PINKERTON II, p. 525) confirms the eating of grasshoppers, reporting that the West Indians consume a species in great quantities which he calls *Oedipoda corallipes*. And also Sir H. SLOANE (1725, p. 204) quotes that LOPES DE GOMERA found baskets full of 'crickets' smong the provisions of the Indians on Jamaica when the island was first discovered.

But the most famous insect resources of the West Indies were a number of beetle grubs, which were soon adopted as a rare delicacy by the new immigrant Europeans as well as Negroes. These consists primarily of two families: longicorn beetles and weevils. Among the longicorn beetles (apart from *Macrodontia cervicornis* L. which is more common in tropical South America) we have to mention especially the fallow-deer beetle *Stenodontes* (*Prionus*) *damicornis* L., another common species of tropical America and the West Indies. Its larva, a grub about 9 cm. long and thick as a little finger, is in great demand as an article of food, being considered by epicures as one of the greatest delicacies of the New World. Authors of the highest respectability inform us that some rich people in the West Indies keep Negroes for the sole purpose of going into the woods in quest of these prized larvae, which are dug out of the trees in which they reside. Dr. BROWNE, in his History of Jamaica, states that they are chiefly found in the plum and silk-cotton trees (*Bombax*), their native name being *Macauco* or *Macokkos*. The mode of dressing them is first to open and wash them and then carefully to broil them over a charcoal fire (vide COWAN 1865, p. 73). Sir HANS SLOANE (1725 II, p. 193 f.) mentions that the Indians of Jamaica boil them in their soups, potages, olios, and pepper-pots, and consider them to have a delicious flavour, similar to, but better than, marrow. The Negroes of this island roast them slightly on the fire, and eat them with bread.

THE PÈRE LABAT, who visited the West Indies at the beginning of the 18th century, tried the famous palmworm (1931, p. 100): This insect originates in the heart of trees, when these are felled. The palmworm is as thick as a finger and about two inches long. It may be compared to a lump of fat from a capaun, wrapped into a very tender and transparent pellicle. They are placed in a row on a piece of wood and turned over the fire. When they begin to get hot, they are covered with powdered bread crust, some pepper and muscat. If they are boiled, they are served with a few drops of orange or lemon juice. They are good to eat and very delicate, once the natural repugnance at eating worms has been overcome, especially when one has seen them alive. When these worms are exposed for some time to the sun, they exude an oil which helps excellently against cold pains and especially against haemorrhoids.

The larva of the big black weevil *Rhynchophorus palmarum* L., which is over 5 cm. long, is the *grou-grou* of the West Indies. This large white, oval grub lives in the tenderest parts of smaller palms and, fried or broiled, is considered as a great dainty. MARIA SIBYLLA MERIAN (1771, pl. 48) says: 'The tree grows to the height of a man and is cut off when it begins to be tender, is cooked like cauliflower and tastes better than an artichoke. In the middle of these trees live innumerable quantities of grubs, which at first are as small as a maggot in a nut, but afterwards grow to a very large size, and feed on the marrow of the tree. These grubs are laid on the coals

to roast and are considered as a highly agreable food'. Captain
STEDMAN (1796 II, p. 23, 115) mentions that these larvae are a
delicious treat to many people and that they are regularly sold at
Paramaribo. They are fried in a pan with a very little butter and
salt or are spit on a wooden skewer. Prepared thus they partake in
taste of all the spices of India: mace, cinnamon, cloves, nutmegs,
etc. STEDMAN once found concealed near the trunk of an old tree
a 'case-bottle filled with excellent butter', such as the natives make
by melting and clarifying the fat of this grub.

DOBRITZHOFFER (I, p. 410, vide COWAN 1865, p. 69) refers to
Rhynchophorus palmarum or a related species, when he says: 'The
Spaniards of Santiago in Tucuman, when they search for honey in
the woods, split open certain palmtrees upon their way, and on
their return find large grubs in the wounded trees, which they fry
as a delicious food'. The Guaraunos of the Orinoco (SOUTHEY 1817,
I, p. 110) find similar grubs in great numbers in the palms, which
they cut down for the sake of their juice. After all has been drawn
out that will flow, these grubs breed in the incisions and the trunk
produces, as it were, a second crop.

The Creoles of Barbados (SCHOMBURGK 1847, p. 646) consider
the *grou-grou* a great delicacy when roasted and say it resembles
in taste the marrow of beef-bones. A. DE ULLOA in his Noticias
Americanas (vide COWAN 1865, p. 70) says this grub has the singular
property of producing milk in women.

5. SOUTH AMERICA

In the excellent study of DE WAVRIN on the Habits of the native
Indians of S. America (1937) we find the following notes on insect
food: 'The Indians also like bee-honey very much and the bees
abound in certain parts of the forest. They have no stings and build
their nests in hollow tree-trunks. When the nest is close to the soil
and its entrance hole sufficiently large and not very deep in the
interior, the contents are collected by putting one's arm into the
hive. Otherwise the Indians use a hatchet, but they fell the trees
only when they judge the honey crop to be abundant'. (p. 58).

'One may state as a general rule that all tribes eat almost all mammals, from the largest to the smallest. Only a few are protected by their bad smell or by religious taboos (p. 121). The Piaroa eat many fruits and vegetables, all fishes and all reptiles: from toads to the poisonous spider *arana mona*'. (p. 122).

'One day an old man entered the village with three parcels wrapped in leaves. The third parcel contained some earth from a termite nest full of these small insects. The Indian indicated that they were for eating and offered them to us. He himself put a handful into his mouth, apparently delighting in this and asking the others to participate, which the travellers had to do in order not to offend him. He could not, however, swallow it'. (p. 123). DE WAVRIN concludes that the Indians in general are not strict vegetarians. Many tribes are exclusively carnivorous or ichthio-vorous, others can be regarded as omnivorous (p. 130).

A. R. WALLACE devoted a special paper to 'The Insects used for Food by the Indians of the Amazon' (1853). The local Indians devour almost everything eatable, such as turtles, alligators, lizards, snakes, frogs and many insects.

All over Amazonia the great-headed ant, *Atta cephalotes* Latr., living in hillocks 20 feet square and 1 yard high, are eaten. 'At a certain season the insects come out of their holes in such numbers, that they are caught by basketsfull. When this takes place in the neighbourhood of an Indian village, all is stir and excitement; the young men, women and children go out to catch the *Saubas* with baskets and calabashes, which they soon fill; for though the female ants have wings, they are very sluggish and seldom or never fly. The part eaten is the abdomen, which is very rich and fatty from the mass of undeveloped eggs. They are eaten alive, the insect being held by the head as we hold a strawberry by its stalk, and the abdomen bitten off, the body, wings and legs are thrown down on the floor, where they continue to crawl along, apparently unaware of the loss of their posterior extremities. They are kept in calabashes or bottleshaped baskets, the mouths of which are stopped with a few leaves, and it is rather a singular sight to see for the first time an Indian taking his breakfast in the *Sauba* season. He opens the basket, and as the great-winged ants crawl slowly out, he picks them up carefully and transfers them with alternate handfuls of farine to his mouth. When great quantities are caught, they are slightly roasted or smoked, with a little salt sprinkled among them, and are then generally much liked by Europeans (p. 242 f.)'.

The large South-American termite (*Termes flavicolle* Pty.), also much sought after by the big ant-eater *Myrmecophaga jubata*, is eaten on the Upper Amazon by the Indians. 'The great-headed, hard-biting worker is entrapped by means of its claws. An Indian boy going after '*Cupim*' takes with him a calabash or a bottle-basket, and searches about for a nest. He then scrapes away some of the earth and taking a large piece of grass inserts it as far as it will go, and on withdrawing it finds a row of 10 or 12 termites holding tightly on to it. He repeats this operation till the basket is filled. These insects are also eaten alive or roasted; but in this case it is not the abdomen but the enormous head and thorax which is eaten, as those parts contain a considerable mass of muscular matter. These insects have generally a bitter taste and are not much esteemed, except by the Indians themselves'. (p. 243).

'The Homopteron *Umbonia spinosa* swarms at certain seasons on the *Inga*-trees, which are universally planted by the Indians near their cottages. The insects fall on the ground in great numbers and the sharp spine on their thorax renders walking barefoot very disagreable. This spine seems to render them very ill adapted for food, but when they first appear the whole body is soft and flaccid and they àre then collected and roasted in a flat earthen pan. They are not, however, so much esteemed as the other insects I have mentioned'. (p. 243).

'Headlice of men are probably more a delicacy than an article of food and they are caught exactly in the same way the monkeys catch them in the zoo. A couple of Indian belles will often devote a spare half hour to entomological researches in each others' glossy tresses, every capture being immediately transferred with much gusto to the mouth of the operator'. (p. 244).

WALLACE was not aware of any other insects eaten by the Indians of the Amazon, yet earthworms were also used as food.

A number of early South American references to ants may in some part properly refer to termites. G. PISO (1658, pp. 9, 291) writes: 'Alia praeterea datur grandis species *Tama-ioura* dicta digiti articulum adaequans. Quarum etiam clunes dessecantur et friguntur pro bono alimento'. The *Tama-ioura* is the ant *Atta cephalotes*.

G. Marcgrave (1648, p. 56) writes: 'Denique formicae hic visuntur grandissimae, quas indigenae vulgo comedunt et in foris venales habent'.

T. de Laet (1630, pp. 333, 379) adds: 'Formicis vescebantur, easque studiose ad victum educabant'.

The Indians of Amazonia eat only the soldiers. They introduce a stem of grass into the termite hills, draw them back with about a dozen soldiers holding fast to the intruding stem and eat them raw (E. Hegh, 1922, p. 670).

In the narratives of Barrère, de Azara, A. v. Humboldt (1822, pp. 443 ff., vol. VII, Rengger (1835, p. 253), R. Schomburgk 1848, p. 112), J. Orton (1876, p. 301), R. Spruce (1908, p. 484) and others, there are many references to the esteem in which the large sexuals of the leaf-cutting ants (*Atta sexdens*, *A. cephalotes*) and the honeypot ants (*Myrmecocystus*) are held by many S. American Indians. P. Barrère (1741, p. 198) mentions a big, winged, edible ant from British Guiana appearing in great numbers at the beginning of the rains. Negroes and creoles eat its abdomen which is the size of a chick-pea and is full of a whitish, honey-like liquid which apparently is nothing else than its eggs. And F. de Azara (1809, p. 199) describes how the inhabitants of Santa Fé go hunting the winged ants. The fat abdomens are eaten raw, or they are fried, passed through syrup and eaten as sweets. Gallendo (1916, p. 344) reports even recently about the survival of the eating of the abdomens of *Atta sexdens*, the *tanajura*, in Brazil. Schomburgk (1848, II) vividly described how the Indians in British Guiana collect these ants, when they swarm out of their mound-shaped nests with the first rainstorms. Their heads are pulled off as soon as they are caught and the swollen gaster, filled with fatty tissue, is roasted or otherwise cooked. Thus prepared these insects are considered even daintier than the palmworms (Cf. J. Bequaert 1921, p. 330). The Galibi of Guiana call them *koumaka*, and they are eaten by Negroes, creoles and Indians fried with flour in fat. The various castes are well distinguished and the females with eggs are the most appreciated (Cf. E. Daguin 1900, p. 17).

The S. American Indians have a great passion for the large queens of the *Atta*-ants. At certain seasons the winged queens fly in thousands and are then collected. A. Hyatt Verill (1938,

306

p. 161) saw Indians wading in ant hills up to their knees and bleeding from their bites, yet still quietly collecting them, and when leaving eagerly cracking the queens. When full of eggs, these queens taste like condensed milk.

The pleasant acid taste of ants tempts many races of Brazil and India. The termites in India are caught by the thousands and baked in pies which are brought to the markets, as in S. America.

W. R. PHILIPSON (Brit. Mus. London) told us that in the primeval forest of Columbia the mestizos roast *suaba* ants (*Atta spp.*) in great quantities. They are generally eaten in the villages. This food is quite a national dish all over the Andine region.

The eating of caterpillars is widespread in South America, while grubs of beetles, apart of course from those of palmworms, are less commonly eaten. The following random observations will exemplify this statement.

PÈRE CALANCHE (vide DAGUIN 1900, p. 7) observed that in the Peruvian Andes a caterpillar, called *sustillo*, resembling the silkworm, is collected in large quantities from *Mimosa nigra*, as a most delicate dish.

AUGUSTE DE ST. HILAIRE (vide DAGUIN 1900, p. 7) saw natives in Brazil eating the big caterpillars of a Hepialid moth which lives on bamboo. The Indians take great care to remove the head and intestines before consuming the caterpillars.

MAINE-RAID has the following story about the Mundracos of Central America (vide DAGUIN, 1900, p. 8): 'When arriving at a lake whose shores were covered with reeds, these Indians started to cry joyously, to break the reeds and to extract from each knot fat, white larvae, 8 to 10 cm. long, called *Maquara*, which they ate eagerly, and they fell like dead-drunk soon afterwards. These caterpillars are really excellent, and to prevent intoxication it is sufficient to remove their head before consumption'.

In South America a great hawk-moth caterpillar, which congregates, at a certain season by hundreds before descending to pupate in the soil, is regarded as a delicacy. In British Guiana HYATT VERRILL (1938, p. 160) tasted them and compared them to soft-skinned shrimps and he asked for more.

SIMOUIN WRITES: 'During a halt . . . . OYER took out from some decaying trees some big worms, which we roasted and spread on our bread in place of butter. First I cut some faces, yet soon changed my opinion . . . . The worms of Cilaos should be regarded as true delicacies. LABLAND also (TOUR DE MONDE 1862, II, p. 171) calls these *gusanos* (worms) excellent (vide DAGUIN 1900, p. 13).

RICHARD SCHOMBURGK (vide BEQUAERT 1921, p. 197) saw the Indians of British Guiana actively gathering caterpillars and their pupae in the rainy season.

Beetles of the family *Dryopidae* have repeatedly been reported as being used in Chile and Peru in the form of a paste for the preparation of a national soup called *chichi*. The species mentioned are *Elmis chilensis* Phil. (NETOLITZKY) and *E. condimentaria* Phil. (BRYGOO). The larvae of *Passalus interruptus* L., feeding on the roots of batatas, are collected for food in Surinam (BRYGOO). And VELLARD (1939, p. 85) in his 'Civilisation du Miel' writes of the Guayakis of Paraguay as follows: 'The stems of the pindo-palm (*Cocos romanzoffiana*) when felled and decaying, harbour in abundance the white, thick, soft and up to 12 cm. long larvae of a Passalid beetle. The Guayakis like them very much as food, as well as other, similar larvae from other tree-trunks. They put them on small spits above the fire, until a fluid oil emerges from them. When they are brown, they melt in the mouth. Their taste is like that of brain, and slightly salted they are quite good. Yet the Guayakis usually take them by their heads and bite off the body, which they eat raw'.

The grubs of *Megasoma hector* Gory are roasted over burning coals in Brazil (NETOLITZKY) and those of *Lamia tribulus* F. in South America (BRYGOO). The grubs of *Macrodontia cervicornis* L., growing up to 21 cm. in length in the stems of *Bombax*-trees, are roasted in Brazil and Guiana. The *moutac* or *macoco* of Guinea and the Antilles is the larva of the related *Stenodontes damicornis* L. (NETOLITZKY, BRYGOO, BEQUAERT, BLANCHARD). Palmworms are not only appreciated in the Antilles. MERIAN and LINNÉ had previously mentioned the *Rhynchophorus palmarum* L. from Surinam. *R. cruentatus* F. is another palmworm developing in various palms in tropical South America, where it is an appreciated delicacy (GHESQUIÈRE 1947).

J. CREVAUX (1877, II, p. 144) was stung by a wasp in the Andes. The combs of its nest, which never have honey, are full of maggots, and these CONASSI hurried to eat with his cassave. This wasp is greatly appreciated all over Guiana, where the Roucouyen Indians call them *ocomo*. When the Negroes of Guiana are stung by bees, they in revenge eat as many as they can catch (E. BANCROFT 1769, p. 230).

Father OVALLE (vide SIMMONDS 1885, p. 357) observed in 1649 that the Indians of Chile, in the absence of grain, make bread from locusts. They watch where the insects alight to rest at night, then setting fire to the bushes, reduce everything to ashes, which are gathered and baked into cakes.

TSCHUDI saw the natives of Peru hunt and eat the lice from their children's hair (SIMMONDS 1885, p. 350).

TH. KOCH-GRÜNBERG (1909) made a number of observations during his two years' stay with the Indians of N. W. Brazil. 'The Indians consume the lice which densely populate their hair. In Yurupary-Cocheira we saw every morning and every evening how the women eagerly collected them from each other, and I gained the impression that the lice are regarded less as vermin than as a delicacy. Another Indian delicacy, which also tastes not bad to our European palate, is the large winged ant, roasted on the hearth and tasting like fine Christmas cookies. The head and remnants of the wings are detached and the fat abdomen only is eaten. Gourmets even consumed them alive. The swarming season of these ants, early in the rainy period, is a period of festival for the entire village. At the first announcement of their swarming everybody, old and young, joyously collects baskets and jars and runs to the ants' nests to gather as many as possible of this appreciated delicacy' (1909 I, p. 141 f.).

'After the rich food of the Tukano and Tuyuka, we now had to manage with mandioca in abundance from the *Inga*-shrubs . . . . Still in darkness, we roused one another, in order not to lose our first breakfast, the only substantial meal of the day, consisting of a soup of *biju* and roasted ants. In Pinokoalisa the women answered our request for food with the words: 'No chicken, no pineapples, no bananas! But beetles and ants!' ' (pp. 332 ff.).

'One evening an Indian ran into the house calling: 'The *sauba*-ants are swarming'. This event had been expected for some days. Early before sunrise all inhabitants departed for the ant-hunt. A low scaffold was erected over the nest, on which the Indians posted themselves, in order to avoid the bites of the ants. They burned the wings of the swarming sexuals with torches and then gathered them as speedily as possible into baskets and leaf bags. During the following days we had for lunch highly flavoured sandwiches: *biju* with finely pounded ants, which were roasted and seasoned with pepper and salt and which we enjoyed very much' (p. 96).

Some other observations are taken from a later edition (KOCH-GRÜNBERG 1921, pp. 86, 241, 380, 385, 397). 'The Cayara Indians also eat the *pinos*, tiny gnats. Often a woman is seen squatting behind he husband and collecting the tiny bloody beasts from his back. In Yauarté the women brought us four containers and one bag full of cockchafers, which now abounded and covered the water to the great delight of the *aracu*-fish. These beetles are very fat. The Indians eat them roasted, and as I found out myself, they have a rather good taste. On the Yapura thick white grubs of beetles from palm stems were consumed as delicacies raw and alive. At the Makunas, after the evening toilet, the mothers collect the lice from the hair of their children; they comb their hair over a chair and, catching the vermin with their moistened forefingers, eat them with delight'.

A rare domestic animal was observed among the Makunas: a piece of hollow tree-trunk, a bee-hive, was hung on the pole of the huts at a height of about two meters.

In South America also insects are included in magic performances. KOCH-GRÜNBERG (1921, pp. 83 ff., pp. 328 ff.) demonstrates this well for the Indians of Amazonia. Their magic mask-dances and ceremonies aim at obtaining the favour of evil and dangerous spirits, which have caused the death of a parent and to prevent their causing any further harm. The great carnivores of the forests and the pests of the field are also included in these masks, while – probably by chance – none of the important insects which are eaten, such as honey-bees and ants, are among KOCH-GRÜN-BERG's lists. The secondary aim of these magic mask ceremonies is to increase the fertility of game and crops. The dances imitate the movements and actions of the spirits or animals, with the aim of exerting a magic influence on them, just as the Palaeolithic hunter aimed to further the success of his hunt by magic drawings of his game animals. Living creatures, such as butterflies, vultures, jaguars, fish, caterpillars, grubs, etc., are represented by these masks in

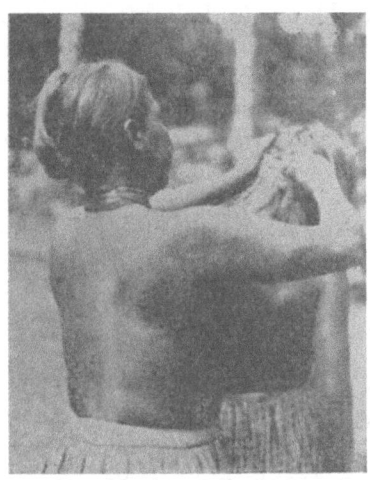

Fig. 40. Indian mother collecting lice from her daughter's head. From KOCH-GRÜNBERG, 1909, Fig. 188.

Fig. 41. Indian drawing of a butterfly-mask dancer with a drinking calabash on the edge of which he drums with a stick. From T. KOCH-GRÜNBERG 1921, p. 332.

addition to demons in human shape. No doubt the ceremonies for insects which are eaten are very similar to those performed for other insects. We give a short description of their dances based on accounts by KOCH-GRÜNBERG:

The dance of the dung-beetles illustrates the cleansing work performed by this industrious insect, which constructs balls of dung and buries them in the soil for future consumption. Two dancers, chanting, advance and retreat hand in hand. They press the dancing sticks to their sides under their arms, and with them roll another stick, representing the dung ball.

The azure-blue *Morpho*-butterfly, despite its harmless beauty, is regarded as one of the most dangerous demons, often even as the 'master of all masks'. Fig. 41 shows its mask in the Yurupary-Cachoeira, where it 'induces malaria'. The mask holds the malaria-spreading drinking calabash and beats it with a small stick. Usually two butterfly masks appear together, now speedily advancing, again advancing and retreating, in imitation of the fluttering flight of the

Morpho and its congregation on rocks. Dance and chants are accompanied by rattling with the calabashes, ending with a long call of the dancer: 'e - - he - - -'. Another greatly feared demon is a tiny plant bug of the Indians' plantations. It is portrayed as pounding roasted pepper in a wooden mortar and spreading this fine dust into the air, to infect the eyes of the workers in the plantations with diseases. Other dances of insect masks described by KOCH-GRÜNBERG are those of dragonflies (pp. 330, 398) and sandfleas (p. 330). These few lines are sufficient to show that in South America, as everywhere in the tropics, insects are a living component of the magic ceremonies.

6. HONEY-HUNTING AND BEE-KEEPING IN SOUTH AMERICA

The following notes by K. SAPPER (1936, pp. 24, 83, 95) may serve as a first introduction to the complicated situation of honey-bee utilization in tropical America, which has undergone all stages of development from bee-killing honeyhunts to true primitive bee-keeping. SAPPER states that the domestication of the stingless honey-bees (*Meliponidae*) was apparently relatively easy. Beekeeping was restricted at the time of the conquest to Central and the northern areas of South America. There it still survives today in a primitive form. The Indian cuts the trunk or a branch of a tree, in which a wild bee swarm is located, above and below the nest of the swarm, covers its entrance opening, and transports it thus to his home. The swarm remains at his home, protected by the roof of the house, provided that the distance from the earlier location is sufficiently large for the bees to be unable to find their way back to it. When the Spaniards first came to America, they found at Yucatan, on the Cosumel, on the Sierra de Santa Marta, and elsewhere a highly developed bee-keeping in artificial hives, such as calabashes, pots, barrels and hollowed tree trunks. Probably after careful observations on the habits of the wild honeybees, and apparently first at Yucatan, the Indians made artificial hives and put them near their houses, when young swarms left the domestic apiary. At present, some native bee-keeping is still maintained in Mexico, El Salvador and in S. America, for instance by the Meniheme-Indians on the Rio Japura. IHERING reports that occasionally, when some hives are kept near houses, the swarms select as their home a wooden box close to the apiary.

European bees are now kept by many whites, Indians and mestizos, while a few local bees (*Melipona, Trigona*) are still kept by Indians (1936, p. 83). The once flourishing bee-keeping of the Indians is today much reduced. Where bees are kept at present, as in the coffee-plantations of Guatemala for the improvement of the fertilization of the coffee-flowers, or as an actual profession as in Costa Rica, European bees are used and kept by European bee-keepers (1936, p. 95).

The most important single paper on Indian apiculture is that of E. NORDENSKIÖLD (1929, pp. 169 ff.) who states that even now the modern Indians gain almost all their honey and wax from wild bees. He repeatedly accompanied them on their honey-gathering trips. Their importance for the Indian economy is great, especially in the Gran Chaco. The Indian name for the bee, *iramanha,* means honey vigilance. The honey of the stingless bees has a particularly agreable taste. It is often served as a drink mixed with water. Yet the honey of a few species has an outright bad taste or may even be poisonous. The usual honey crop is from half to two liters, while from a nest of *Trigona nigra* Cress. up to ten or fifteen liters may be taken. The Indians often have to migrate over long distances to find wild bees' nests and they are ready to undergo great pains to get at them.

The first Spanish travellers used to report on the bee-cultivation by the Indians. Yet C. THOMAS' (1882) interpretation of a chapter of the Maya Codex Troano as an apicultural calendar is certainly wrong. Mixtec apiaries in Mexico (STARR 1899, p. 58) show that bee-domestication spread from its Central American origin northwards as well as southwards.

NORDENSKIÖLD (1929, p. 285) concludes from the fact that all stages of apicultural development are met with in America, that it has developed autochthonously in this continent, and this conclusion is, of course, entirely confirmed by the eye-witness reports of the first European travellers.

N. E. H. NORDENSKIÖLD (1929) thought that tribes like the Guayaqui Indians of Paraguay which possess stone axes, notwith-

Fig. 42. American apiculture: Dots: fully
established Indian bee-keeping. Circles: tem-
porary hive-keeping. From NORDENSKIÖLD
1929.

standing the fact that they neither cultivate the ground nor make
hollowed-out canoes, find these implements primarily of use in
cutting into tree trunks where bees'-nests are obtainable. Others
improvise crude chisels. The Guayaqui also employ a more ingenious device for honey extraction. A long-handled brush is thrust
into the nest and then brought to the surface, impregnated with
honey. NORDENSKIÖLD saw such brushes in use among the Ashluslay
Indians in the Gran Chaco.

At Yadoconno and other towns the Mixtec Indians keep many
bees (STARR 1899). The bee-hives are cylindrical, made of tied-
together sticks, which are wrapped in matting and hung to the sides

314

of the houses or arranged upon racks over which protecting thatches are constructed.

The introduction of *Apis mellifica* L. into S. America has greatly reduced the domestication of wild stingless bees (SCHWARZ 1948, p. 160). The history of the attempts to export the *Meliponidae* into the eastern hemisphere for domestication have been fully reported by the same author (pp. 160 ff.), none of which however proved itself to be a success.

GÜNTHER (1931, p. 298 f.) confirms that the wild stingless bees are known to the Brazilian Indians under many names. Like the Indians, travellers in that country have found that honey forms an important article of diet. Many species produce from 3 to 25 pints of honey in a single nest, which according to the species may contain from a few hundred to 70,000 inhabitants. It is more savoury and aromatic than European honey and is thin and liquid, setting only after it has been boiled. The honey may occasionally be poisonous owing to collection from poisonous flowers. The cells are closed after provisions for the full development of the grubs have been filled into them. Above and below the combs the bees build a great heap of waxen honey-jars, which in some species are as big as the cells of a hen's egg, and which are filled with either honey or pollen. The pots in the interior of the ball cannot be got at until the outer pots are emptied, when the bees bite through them.

A few of the social wasps in Brazil also store honey (GÜNTHER 1931, p. 294). The wasps feed their maggots on the flesh of masticated insects. The adults are fond of sweet food. *Nectarinia mellifica* And. and *Polybia occidentalis* are the main wasps which store honey in unfavourable weather, the former species making a regular habit of this. Their honey also is sometimes poisonous or intoxicating. Nevertheless, it is readily collected and consumed.

Full information on the 'Stingless bees (*Meliponidae*) of the Western Hemisphere' is found in the stupendous monograph of that name by H. F. SCHWARZ (1948). Before the introduction of sugar cane and the European honey bee, i.e. in the pre-Columbian era, the *Meliponidae* were almost the only source of sweet food in America. COLUMBUS noted on his first voyage in 1493 a variety of wild honey in Cuba (Juana), and SCHWARZ also (p. 123) comments on the honey of *Melipona beechei fulvipes* Gu. as among the natural assets of the island. In nearby Yucatan, where agriculture had a long tradition, F. L. DE GOMERA (1578, p. 200) mentions the sale of honey of various kinds on the great market of Mexico City. The Indians of Central and South America likewise enjoyed wild honey.

The wild honey of the stingless bees resists crystallization and often remains indefinitely in a fluid state (P. FERMIN 1769 II, p, 300 f.; O. GOLDSMITH V, p. 146). And T. PECKOLT (1893/4) kept this honey for thirty years without observing any crystallized sediment. However, A. de BERTONI (1911) and H. v. IHERING (1904) observed full crystallization with the honey of *Trigona jaty* Sm., and IHERING noticed the same with that of *Melipona marginata* Lep. FERMIN, BERTONI, and others stated long ago that the Meliponid honey does not keep and sooner or later turns sour; yet this sour taste can be made to disappear by boiling. When in a more concentrated form, as after boiling for a considerable time, a practice found all over Brazil (K. GÜNTHER 1931, pp. 296, 376), it may keep

Fig. 43. Trees with nests of wild bees and wasps from the South American primeval forest (from HERNANDEZ and RECCHI).

for a lengthy period. It is widely used for the preparation of intoxi-cating drinks, such as the *Balché* of the Maya, and has also many medicinal uses. PECKOLT's analysis (loc. cit) showed 50% levulose plus dextrose and 50% water. The amount of honey per nest varies considerably according to the species, size of colony, season, etc. H. v. IHERING (1903, p. 273) took as a rule half to two liters, in exceptional cases up to 50 liters from a single nest in S. America; P. R. HENDRICHS (1941) one to two liters in Mexico. In Paraguay it was customary to store and transport honey of stingless bees in leather sacks. For this purpose a hide was cut into three pieces, tanned, and all hairs were removed, after which it was sewn to-gether with rawhide cords; a tube was then fixed to one end through which the honey was poured (A. HANNEMANN 1872). In Bogota honey was sold on the market in bamboo containers (GOUDOT 1848). The bees'-wax had, of course, the most manifold uses (see SCHWARZ 1948, pp. 133–143).

A number of stingless bees, mainly *Melipona beechi* Benn., the *yilkil cab* (honey bee), *colel cab* (lady-bee), *xunen cab* (hove-bee) have been on a number of occasions extracted from their forest hives and brought into artificial hives. BENNETT (1868) describes such an attempt: 'A hollow tree is selected and a section of it, 60 to 90 cm. long, cut off. At about the middle of this hollow log the future entrance hole is bored and the log is plugged at each end with clay or with stones and clay or, according to HALL (1824) with circular doors, cemented closely to the wood, but capable of being removed at pleasure. The future hive is thereupon suspended horizontally on a tree, and it is not long before a swarm takes possession of it'. Offering the swarming bees suitable empty hives close to their original home was doubtless the earliest step towards domestication, by imitation of the natural production of new hives in stingless as well as in European bees. J. P. HUBER (1839) was informed by Mexican bee-keepers of a more effective method for establishing colonies. A portion of the brood comb of the parent hive is sepa-rated and set up with 'a handful of old bees' as the nucleus of a daughter colony. Such a new nest must, of course, include also a member or members of the royal caste, either as brood or as adults. This was described as a normal practice from Mexico and far into S. America.

BENNETT also states that the collection of the honey does not necessitate stupefying or destroying the bees as with *Apis mellifica*. The plug at each end of the hive was merely withdrawn and the adjacent honey-combs were removed. One such artificial hive would yield at least two honey crops during the summer, while honey could also be extracted in any other month.

SCHWARZ (1948, p. 114) was told by a Maya apiarist in Yucatan, who had no less than than 110 log hives of *Melipona beechei* carefully banked on a scaffolding of wood under a sloping thatched roof, that he gathered honey only in April and December. G. M. FORSTER (1942) writes: 'A Popoluca Indian is required to practice sexual continence for seven nights before opening a hive, seven being the mystic number of this people. At the end of this period, the honey-gatherer goes forth before sunrise, smokes the hive with copal, which he burns in a small clay pot, and thereupon removes the honey. After a second smoking, the ends of the hive are sealed, and are kept sealed until the honey-gathering of the next year'. Similar sex-inhibitions are applied for the apiarist and for his assistant among the Akamba in Kenya lasting for ten days (J. K. R. THORP, 1943; G. LINDBLOM 1920), HUBER (1839) was informed that at Tampico the normal crop of artificial hives is three gallons of honey. ALEXANDER VON HUMBOLDT (1811, p. 204) found as many as 600 to 700 hives around Campeche. At the same time bee-hives formed an important item in legal testaments in Yucatan (ROYS 1939).

In addition to these hollowed wooden logs, B. HALL (1824, II, pp. 224 ff.) mentions the use of cylindrical earthenware hives, bearing ornaments of raised figures and circular rings. The little entrance hole in the middle of the cylinder was so moulded as to suggest the mouth of a man or of some monster, the head of which was represented in the earthenware of the hive. Above the hole was a small projection serving as a shield against rain. The wooden bee-logs were also sometimes ornamented 'to show the bees where the door is'. The hives were arranged on racks in rows, and sheltered by a roof of palm leaves, pointing from east to west, in a corner of the farmyard.

In other cases the section of the hollow part of the tree with the hive and its bees was transported from the forest and suspended beneath the eaves of the owner's house or in some other shady spot.

318

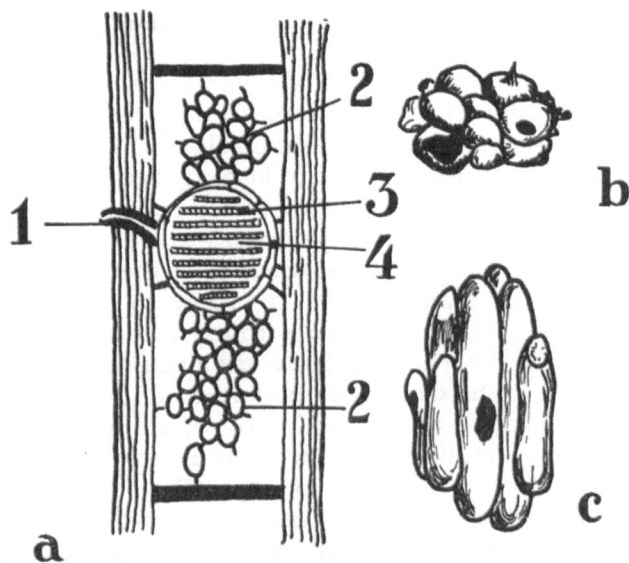

Fig. 45. Nest of a stingless bee (*Melipona ap.*). a. Schematic cross-section: 1. entrance tube; 2. storage cells; 3. brood; 4 combs. b. Honey cells. c. Pollen cells. (From SCHOLZ 1913, p. 158 f.).

The Popoluca sprinkle water over the cut ends, before taking the wild hive from the forest, to assure good honey production (FORSTER 1942, p. 538). In Guerrero in Mexico, which was one of the regions from which the Aztec king collected his tribute of honey, bee-keeping is now more restricted than of old (HENDRICHS 1941, p. 367). One of the bees kept there is *Melipona beechei*. The hives are taken from the forest and hung on the outer wall of the houses. The honey is usually taken in April, and in good years again in October. Bee-culture in the ancient manner is still maintained in many places in Mexico (HENDRICHS loc. cit; SAPPER 1935, p. 191). The old bee-centre of Yucatan is still an important place for honey production among the Maya. Apart from *M. beechei*, *Trigona pectoralis* D. T. is also kept in isolated hives, well removed from the large apiaries of the other species (BEQUAERT 1932, p. 17), as are also those of *Trigona jaty* F., *Partamona sp.*, *Trigona nigra nigra* Cress,

Lestrimelitta limao Sm. with its unpalatable honey, *Melipona fasciata guerreroensis* Schw., and others.

In Central America also *Melipona beechei* is the most common domesticated bee, and *Trigona fulviventris* Geur. is also kept. In British Honduras bee-keeping is as old as the colony. The bees were kept in the logs in which they originally formed their hives, the honey being drawn at intervals by probing into the log with a sharpened stick (H. GAHNE 1902, p. 947). Yucatan was a great export centre for honey. In Honduras honey-gathering in the forests was more common than bee-keeping. SCHWARZ (1948, pp. 149 ff.) describes a number of the old ceremonial rites from Yucatan, relating to bees where Hobnil was the special patron of the apiculturists. Ceremonial honey offerings to the Gods were performed on harvest day.

As bees were under the protection of certain deities, they were treated with consideration so as not to arouse the anger of the gods (REDFIELD and VILLA 1934, pp. 50, 117). War among bees was an evil omen for their owner. SCHWARZ points out (loc. cit., p. 153) that the so-called bee-figures in the old Maya Codex Tro-Cortesianus do not represent bees but the star Venus. But in the 'Book of Chilam Balam of Chumayal' (ROYS 1933), a sacred Maya-text dating from about 1500 A.D., many references are found to bees, bees'-wax and honey. H. M. RANSOME (The Sacred Bee 1937, p. 262 f.), who gives much additional data on ancient bee-keeping in Mexico, quotes from the Popol Vuh, the sacred book of the Quiche Mayas, that when the heroes conquer their enemies, the only occupations which they allow them to retain are the making of pots and bee-keeping (T. A. JOYCE 1914, p. 243). And in a certain story of the Popol Vuh enemies are defeated with the help of bees, wasps and hornets (LIEBRECHT 1879, p. 75). Yucatan remained from pre-Columbian days the great centre of the Mayas for bee-keeping (SAPPER 1935), to the delight of the invading Spaniards.

In S. America also bee-keeping was practised. Although most wild nests were pillaged so ruthlessly that no opportunity was given to the bees to recuperate, there are other cases where the bees were treated more considerately. When the Caingua Indians of the Alto Parana find a bees' nest in the forest, they do not utterly destroy it, but leave enough of it to induce the bees to rebuild the hive, so that

Fig. 46. Hives of domesticated *Melipona beecheii* Benn. in Mexico. Below: Terracotta hive from Tampico. The entrance of the bees is the open mouth of the figure. The hive was plugged at each side, and the plugs are taken out for collecting the honey crop. From P. HUBER. Above: Maya-apiary in Yucatan, in August 1946, composed of 110 log-hives, stored beneath a protective roof. From H. F. SCHWARZ 1948, p. 44.

when they again visit the spot, they may obtain another honey crop (J. B. AMBROSETTI, 1894). N. E. NORDENSKIÖLD (1929, p. 172) remarks that such an act of foresight may be regarded as the initial step towards bee-domestication. This step, actual domestication of stingless bees, was taken by some tribes. Thus, the Makino Indians of Brazil bring in entire nests from the forests and suspend them near their huts. The Paressi of the Matto Grosso, the Menimehé of the Rio Japura and the Mixtec of Central America construct true hives for raising the bees. NORDENSKIÖLD (1922, p. 285) assumes that these practices are derived from pre-Columbian apiculture, spreading from its Central American home southwards, whilst FORSTER (1942, p. 542) believes in the independent polytopic origin of these habits of hive transportation and of hive construction. The Chorti Indians of Guatemala, after collecting a wild hive from the forest, even plant nearby a number of flowering plants. Certainly bee-domestication had taken place in Venezuela before the arrival of the Europeans (SCHWARZ loc. cit., p. 155 f.). Bee-hives of clay were noted by a soldier in the 16th century in the Valle de Caldera in Columbia, who saw altogether 80,000 hives. ROQUETTO-PINTO (1919, pp. 13, 341, fig. 3) mentions another old type of hive, calabashes made from the fruit of *Crescentia cujete*, in use by the Paressi Indians of the Matto Grasso. Thus we find calabashes, clay vessels and hollow logs in use as hives in S. America.

More than a hundred years ago, the inhabitants of Sabara in Brazil had hit upon a scheme 'which has succeeded perfectly' for multiplying the colonies of stingless bees by removing from a hive some of the combs and establishing these in a new hive, which was first perfumed with incense (ST. HILAIRE 1830, I, vol. 2, pp. 370 ff.). The honey of *Trigona jaty* Sm. was much esteemed in S. Brazil, where the species was domesticated. In Northern Brazil *Melipona fasciata scutellaris* Latr. is the bee most favoured for domestication (MARIANNO 1910). J. NIEUHOFF (in PINKERTON 1893, vol. 14, pp. 732 ff.) noticed around 1650 the use of a hollow pipe for the removal of the combs of the bees which nest in the cavities of trees in Brazil.

In addition to domestication, the periodic exploitation of accessible hives in the forests is often habitual. In New Granada parties of honey-gatherers go out in April/May and October/November (J. GOUDOT, 1846). D'ORBIGNY (1839/43 II, pp. 614 ff.) comments on the systematic manner in which the Indians of Chiquitos in Bolivia explored the primeval forest for bees'-nests. Organized bands from each mission would set forth each year from June to Septem-

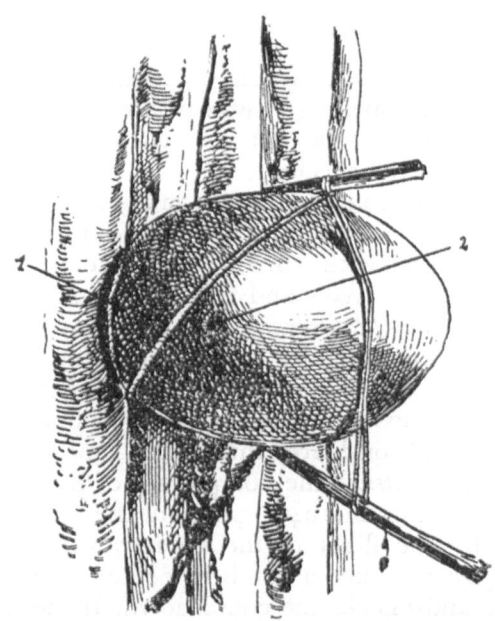

Fig. 47. Hive of the Paressi Indians. 1. Entrance for the bees; 2. Opening for honey extraction. After ROQUETTO-PINTO.

ber, carefully combing a given area for honey and wax. So successful were these foragers that only rarely did such a band of ten to twenty natives fail to bring home the 75 pounds of wax exacted from them as the state tax. The Chiquito Indians, apart from plundering many a nest of stingless bees and destroying them, would often carry off a whole colony for domestication. D'ORBIGNY saw such colonies in several houses in Santa Cruz (Bolivia) housed in jars, GARDENER (1846, pp. 327 ff.) passed by a small village in Goyaz (Brazil) at a season in which the people go to the woods in search of honey. This is so common there, that after passing Duro a portion of honeycomb was presented to him at almost every house where his company stopped. P. SPEGAZZINI (1909) collected 14 species of honeybees in Missiones, where honey gathering is the keenest pleasure enjoyed by the rural peon. For a spoonful of honey he is ready to work an entire day on a tree-trunk and often endangers his life. One cannot estimate the risk which people are prepared to take in the mountains for the sake of honey. All that is needed is that a peon observe a small opening of wax or a cleft in a trunk to make him go immediately for a hatchet and to fell or destroy a beautiful tree of the most valued species.

And around Guerrero in Mexico as many as ten men are known as *mieleros* or honey-gatherers. These band together in the dry season and, equipped with burros (asses) laden down with food supplies, empty bottles, axes and other equipment, scour the un-inhabited mountains for weeks, chopping down bee trees which they have been lucky enough to lacate, and filling their bottles with the honey flowing out of the wrecked nests (HENDRICHS 1941, p. 366). In Cheran (Mexico) gathering of wild honey is the one col-lecting technique still considered important. Especially in late spring and autumn the *panaleros*, the conquerors of honey, spend many weeks on this task. BEALS (1946) speaks of the skill of these men in locating nests by the flight of the bees (possibly wasps such as *Nectarinia lecheguana*): Once a comb is located, the *panalero* usually climbs the tree and knocks the comb down. If the tree is large he cuts notches in the trunk with a small special hatchet, tying himself with a riata or a length of rope, while he chops. The rope is not tied, but passes around the tree trunk and the two ends are held in the hands. The *panalero* holds the rope tightly, while he edges up the tree. Then, with a skillful motion, he throws the loop of rope higher up the tree trunk. Once the comb is reached, the panalero covers his face and hand with a blanket and taps the comb until the bees leave. This procedure is regarded as highly dangerous. *Panaleros* are always careful to take a good rest before climbing a tree. They are also particularly attentive to their saint, San Anselmo. The danger of the occupation is increased by the fact that the *panalero* always works alone and in case of accident could expect no assistance. Among the Chorti of Guatemala also a number of families are almost professional honey-hunters.

DE WAVRIN (1926, p. 19 f.) refers to recognized property rights in wild honey in the Paraguay basin. When a man has located a wild swarm that has not yet reached the peak of honey production, he marks the site and this indication of his claim is sufficient to restrain others from gathering the honey. In due time, accompanied by his wife, he would chop down the tree and collect his property.

Many tribes of S. American Indians were very dependent on the wild honey of stingless bees. The almost monotonous repetition of honey mentioned by G. TESSMANN (1930, pp. 52, 71, 89, 110, 132, etc.) as constituting a food of tribe after tribe of the Indians of N.E. Peru is sufficient indication of its importance as an article of food to this very day.

Some species of *Meliponidae* are sufficiently abundant at New Granada to induce a certain number of natives each year to search for their nests (GOUDOT 1846). At two different seasons they search the forests for the sole purpose of collecting honey and wax, which is hard work to gather. The most abundant stingless bees nest in cavities of hollow trees and to reach them, the trees usually have to be felled. The honey sometimes appears on the market at Bogota, packed into the internodes of bamboo. Within one such vessel the honey of various species is often mixed, as is also the wax offered for sale. The wax is dark and apparently it is impossible to bleach it. The nests are maintained for some years, when they are abandoned for unknown reasons. The large *Melipona fasciata* Latr. produces most of the honey and wax available. Its honey is like a thick greenish-yellow syrup of agreable taste. A normal hive yields 3 liters of honey, in addition to the waste, and 1 kg. wax. This small quantity of honey which is kept in store is due to the absence of a winter in New Granada; the time between the two seasons when the plants are in flower is much reduced, and then the bees are in need of their reserves. Just before these two seasons the hives contain their largest stock of honey. Hence the honey-hunters start on their search in April/May and October/November.

M. REVERET-WATTEL (1875) made various enquiries and collected a number of data on stingless bees for the French Society of Acclimatization. MAXIMILIAN DE WIED-NEUWIED (1820/21. II, p. 49) mentions intensive honey-hunting from stingless bees at Ponte de Gentio (Brazil). A refreshing drink is obtained from this aromatic honey by mixing it with water. Around Arragal da Conquesta the wild Camacan Indians sell honey, which they gather in quantities in the forests. It is one of their favourite foods. The pavillion of the Empire of Brazil at the International World Exhibition at Vienna in 1873 offered a delicious honey of stingless bees and also their soft wax from which Brazilian industry derives great benefit. M.

325

BRUNET informed REVERET-WATTEL that in the interior of Rio Grande du Nord he found an ingenious farmer, who planted around his house a big plantation of papayas. In the trunk of the largest of these trees he had settled in holes many swarms of *urucu* (*Melipona scutellaris*), which thrived in these artificial hives. The *urucu* is almost the only stingless bee which is domesticated in Northern Brazil. Near Bahia the following species were all domesticated by BRUNET who also sent specimens to the Société d'Acclimatisation: *Melipona scutellaris* Latr., *M. marginata*, *M. bilineata* Sm., *M. dorsalis* Sm., *M. atratula* Ill., *Trigona muscaria* G., *T. geniculata*, *T. angustula*, *T. flaveola*.

While little is known of the honey ceremonials of the Indians of the primeval forests, information has been preserved about those of the primitive beekeeping agricultural communities of the ancient Mayas in Yucatan (SCHWARZ 1948, pp. 149 ff.). The priest and his assistants had to prepare themselves for the great bee-rituals in the month of Tzec by fasting, while fasting was merely optional for the other participants. To Hobeil, a special patron of the bee-keepers, and to Bacab, Kanonholkan and other deities offerings were made representing honey. The original purpose of the ceremonies, to bring about abundance of honey, is still conspicuously obvious. Another minor honey ceremony took place in the month of Mol, to appeal to the gods to provide ample flowers for the bees, so that the connection between nectar gathering and honey production was recognized. This is, of course, not surprising in a people generally excelling in keen observation of nature.

REDFIELD and VILLA (1934, pp. 117, 145) give the following information about the Maya village Chom Kon: As with the harvest, so with honey from the hives. For each year's yield man owes a ceremonial offering to the gods, the *u hanli cab* ritual, while the *u hedz luumil cab* ritual is applied when the hives are moved or a new hive is established. The patrons of apiculture will punish with sickness the bee-keeper who does not fulfil them. The *u hanli cab* ceremony or the 'dinner of the bees' as a rule takes place at intervals of four years. On the evening preceding the performance of the rites, the leader of the ritual invites the lords of the bees for the following day. On the altar erected beside the hives, he places a gourd vessel and a maize-dish as offerings. An invocation is then delivered. The public waits for 1 or 2 hours while the gods accept the offering, and the leader drinks much rum. The ceremony on the following morning is much more elaborate and the rituals end with a banquet.

These are all shadowy remnants of older ceremonies. Many other honey ceremonials, including processions with hives, are still preserved among the Christian Indians. These, however, would not have deserved mention here, if they did not go back to much older and more important rituals after the Australian style, which have now disappeared. Certain habits, such as deep incisions above the eyes with a sharp stick until blood flowed before leaving for the honey-hunt, habits of sexual abstention when putting out hives in the forest, etc., indicate such more primitive stages of ceremonial development.

7. THE HONEY-CIVILIZATION OF THE GUAYAKIS

The Guayaki Indians of the primeval forest of Paraguay are still in the stage of food gathering. Their ways of hunting and fishing are still so primitive that they cannot rely on their results for their living. Even trapping is unknown to them. The primeval forest is very poor in nutritive plant products except to some degree for the pindo palm (*Cocos romanzoffiana*). Yet the real base of their nutrition is honey. J. VELLET, in a remarkable monograph (1939, pp. 79–83) goes as far to describe their civilization as a honey-civilization. Apart from bees, this honey is also produced by a number of social wasps (*Polybia, Nectarinia*) which are much feared because of their stings.

European bees, which were introduced into Paraguay long ago, entered the forests with the oranges, and are nesting wild in tree trunks. Their honey is sweeter, less aromatic than that of the native bees. The Guayakis show little appreciation for their yellow wax, and it is not eaten. The honey of the European bees and of the wild wasps, although it is collected, is less appreciated and much less important than the abundant honey and wax of the *Melipona*-bees. The *Melipona* are very small, velvety black bees of plump body and with an atrophied sting, thus making them stingless. When disturbed, they leave their hives in masses, and congregate on the clothes, the ears, the eyes and the hair of the intruder. Some of their species bite and their toxic saliva produces a small red mark for several days. In the cold days of winter they are especially nervous, sometimes attacking even without provocation. Perspiration is rather attractive for them.

Most species of *Melipona*, those which produce the best honey, such as the *yatei*, the *tapesuha* or the *eyra-apua*, build their nests in the trunks of trees. A small, soft wax-tube protrudes over the one opening of the nest, and together with a light oozing of honey, announces to trained eyes the existence of the hives, which may contain even more then ten liters of a sourish, good-tasting honey. This honey crystallizes very badly, in contrast to the wasp-honey. According to the species, this honey varies light brown to yellow, and is contained in the wax cells of the combs. The wax, black and mixed with resin, differs much from the yellow wax'of the European bees. Other *Melipona*-species, such as the *eyra-ybyguy* (soil-bee), nest in the soil, especially on slopes. A small collar of black wax mixed with soil indicates the entrance to the nest. The honey of some of these soil-bees is edible, whilst other species produce a more or less alkaline honey, which is a laxative and believed to be dangerous.

The Paraguayans and the Indians claim that the *Melipona*-honey, and still more that of the wasps, is inebriating; it would be more exact to say, that under seasonal and other, less known influences the slightly stupefactive character of some of these honeys is strenghtened. Certain honeys are always regarded as very poisonous and the Indians never consume them. One spoonful of the honey of a certain wasp, the *lechiguana colorada* (probably a *Nectarinia sp.*), is sufficient to produce a deep coma in an adult man and often even to induce death. The mestizos and the Indians throughout tropical America know these toxic honeys. The botanist SAINT-HILAIRE died in consequence of such a poisonous honey in Central Brazil.

These different honeys, especially those of the *Melipona*, offer an abundant and highly nutritive food to the Guayakis, and are much more important to them than hunting or even vegetable products. The honey forms the basic food of these Indians. Other provisions may be lacking in their camps, but they are always stocked with honey and wax. One group of 15 individuals had seven great jars

328

full of honey, at least 40 liters. The Guayakis do not cease searching for bees for one moment during their ceaseless marches through the forests, and their paths are marked by broken hives.

The usual tools of the Guayakis are few in number and vary but little; they almost all serve for the collection and preservation of honey or for the working of wax. The stone hatchet, their main tool, is used almost solely for opening up the bees'-nests in the tree-trunks. Long ropes are woven from vegetable fibres and animal hairs for climbing up the enormous trunks of the trees and for supporting them at the height of the nests. During a march these ropes are kept rolled around the left arm. To absorb the very liquid honey of the bees they crush palm twigs and transform them into brushes. Their great basket-jars, which are made impermeable by a thick wax layer, are used mainly for transporting and storing the honey. The use of the wax is manifold. It largely replaces clay, as a covering for their basket-jars. Mixed with a coarse clay, a black soil of the swamps rich in organic matter, the wax is used for the fabrication of primitive small pottery ware, the only kind known among the Guayakis. The wax is melted over a low fire, and this molten wax is worked with the aid of wooden spatulas and polished with the shells of *Anodonta*, collected from the forest lagoons. This short lis of tools connected directly with the collecting and the working of honey and wax represents almost the entire material culture of the Guayakis. The wax also protects the bows and the arrows against humidity, it keep the ropes in good condition and it is a component of the rare body dyes of these Indians.

The search for bees'-nests does not depend entirely upon chance. They follow the bees returning to their nest with their crop, or they observe the direction from which a number of bees arrive, in order to calculate the approximate position of a nest. Many *Melipona*-nests which have been robbed are not abandoned by the bees. They are closed up again after the honey has been removed and can be visited periodically. The nests in the tree-trunks are opened with the stone hatchet. Whether or not they are previously fumigated depends on the agressiveness of the various species. The Guayakis climb up the heavy trunks with the greatest agility, helped by the above mentioned 10 to 20 meters long ropes, which they utilize in many ways, the hatchet being kept on the wrist by a lash of hide . . . . These scalings are not without danger, especially for isolated men. A story is often heard in Paraguay, that a Guayaki was found dead, high up on a tree, his hand fixed in the nest and his arm broken. NORDENSKIÖLD has reported a similar tale from the Cavina Indians of Bolivia. Such accidents can occur in all forests where the Indians are honey-hunting.

The honey is separated from the wax by manual pressure and kept in the basket-jars we have described. It is then called '*bwe eyetükwé*', i.e. that which comes from the bees. The honey is eaten with the help of a brush made from crushed palm twigs. It has been suggested that this habit, which is widespread amongst forest Indians may also be used for collecting honey from the nests, without opening them. The Guayakis, however, use it only for taking the honey from their jars.

The wax of the *Melipona*, blackish and rich in resin, is kneaded or chewed, and then melted at a low fire, in small black earthenware pots, after which it is made into fist-sized balls. When used, it is melted again, and, according to the purpose, is mixed with soil or with vegetable resins. This wax is very valuable for the poor Guayakis, as well as for the farmers of Paraguay, who use it in many ways. When the latter surprise a group of Guayakis, they always take their honey and their wax.

Other Indian tribes also eat honey and use wax on a large scale, but their existence and their material culture are not based almost entirely on these products. The Guayakis almost form a honey-civilization. This very old and very primitive form of forest civilization has been developed and maintained in the vast forests of Eastern Paraguay owing to the isolation of these Indians, who have been refused entrance into richer regions, first by Indians of higher culture, later by the Spaniards and the Paraguayans. This type of civilization exists in America only among a few forest tribes, such as the Guayakis, who are little known and have remained nomads.

VII. BIBLIOGRAPHY

K. E. Abbot, On the poisonous honey from Trebizond, described by Xenophon. Proc. Zool. Soc. London. 1834. p. 50.

M. Adanson, Histoire Naturelle du Sénégal. Relation d'un Voyage fait en 1749/53. Paris. 1757. p. 82, 88.

E. Adler, Nutritive Value of Locusts. Farming in S. Africa. 9. 1934. p. 232.

J. Leo Africanus, Descriptio totius Africae. Anvers. 1556. German Transl. Herford. 1805.

J. d'Aguilar, L'Entomophagie. La Nature. 1941. p. 114–117.

J. d'Aguilar, La Gastronomie orientale fait grand cas des Insectes. Science et vie (Paris) 79. Febr 1951 no. 401 p. 141–144.

J. M. Aldrich, Flies of the Leptiid genus *Atherix* used as food by the California Indians. Entom. News 23. 1912. p. 159–163.

J. M. Aldrich, Larvae of a Saturniid moth used as food by California Indians. Journ. New York Entom. Soc. 20. 1920. p. 28–31. pl. I.

Ulysse Aldrovandi, De animalibus Insectis libri septem. Bologna. 1602.

W. W. Alpatov, Bull. Soc. Nat. Moscou, Sect. Biol. 44. 1935. p. 284–292.

J. B. Ambrosetti, Los Indios Caingua del Alto Parana (Misiones). Bol. Inst. Geogr. Argentino. 15. 1894. p. 699 f.

L. Ancona, Los Jumiles de Taxco, *Atizies taxcoensis n. sp.* Anal. Inst. Biol. Univ. Mexico. 3. 1932. p. 149–162.

L. Ancona, El Ahuautl del Texcocco. Anal. Inst. Biol. Univ. Mexico. 4. 1933. p. 51–69.

L. Ancona, Los Jumiles de Cuautla, *Euschistus zopilotensis* Dist. Anal. Inst. Biol. Univ. Mexico. 4. 1933. p. 103–108.

C. R. Anderson, Lake Ngami; or Explorations and Discoveries during four years wandering in S.W. Africa. New York. 1856. p. 284.

G. F. Angas, Savage Life and Scenes in Australia. 2 vol. London. 1847 I, p. 57.

G. F. Angas, South Australia illustrated. London. 1847.

G. F. Angas, The Kafirs. London. 1849.

N. Annandale, Proc. Zool. Soc. 1900. p. 859.

L. Armbruster, Die Biene im Orient. I. Der über 5000 Jahre alte Bienenstand Aegyptens. Arch. f. Bienenkde. 12. 1931.

L. Armbruster, Die Biene im Orient II. Bibel und Biene. Arch. f. Bienenkde. 13. 1932. p. 1–43.

W. Arndt, Bemerkungen über die Rolle der Insekten im Arzneischatz der alten Kulturvölker. D. Ent. Zeitschr. 1923. p. 553.

Ch. Auffret et F. Tanguy, Note sur la valeur alimentaire des termites. Bull. Médic. d'A. O. F. 1947/8 p. 395–396.

F. DE AZARA, Voyages dans l'Amérique méridionale. 4 vol. Paris. 1809.

H. AUHAGEN, Beitraege zur Kenntnis der Landesnatur und Landwirtschaft Syriens. Berlin. 1907.

J. BACKHOUSE, Narrative of a Visit to Mauritius and S. Africa. London. 1844. p. 584.

S. W. BAKER, Eight years' wandering in Ceylon. New York. 1881.

A. BALDANUS, Locustae majores, quibus Johannes in deserto vitam tolerare dicitur Commentar. Bononienses 5. 1767 p. 53–54.

E. BANCROFT, Essay on the Natural History of Guiana in S. America. London. 1769. p. 230.

P. BARGAGLI, Insetti Commestibili. Lettura nell. Soc. Ent. Ital. Rivista Europaea-Riv. Internaz. Fasc. 16. 6. 1877. Reprint 11 pp.

H. BARTH, Travels and Discoveries in North and Central Africa 1849/55. 5 vol. London. 1857/58. II, p. 30.

B. S. BARTON, An inquiry into the question, wether the Apis mellifica or true honey bee is anative of America. Trans. Amer. Soc. 3. 1793. p. 241–261.

H. BASEDOW, The Australian Aboriginal . Adelaide, 1925, p. 122 ff., 153, 145 ff., 283 f.

A. BASTIAN, Indonesien, Timor und umliegende Inseln. Berlin. 1885. p. 4.

O. BAUMANN, Beitraege zur Ethnographie des Congo. Mittheil. Anthrop. Ges. Wien. 17. 1887. p. 160–181.

R. L. BEALS, Cheran, a Sierra Tarascan village. Inst. Soc. Anthrop. Smithsonian Inst. Publ. no. 2. 1946. p. 13, 126, 141.

O. BECCARI, Wanderings in the great forests of Borneo. London. 1904. p. 106, 161.

B. F. BECK and D. SMEDLEY, Honey and your health. London. 1947.

E. T. BENNETT, Mexican Bees: Some account of the habits of a Mexican bee. In: Beechey, F. W., Narrative of a voyage to the Pacific and Berings Strait. London. vol. 2. 1831. p. 357 ff, 613 ff. Partial reprint in Amer. Bee Journ. 4. 1868. p. 24–26.

G. BENNETT, Wanderings in New South Wales, Batavia, Pedis Coast, Singapore and China, being a journal of a naturalist. 2 vol. London .1834.

G. BENNETT, Gatherings of a Naturalist in Australasia. London. 1860.

G. BENNETT, On the physiology, utility and importance of the acclimatization or naturalisation of animals and plants to Australia. Ann. Rept. Acclim. Soc. N.S. Wales 3. 1864. p. 18–36.

J. BEQUAERT, Insect as Food. How they have augmented the food supply of Mankind in early and recent times. Nat. Hist. Journ. Amer. Mus. Nat. Hist. 21. 1921. p. 191–200, 8 fig.

J. BEQUAERT, The predacious enemies of Ants. Bull. Amer. Mus. Nat. Hist. 45. 1922, p. 329–331.

J. C. BEQUAERT, Appendix to T. D. A. COCKERELL, Bull. Brooklyn Ent. Soc. N.S. 27. 1932. p. 17.

H. P. BERENSBERG, The uses of insects as food delicacies, medicines, or in manufactures. Natal Journ. and Min. Rec. 10. 1907. p. 757–762, 1 pl.

L. BERLAND, Les Insectes et l'homme. Paris. 1942. p. 112.

H. A. BERNATZIK, Owa Raha. Wien-Leipzig-Olten. 1936. p. 71 ff., 91.

332

M. Bertaut, Contribution a l'étude des Negrilles de la Région du Haut-Nyong. Bull. Soc. Etud. Cameroun. 1943, no. 4. p. 73–91.

A. de W. Bertoni, Contribucion a la biologia de las avispas y abejas del Paraguay. An. Mus. Nac. Hist. Nat. Buenos Aires. 22. 1911. p. 99, 138 ff.

J. G. Bessler, Geschichte der Bienenzucht. Stuttgart. 1886.

H. Bigg, On a species of Bee from the Brazils, found living on splitting a log of peach wood containing its comb. Proc. Zool. Soc. London 2. 1834. p. 118.

C. T. Bingham, The Aculeate Hymenoptera procured at Aden by Yerbury and Nurse. J. Bombay Nat. Hist. Soc. 12. 1898. p. 114.

C. I. Bingham, Fauna of British India: Hymenoptera II. London. 1903. p. 311.

G. M. de Bione, Des Abeilles dans l'Inde. L'Apiculteur. 4. 1860. p. 141–142.

L. Biro, Beschreibung der ethnographischen Sammlung Ludwig Biro's aus Deutsch Neu Guinea. Budapest. 1899. p. 247.

R. Blanchard, Traité de Zoologie Médicale. Paris. 1890. vol. II, p. 545 ff.

W. H. Bleek and C. L. Lloyd, Bushman Folklore. London. 1911. p. 353.

A Blunt, A Pilgrimage to Nejd. London. 1881. 2 vol.

S. Bochart, Hierozoicon sive bipartitum opus de animalibus S. Scripturae. 2 vol. in one. Frankfurt. 1675.

F. S. Bodenheimer, Über das Tamariskenmanna des Sinai. In: Bodenheimer und Theodor, Ergebnisse der Sinai-Expedition 1927. Leipzig. 1929. p. 45–89.

F. S. Bodenheimer, Die Schaedlingsfauna Palaestinas. Berlin. 1930. p.

F. S. Bodenheimer, The Honeybee in Ancient Palestine. The Bee World. 15. 1934. p. 134–135.

F. S. Bodenheimer, Studies on the Honey-Bee and Bee-Keeping in Turkey. Ankara. 1942. p. 45–50.

F. S. Bodenheimer, The Manna of Sinai. The Bibl. Archaeol. 10. 1947. p. 2–6.

J. Bontius, Historiae naturalis et medicae Indiae orientalis libri VI. Amsterdam. 1658.

J. Bonwick, Daily Life and Origin of the Tasmanians. 2nd ed. 1898. p. 17.

de Bourboulon, Voyage du Schang-Hai à Moscou.

G. Bouvier, Quelques questions d'entomologie vétérinaire en Afrique tropicale. Acta Tropica 2. 1945. p. 42–59.

A. Bréhion, Utilisation des Insectes en Indochine. Préjugés et moyens de défense contre quelques-uns d'entre eux. Bull. Mus. Nat. d'Hist. Nat. Paris. 19. 1913. p. 277–281.

A. E. Brehm, Reiseskizzen aus Nord-Ost-Afrika .... in 1847–1852. 3 vol. Jena. 1853/55.

H. Breuil, P. S. Gomez et J. G. Aguilo, Les peintures rupestres d'Espagne. Les abris del Bosque à Alpera (Albacete). L'Anthropologie. Paris. 23. 1912. p. 529.

M. Briault, Les Sauvages d'Afrique. Paris. 1943.

W. S. Bristowe, Insects and other Invertebrates for Human Consumption in Siam. Trans. Ent. Soc. London. 80, II. 1932. p. 387–404.

333

R. Brooks, On the properties and uses of Insects. Vol. IV. of: A new and accurate system of Natural History. Vol. 4. 2nd ed. 1772.

T. D. Broughton, Letters written in a Mahratta Camp during 1809. London. 1813.

B. Brown, A caterpillar fungus from New Zealand. Proc. Boston Soc. Nat. Hist. 3. 1850. p. 341.

A. R. Brown, The Andaman Islanders. London. 1922. p. 183, 152 f., 357, 104 ff..

W. G. Browne, Travels in Africa, Egypt and Syria. in 1792/98. London. 1799.

W. G. Browne, Nouveau voyage dans la Haute et Basse Egypte, La Syrie, Le Dar-Four (1792–1798). Paris. 2nd ed. 1800. 2 vol. II, p. 31 f.

J. Bruce, Travels to discover the source of the Nile in 1768/73. 5 vol. Edinburgh. 1790.

C. T. Brues, Insect Dietary. Cambridge. 1946. p. 419–422.

H. A. Bryden, Wild Life in South Africa. London. 1936. p. 215.

E. Brygoo, Essai de Bromatologie Entomologique: Les Insectes Commestibles. Thèse de Doctorat. Bergerac. 1946.

F. Bryk, Termitenfang am Fusse des Mount Elgon. Entom. Rundschau 44. 1927. p. 1–3 .

F. Bryk, Termiten und Negerleben. Völkerkunde, Wien. 1929. 5. Heft 7–9. p. 150–163 Heft 10.

F. Buchanan, A. journey from Madras through the countries of Mysore, Canara and Malabar. 3 vol. London. 1807.

Buchner, Contributions to the Ethnography of the Bantus.

W. J. Burchell, Travels in the interior of Southern Africa. London. 1822.

C. L. van der Burg, De Voeding in Nederlandsch-Indië. Amsterdam. 1904.

M. Burr, The Insect Legion. London. 1939. p. 208–217, 263–266.

G. W. Bury, The Land of Uz. London. 1911. p. 311 ff.

G. W. Bury, Arabia Infelix. London. 1915. p. 44.

H. v. Buttel-Reepen, Apistica. Beitraege zur Systematik, Biologie, sowie zur geschichtlichen und geographischen Verbreitung der Honigbiene. Mitteil. Zool. Mus. Berlin. 3. 1906. p. 117–201.

A. Caldcleugh, Travels in South America. 2 vol. London. 1825.

A. F. Calvert, The Aborigines of Western Australia. London. 1894. p. 17, 24, 28 f.

R. P. Camboué, Les sauterelles à Madagascar sur le riz malgache. Bull. Soc. Nat. d'Acclimat. de France 33. 1886. p. 168–172.

R. P. Camboué, Les sauterelles à Madagascar. Bull. Soc. Nat. d'Acclimat. de France 35. 1888. p. 794–797.

V. L. Cameron, Across Africa. 2 vol. London. 1877.

J. Campbell, Travels in South Africa, undertaken at the request of the Missionary Society. 3 vol. London. 1815–1822.

T. G. Campbell, Insect Foods of the Aborigines. The Australian Museum Magazine. 2. 1926. p. 407–410.

D. W. Carnegie, Spinifex and Sand. A narrative of five years pioneering and exploration in Western Australia. London. 1898. p. 308.

G. D. H. CARPENTER, Experiments on the Relative Edibility of Insects, with special reference to their Colouration. Trans. Ent. Soc. London. 1921. p. 1–105.

G. CASATI, Ten years in Equatoria. (Germ. ed. 2 vol. Bamberg, 1891).

J. CASTLES, Observations on the Sugar Ants. Philos. Transact. 80. 1790. p. 346–358.

M. CHANIER, Account of the Empire of Morocco. London. 1788.

J. P. CHAPIN, Profiteers of the busy bee. Observations on the honey guides of Africa. Nat. Hist. New York. 24. 1924. p. 328–336.

J. P. CHAPIN, The Birds of the Belgian Congo. Part II. Bull. Amer. Mus. Nat. Hist. New York. 1939. p. 549 ff.

R. E. CHEESMAN, Lake Tana and the Blue Nile. London. 1936. p. 64, 355.

E. CHINA, An interesting relationship between a crayfish and a waterbug. Nat. Hist. Mag. London. 1931. p. 57–62.

LE CHEVALIER CHARDIN, Voyages en Perse. Amsterdam. 10 vol. 1711. IX, p. 227.

E. COBBOLD, Kenya: The Land of Illusion. London, 1935. p. 36 ff., 54.

P. COLLINSON, Some observations on the Cicada of North America. Philos. Transact. 54. 1764. p. 65-69, 1 pl.

C. COLUMBUS, The discovered islands. A letter from Columbus. Encore. 9. 1946. p. 355.

E. CONZEMIUS, Die Rama-Indianer von Nicaragua. Zeitschr. Ethnol. 59. p. 314 f.

H. B. COTT, Adaptive Coloration in Animals. London. 1940.

CONSETT, Travels in Sweden, London. 1789. p. 118.

F. COWAN, Curious Facts in the History of Insects. Philadelphia. 1865.

C. E. V. CRAUFURD, Land of Ophir. London. 19 n.d. p. 110.

R. G. CUMMING, Five years of a hunter's life in the far Interior of South Africa. Paris. 1850.

P. M. CUNNINGHAM, Two years in New South Wales. 2 vol. London. 1827. I, p. 329.

C. H. CURRAN, Insect Lore of the Aztecs in Revealing early acquaintanceship with many of our agricultural pests and therapeutic measures. Nat. Hist. (New York) 39. 1937. p. 196–203.

C. H. CURRAN, On eating insects. Nat. Hist. (New York). 43. 1939. p. 84–89.

A. CUTHBERTSON, Note on the swarming of Pentatomid-bugs. Nada, Ann. Native Aff. Dept. S. Rhodesia. 1934. p. 38.

G. L. C. F. CUVIER, Animal Kingdom. 16 vol. London. 1827/35. Insecta II, p 205.

E. DAGUIN, Les Insectes comestibles dans l'antiquité et de nos jours. Le Naturaliste. 1900, Reprint 27 pp.

AD-DAMIRI, Hayat al-Hayawan. Transl. by A. S. G. Jayakar. Vol. I. London. 1906.

W. DAMPIER, An account of a new Voyage round the world. London. 1697.

E. DARWIN, Phytologia, or the Philosophy of Agriculture and Gardening. London. 1800. p. 364.

S. DAS, Locusts as food and manure. Indian Farming. 6. 1945. p. 412.

E. DAUMAS et A. de CHANCEL, Le Grand Désert. Paris. 1848.

R. H. DAVIES, On the Aborigines of van Diemen's Land. Tasmanian Journ. Nat. Sci., Agric, Statistics, etc. 1846. p. 409–420.

J. DAWSON, Australian Aborigines. 1881. p. 20.

G. DEBATISSE, L'entomophagie. Natur. Amat. Verviers. 3,6. 1946. 3 pp.

R. DECARY, L'entomophagie chez les indigènes de Madagascar. Bull. Soc. Ent. France 42. 1937. p. 168–171.

R. DECARY, La Fauna Malgache. Paris. 1950. p. 151 ,f., 160, 172 f.

[J .H. DEGNER,] Von einem Knaben, der durch den Genuss eines ganzen May-kaefers getötet worden. Götting. Gelehrt. Anz. 1778 p. 721.

J. DEMOLE, Les Sauterelles dans la Russie Méridionale. Bibl. Univ. Genève 31. 1856. p. 218–229.

H. R. P. DICKSON, The Arab of the Desert. A glimpse into badawin life in Kuweit and Saudi Arabia. London. 1949.

DIODORUS THE SICILIAN, Historical Library in fifteen books. Transl. G. Booth. 2 vol. London. 1814.

W. L. DISTANT, Tenasserim. 1852. p. 385.

W. L. DISTANT, A monograph of Oriental Cicadidae. London. 1889/92. p. 22.

W. L. DISTANT, Fauna of British India: Rhynchota I. London. 1902. p. 283.

F. DOFLEIN, Ostasienfahrt, Leipzig. 1906.

E. DONOVAN, Natural History of the Insects of China. London. 1842. p. 6.

S. S. DORNAN, Pygmies and Bushmen of the Kalahari. London. 1925.

C. M. DOUGHTY, Travels in Arabia Deserta. London. 1926. Two vol. in one.

P. DUDLEY, An account of a method lately found out in New-England for dis-covery where the Bees hive in the woods in order to get their honey. Philos. Transact. 31. 1721 p. 148–150.

C. A. EALAND, Insects and Man. London. 1915. p. 203–214.

J. E. ECKERT, Beekeeping in California. Cal. Agric. Extens. Serv. Circ. 100. 1936.

C. C. EDGAR, Zenon Papyri. Vol. III. Cairo. 1928.

H. ELTRINGHAM, An account of some experiments on the edibility of certain lepidopterous larvae. Trans. Ent. Soc. London. (4). 1909.

G. P. ENGELHARDT, The Saturniid moth *Coloradia pandora*, a menace to pine forests and a source of food to Indians in Eastern Oregon. Bull. Brooklyn Ent. Soc. 19. 1924. p. 35–37.

T. ESAKI, Notes on entomophagous habits (in Japanese). Bull. Takarazuke In-sectarium. 27. 1942. p. 1–8.

K. ESCHERICH, Die Ameise. 2nd ed. Braunschweig. 1917. p. 150 ff.

E. O. ESSIG, The value of insects to the Californian Indians. Scient. Mo. 38 (2). 1934. p. 181-186.

I. H. N. EVANS, Folk stories of the Tempas-uk and Tuatan Districts of British North Borneo. Journ. R. Anthrop. Instit. 43. 1913.

I. H. H. N. EVANS, Among primitive People in Borneo. London. 1922. p. 99, 165.

E. EYLMANN, Die Eingeborenen der Kolonie Südaustralien. Berlin. 1908.

FIDE EYLMANN: G. Taplin, The Narrinyeri.

J. H. FABRE, Souvenirs Entomologiques. 10 vol. Paris. 1922/24. V, p. 262 ff; X, p. 103 ff.

M. M. FAGAN, The uses of insect galls. Americ. Nat. 52. 1918. pp. 157–176.

336

J. C. Faure, Pentatomid bugs as human food. J. Ent. Soc. S. Africa. 7. 1944. p. 110/2.

Rd. Favre, An account of the wild tribes inhabiting the Malayan Peninsula. Paris. 1865.

H. Jaques-Félix, Ignames sauvages et cultivées. Bull. Soc. Etud. Cameroun. 1948, no. 21–22. p. 13–18.

P. Fermin, Déscription générale, historique, geographique et physique de la colonie de Surinam. Amsterdam. 2 vol. 1769. II, p. 300 f.

E. B. Fladung, Insects as food. Maryland Acad. of Sci. Bull. October 1924 .p. 5–8.

F. Fleming, Kaffraria. London. 1853. p. 80.

J. Forbes, Oriental Memoirs. 4 vol. London. 1813. I, p. 46.

J. Ford, Borak and belles. In: T. Harrison, Borneo Jungle. London. 1938. p. 67, 239.

A. Forel, Une nouvelle fourmis melligère. Ann. Soc. Ent. Belg. 39. 1895. p. 429.

A. Forel, The social world of the ants, compared with that of man. 2 vol. London and New York. 1928. I, p. 458 ff.

G. M. Foster Jr., Indigenous apiculture among the Popoluca of Mexica. Amer. Anthrop. N. S. 44. 1942. p. 538–542.

Foucher d'Obsonville, Essais philosophiques sur les moeurs de divers animaux étrangers. Paris. 1783. p. 43 ff.

J. Fraser, The Aborigines of New South Wales. 1892. p. 53.

J. G. Frazer, Totemism and Exogamy. 2 vol. London. 1910. I, p. 24, 228. II, p. 242, 292, 315, 428.

F. Freeman, History of Cape Cod. Vol. II. p. 524.

W. W. Froggatt, Honey Ants. In: Horn Scient. Exped. Centr. Australia. 1896. vol. II.

W. W. Froggatt, Forest Insects of Australia. Sydney, 1923. p. 9.

S. W. Frost, General Entomology. New York. 1942. p. 63 f.

A. J. Fynn, The American Indian as a product of environment. Boston. 1907. p. 87 ff.

Th. Gage, Nouvelle Rélation autant les Voyages dans la Nouvelle Espagne. 2 vol. Amsterdam. 1721. I, p. 144.

H. Gahne, Stingless bees; some of the crude ways in which bees were kept in British Honduras. Gleanings in Bee Culture. 30. 1902. p. 947.

G. Gardner, Travels in the interior of Brazil. London. 1846. p. 327 ff.

Y. Gardon, Le dernier blanc. Paris. (about 1949).

G. Garniere, Viaggio alla Nuova Caledonia. Giro del Mondo. vol. 17.

J. de Gaulle, Les Insectes Comestibles. Feuille des Jeunes Naturalistes .3. 1873. p. 125–127.

W. R. Gerard, Insetcs as articles of food. Proceed. Poughkeepsie Soc. Nat. Hist. 1. 1876. p. 17–31.

J. Ghesquière, Les insectes palmicoles comestibles. In P. Lepesme, Les Insectes des Palmiers. Paris. 1947.

C. C. Ghosh, A few insects used as food in Burma. Rept. Proceed. 5th Entom. Meeting, Pusa, 1923. Calcutta. 1924. p. 403/5, Pl. 36.

K. Giesenhagen, Auf Java und Sumatra. Leipzig. 1902. p. 78.

F. GIRAL, Sobre aceites de insectos I. Rev. Soc. Mex. Hist. Nat. 2. 1941. p. 243–250.

J. F. and M. L. GIRAL, Sobre aceites de insectos II. Ciencia, Mexico. 4. 1943. p. 155–156.

M. J. A. GIRARD, Note sur les moeurs des Mélipones et Trigones du Brésil. Ann. Soc. Ent. Fr. 1874. p. 567–573.

ABBÉ GODARD, Déscription et Histoire de Maroc. Paris. 1860.

O. GOLDSMITH, A history of the earth and animated nature. 2nd ed. Philadelphia. vol. 5 1824. p. 146.

F. L. DE GOMARA, The pleasant history of the conquest of the West India, now called New Spayne. London. 1578. Quoted from Facsimile-edition New, York 1940. p. 200.

J. GOUDOT, Observations relatives à l'histoire des Méliponites. C. R. Acad. Sci. Paris. 22. 1846. p. 710-713.

P. GOUROU, Les Paysans du Delta Tonkinois. Paris. 1936.

P. GOUROU, La Terre et l'Homme en Extrème Orient. Paris. 1940.

P. GOUROU, Les Pays Tropicaux. Principes d'une géographie humaine et économique. Paris. 2nd ed. 1948. p. 76–88.

A. et G. GRANDIDIER, Collection des Ouvrages anciens concernant Madagascar. Paris. A series of compendious volumes, beginning about 1902.

J. GRAY, Notes on a native tribe formerly resident at Orroroo, S. Australia. S. Austral. Natur. 12. 1930. p. 4–6.

F. E. GUÉRIN-MÉNEVILLE, Mémoire sur trois espèces d'insectes hémiptères dont les oeufs servent à faire une sorte de pain nommé Hautlé au Mexique. Bull. Soc. Imp. Zool. d'Acclimat. 4. 1857. p. 578–581.

N. J. B. G. GUIBOURT, Notice sur une matiere pharmaceutique nommé Tréhalose, produite par un charançon, avec une notice sur cette nouvelle espèce de sucre par M. BERTHELOT. C. R. Acad. Paris. 46. 1858. p. 1213–1217.

K. GÜNTHER, A naturalist in Brazil. London. 1931. p. 296 ff., 376.

H. B. GUPPY, The Solomon Islands and their natives. London. 1887. p. 93 f.

B. GUTMANN, Bienenzucht bei den Wadschagga. Globus. 96. 1909. p. 205–207.

H. HAGEN, Die Insekten Namen der Tupi Sprache. Stett. Ent. Ztg. 24. 1863. p. 252–259.

H. H. A. HAGEN, Bibliotheca Entomologica. Leipzig 1862/63.

B. HAGEN, Unter den Papuas. Wiesbaden. 1899.

HAKLUYTUS see: S. PURCHAS.

A. HALE, J. Anthrop. Inst. Gr. Brit. and Ireland. 15. 1886. p. 295, 298.

B. HALL, Extracts from a journal written on the coasts of Chili, Peru, and Mexico in the years 1820/22. 2nd ed. Edinburgh. 2 vol. 1824. II. p. 224 ff.

P. HAMBRUCH, Malaiische Maerchen. Jena. 1922.

H. HAMET, Manière de gouverner les abeilles dans l'Inde. C. R. Acad. Paris. 2. 1858. p. 37–38.

A. HANNEMAN, Die stachellosen Bienen. Bienen Zeitg. 28. 1872. p. 206–208.

G. HARDY et CH. RICHET, fils, L'Alimentation indigène dans les Colonies Françaises. Paris. 1933.

T. Hardwicke, Description of a substance called Gez or Manna, and the Insect producing it. Asiat. Research. 14. 1822. p. 182–186.

P. W. Harrison, The Arab at Home. London. 1924. p. 289.

Hartman, The Indians of North-western Mexico. Congrès Internat. Américanistes. Stockholm. 1894. p. 128.

F. Hasselquist, Om graesshoppers nytta till föda hos Araberne. Svensk. Acad. Handl. 13. 1752. p. 76–79.

F. Hasselquist, Voyages and Travels in the Levant in the years 1749/52. London. 1766.

R. Heber, Narrative of a journey through the upper provinces of India from Calcutta to Bombay. 2 vol. London. 1828.

S. Hedin, Jerusalem. Leipzig. 1918. p. 270–279.

E. Hegh, Les Termites. Bruxelles. 1922.

J. W. Helfer, Schriften über die Tenasserim-Provinzen, den Mergui Archipel und die Andamanen. Mittheil. Geogr. Ges. Wien. 3. 1859. p. 167–390 (p. 266, 356).

W. R. T. Hemsted, Locusts as a Protein Supplement for Pigs. East Afric. Agric. Journ. Apl. 1947. p.225-226.

P. R. Hendrichs, E. cultivo de abejas indigenas en el Estado de Guerrero. Mexico Antiguo. 5. 1941. p. 365–373, figs. 1–4.

F. Hernandez-Pacheco, Escena pictoria con Representaciones de Insectos da Epoca Paleolitica. Bull. R. Sc. Espan. Hist. Nat. Madrid. 50. 1921. p. 62–77.

F. G. Hernandez, Nova plantarum, animalium et mineralium Mexicanorum historia. Rome 1648–1651.

A. Herrera, Observacions acera de la Hormiga de Miel Myrmecocystus melliger Wes. La Naturaleza. Mexico. 7. 1885/86. p. 1–10. Transl. from J. Künckel d'Herculais, Merveilles de la Nature.

J. J. Hess, Von den Beduinen des Inneren Arabiens. Zürich und Leipzig. 1938. p. 118.

A. F. C. P. de Saint-Hilaire, Voyage dans les provinces de Rio de Janeiro et de Minas Geraes. In: St. Hilaire, Voyages dans l'intérieur du Brésil. Paris. Pt. 1, vol. 2. 1830. p. 370 ff.

S. P. Hildreth, Notices and observations on the American cicada or locust. Amer. Journ. Sci. and Arts. 18. 1830. p. 47–50.

C. W. Hobley, Bantu Beliefs and Magic. London. 1922. p. 251 ff.

C. W. Hobley, Further Researches into Kikuyu and Kamba Beliefs. Journ. Anthrop. Instit. 41. 19 . p. 412.

H. J. Hockings, Notes on two Australian species of Trigona. Trans. Ent. Soc. London. 1884. p. 149–157.

W. E. Hoffmann, Insects as Human Food. Proceed. Ent. Soc. Washington. 49. 1947. p. 233/7.

A. C. Hollis, The Nandi. London. 1909. p. 7.

J. Holman, Travels in Brazil, Cape Colony, etc. 2nd ed. London. 1840. p. 487.

V. M. Holt, Why not eat insects? London. 1885.

F. W. Hope, Observations respecting various insects which at different times have offered food to Man. Trans. Ent. Soc. London. 1842. p. 129–150.

Y. Horikawa, Animals used as medicine in Formosa. (In Japanese). Contrib. Extra. Dept. Hyg. Res. Inst. Formosa. 84 pp.

W. Horn und S. Schenkling, Index Litteraturae Entomologicae. Serie I: Die Welt-Literatur über die gesamte Entomologie bis inklusive 1863. 4 vol. Berlin-Dahlem. 1928/29.

C. Hose, Natural Man. A record from Borneo. London. 1926. p. 113.

C. Hose, The field book of a Jungle-Wallah. London. 1929.

E. Hovelacque, La Chine. Paris. 1920. p. 33.

L. O. Howard, (Note on edibility of Cicada septemdecim). Proc. Ent. Soc. Washington 1. 1886. p. 29.

L. O. Howard, The edibility of Insects. Journ. Econ. Ent. 8 (6). 1915.

L. O. Howard, Lachnosterna larva as a possible food supply. J. Econ. Ent. 9. 1916. p. 389–392.

A. Hrdlicka, Physiological and Medical Observations among the Indians of S. W. United States and N. Mexico. Smithsonian Inst., Bureau of Ethnol. Bull. 34. Washington. 1908. p. 25, 264.
Contains a number of bibliographical indications, which could not be looked over. We mention the most important ones: J. Baegert (1864/5), C. M. Buchanan (1899), L. Carr (1895), M. Eells (1877), P. Ehrenreich (1891), W. M. Gabb (1875), G. Gibbs (1877), F. E. Grossmann (1873), A. Hrdlicka (1904), J. G. Kohl (1860), F. Krause (1907), W. J. McGee (1898), S. Parker (1844), F. Russell (1908), M. C. Stevenson (1901/2), J. G. Swan (1870), J. R. Swanton (1905), J. Teit (1900), M. L. Turner (1889/90), T. Waitz (1862).

J. P. Huber, Notice sur la mélipone domestique, abeille domestique mexicaine. Mém. Soc. Phys. Hist. Nat. Genève. 8. 1839. p. 1–26, pls. 1–3.

J. E. Hughes, Eighteen years on Lake Bangweulu. London. 1933. p. 19, 158 ff., 184, 335.

F. H. A. von Humboldt, Essai politique sur le royaume de la Nouvelle Espagne. in: Humboldt et Bonpland, Voyage aux régions équinoxiales.., Paris. Pt. 3, vol. 3 1811. p. 240 ff.

J. H. Hutton, The Sama Nagas. London. 1921. p. 72.

J. H. Hutton, The Angami Nagas. London. 1921. p. 236.

C. Ichikawa, Biochemical studies on the locust (note 1 to 4). J. Agric. Chem. Soc. Japan 12. 1936 p. 408–414; 710–712; 14. 1948 p. 43–44.

H. v. Ihering, Biologie der stachellosen Honigbienen Brasiliens. Zool. Jahrb., Syst. 19. 1904. p. 179–287, fig. A-H, pls 10–22.

K. Illiger, Die essbaren Insekten und eine neue Art von Spinnen. (Illiger's) Magaz. für Insektenkunde. 3. 1804. p. 207–214.

Immanuel Kant, Physische Geographie. Leipzig. 1905. p. 238 (II, 6, Ic).

H. Ingrams, Arabia and the Isles. London. 1942. p. 172 ff.

W. Irving, A tour in the prairies. London. 1835.

T. Itikawa, Biochemistry of the locust. Nat. Sci. and Mus. Tokyo 11 (7) 1940. p. 4–11. (In Japanese).

J. G. Jackson, Account of the Empire of Marocco, and Districts of Suse and Tafilelt. 2nd. ed. London. 1809. p. 104.

J. de Joannis, Les insectes comestibles au Tonkin. Révue d'Hist. Nat. 10. 1929. p. 339–343.

T. A. Joyce, Mexican Archaeology. London. 1914. p. 243.

Joyeux et Sicé, Précis de Médicine Coloniale, 2nd. ed. Paris. p. 7.

W. Junker, Travels in Africa during 1879–1883. London. 1891.

H. H. Junod, Life of a South African Tribe. London. 1912. 2 vol. I, p. 239, 363, II, p. 319.

A. Kaiser, Der heutige Stand der Mannafrage. Mittheil. Thurgau. Naturf. Ges. Heft 25. 1924.

A. Kaiser, Neue naturwissenschaftliche Forschungen auf der Sinai-Halbinsel. Zeitschr. D. Pal. Ver. 1930. p. 63–75.

D. and R. Katz, Some Problems concerning the Feeding Behaviour of Monkeys. Proc. Zool. Soc. London. 1936. p. 579–582.

G. A. Keferstein, Über den unmittelbaren Nutzen der Insecten. Erfurt. 1827. 104 pp.

G. A. Keferstein, Vom giftigen Honig. Eine Entomologische Skizze. Wissensch. Ber. Erfurt. Akad. 1. 1853. p. 134–148.

O. Keller, Die Antike Tierwelt. Leipzig. vol. I. 1909. vol. II. 1913.

C. R. Kellogg, Beekeeping in Mexican villages. Amer. Bee. Journ. 85. 1945. p. 356.

Saville Kent, The Naturalist in Australia. London. 1897. p. 253.

A. Kerr, An edible larva (Zeuzera coffeae). Journ. Siam Soc. Nat. Hist., Supplem. 8. 1931. p. 217–218.

H. H. King, A beehive designed for the production of Beeswax suitable for use by the Natives of the Southern Sudan. Entom. Bulletin no. 11. Khartoum. (n.d. 1920).

W. Kirby und W. Spence, Einleitung in die Entomologie. 4. Vol. Stuttgart. 1823. Vol. I, p. 331–371.

J. C. F. Klug, Symbolae Physicae, seu Icones et descriptiones Insectorum, quae ex itinere per Africam borealem et Asiam G. F. Hemprich et C. H. Ehrenberg studio novae aut illustratae redierunt. 5 decades. Berlin. 1829–1845.

R. Knox, An historical relation of the Island of Ceylon. London. 1681; London. 1817. p. 48.

T. Koch-Grünberg, Zwei Jahre bei den Indianern Nordwest-Brasiliens. Stuttgart. 1921.

H. Koch, Proverbes Badjoue et Bikele. Bull. Soc. Etud. Cameroun. 1944, no. 5. p. 39–53.

P. Kolben, The Present State of Cape of Good Hope. 2 vol. London. 1731, 1738. II, p. 179.

J. G. König, Naturgeschichte der sogenannten weissen Ameise. Beschaeft. Berlin. Ges. Naturf. Freunde. 4. 1779. p. 1–28.

K. Korikawa, Chemical studies on the locust. Zyozogaku Zassi, Osaka. 12. 1934. p. 361–365, 445.

H. Koster, Travels in Brazil. London. 1816.

S. P. Krachcheninnikov, The History of Kamtshatka. Glocester. 1764.

J. Künckel d'Herculais, Merveilles de la Nature. Les Insectes. 2 vol. Paris. 1882. I, p. 575 ff., II, p. 130 ff., 148, 366, 473.

J. Künckel d'Herculais, Note sur les populations acridophages en extrème sud de l'Algérie. Bull. Soc. Ent. France. 1891. p. XXIV–XXVI.

Père J. B. Labat, Nouveau Voyage aux Iles de l'Amérique. Paris. 1722. Quoted from ed. Paris. 1931. p. 100.

J. B. Labat, Nouvelle Rélation de l'Afrique Occidentale, 2 vol. Paris. 1728.

J. de Laet, Novis orbis seu descriptio Indiae occidentalis libri XVIII. Leyden. 1625; 1630. p. 333, 379.

C. Lapp et J. Rohmer, Composition et valeur alimentaire du Criquet pélérin. Bull. Soc. Chim. Biol. Paris. 19. 1937. p. 413–416.

J. L. Lassaigne, Examen chimique du miel de la Guèpe Lecheguana. Mém. Mus. Hist. Nat. 2. 1824. p. 319–320.

C. T. Lefebvre, Voyage en Abyssinie. 9 vol. Paris. 1845–1851.

F. W. L. Leichhardt, Journal of an overland expedition in Australia from Moreton Bay to Port Essington. London. 1847.

F. Liebrecht, Zur Volkskunde. Heilbronn. 1879. p. 75.

G. Lindblom, The Akamba in British East Africa. Arch. d'Etudes Orientales. Uppsala. 17. 1920.

De Lisle, Note sur la faune coléoptérologique du Cameroun. Bull. Soc. Etud. Cameroun. 1945. no. 5. p. 55–71.

D. Livingstone, Missionary Travels and Researches in South Africa. London. 1857. p. 164, 464; New York. 1858. p. 48.

D. and C. Livingstone, Explorations dans l'Afrique. Paris. 1868.

F. A. de la Llave, Sur les Busileras ou Fourmis à miel. Rev. Mag. Zool. (2) 14. 1862. p. 457–462.

F. A. de la Llave, Ahuatle. Rev. Mag. Zool. (2) 14. 1862. p. 222–223, 251–255.

A. M. Long, Red ants as article of food. Journ. Bombay Nat. Hist. Soc. 13. 1901. p. 536.

P. H. Lucas, Note sur Acridium peregrinum. Bull. Soc. Ent. France 1845. p. XXXII.

P. H. Lucas, Observations sur les Busileras ou fourmis à miel du Mexique. Rev. Mag. Zool. (2) 12. 1860. p. 271–280.

C. Lumholtz, Among Cannibals. 1890. p. 153 ff., 117, 223, 178, 186, 142.

C. Lumholtz, Au Pays des Cannibales. Paris. 1890.

C. Lumholtz, In Unknown Mexico. 2 vol. New York. 1902.

J. M. D. MacKenzie, Means of Defence in Insects. J. Bombay Nat. Hist. Soc. 33. 1930. p. 1000.

W. S. Mac Leay, Horae Entomologicae, or essais on the annulose animals. Vol. I. London. 1819/21.

A. E. N. Marais, Wild honey; with notes on the Moka-bee. South Afric. Agric. J. 3. 1912 p. 790–795.

G. Marcgraf, Historiae rerum naturalium Brasiliae libri VIII. Amsterdam. 1648. p. 56.

J. Marianno (filho), O cultivo das abelhas indigenas, e um typo de colmeia para o seu desfrutamento industrial. Entom. Brasileiro. 3. 1910. p. 14–18, fig. 1.

342

C. L. Marlatt, The cicadas as an article of food. in: The periodical cicada. U.S. Dept. Agric. Bur. Ent. Bull. 71. 1907.

G. A. K. Marshall and E. B. Poulton, Five Years' Observations and Experiments on the Bionomics of South African Insects, chiefly directed to the Investigation of Mimicry and Warning Colours. Trans. Ent. Soc. London. 1902. p. 287–584.

R. Martin, Die Inlandstaemme der Malayischen Halbinsel. Jena. 1905.

Peter Martyr, De Nouo Orbe; or, The History of the West Indies. Transl. by Eden and Lok. London. 1612. p. 121 f.

J. A. Massam, The cliff dwellers of Kenya. London. 1927.

J. Matthew, Two representative Tribes of Queensland. 1910. p. 86–90.

A. Maurizio, Histoire de l'alimentation végétale. Paris. 1932. p. 30–32.

Brantz Mayer, Mexico as it was and as it is. New York. 1844. p. 218.

H. C. McCook, The Honey ants of the Gardens of the Gods and the Occident ants of the American plains. Philadelphia. 1882. p. 32.

K. C. McKeown, Australian Insects. Sydney. 1944. 2nd ed.

K. C. McKeown, Insect Wonders of Australia. 2nd. ed. Sydney and London. 1944.

B. Meissner, Babylonien und Assyrien. Heidelberg. 1920. Vol. I.

F. H. Melland, In Witch-bound Africa. London. 1923. p. 123, 158, 252.

M. S. Merian, Histoire Générale des Insectes de Surinam. Paris. 1771. pl. 48.

M. Merker, Die Masai. 2nd ed. Berlin. 1910.

J. J. Meyer, Trilogie altindischer Maechte und Feste der Vegetation. III. Indra. Zürich und Leipzig. 1937.

J. Michaud, Enquête aux pays du Levant. Quoted from H. Bordeaux, Voyageurs d'Orient. 2 vol. Paris. 1926. II, p. 180.

N. von Miklucho-Maclay, Ethnologische Bemerkungen über die Papuas der Maclay-Küste in Neu Guinea. Natuurk. Tijdschr. voor Nederlandsch Indië. 35. 1875. p. 70.

J. G. Millais, Far away up the Nile. London. 1924. p. 201, 207 f.

H. B. Mills and J. H. Pepper, The effect on man of the ingestion of the confused flour beetle. Journ. Econ. Ent. 32. 1939. p. 874–875.

Mjele, Haplosterna delagorgei (Hem.). Nada, Ann. Native Affairs Dept. S. Rhodesia. 1934. p. 37–38.

E. Mjöberg, Bland stenaldersmaenniskor i Queenslands vildmarker. Stockholm. 1918. p. 176 ff., 270, 490.

R. Moffat, Missionary Labors and Scenes in S. Africa. London. 1842. p. 448 f.

G. A. De Mol, Collecting wax and honey in the Lake Region of Western Borneo. Landbouw. 9. 1933/34. p. 80–86, 3 fig.

W. Moorcroft, On the management of Bees in Cashmere. Ent. Mag. 5. 1838. p. 119–122.

C. Moore, Psylla eucalypti, forming the manna of Australia. Proc. Ent. Soc. London. 1862. p. 104.

O. Morger, Ein afrikanischer Bienenvater. Schweiz. Bienenztg. 1950. p. 431–433; also in: Jugend-Missions-Kalender Zug. 1950.

H. A. Morstatt, Die stachellosen Bienen (Trigonen) in Ostafrika und das

Hummelwachs. Arb. Biol. Reichsanst. Land- und Forstw. Berlin. 10. 1921. p. 283–305.

M. V. Motschoulsky, Etudes Entomologiques. 5. Helsingfors. 1856. p. 77.

Th. Moufetus, Insectorum sive Minimorum Animalium Theatrum. London. 1634.

C. P. Mountford, Earth's most primitive people. Geograph. Mag. January 1946 p. 89–104

C. Mulsant, Dissertation sur le Cossus des Anciens. Soc. Roy. d'Agric. Lyon. 4. 1841. p. 1–9.

J. Muir, My first summer in the Sierra. Boston and New York. 1916. p. 46.

F. Netolitzky, Kaefer als Nahrung und Heilmittel. Kol. Rundschau 8. 1920. p. 21–26, 47–60.

B. Neumann, Jerusalem. Frankfurt. 1877.

Nguyen-Cong-Tieu, Notes sur les insectes comestibles au Tonkin. Bull. écon. Indochïne, 31. 1928. p. 735–744, 4 pl., 3 fig.

J. E. Nieremberg, Historia Naturae, maxime peregrinae, libris XVI. Anvers. 1635.

J. Nieuhoff, Voyages and travels into Brazil. In: J. Pinkerton, General Collection of Voyages. London. vol. 14. 1813. p. 732 f.

F. Noetling, The Food of the Tasmanian Aborigines. Papers and Proceedings R. Soc. of Tasmania. 1910. p. 281, 290 f.

N. E. H. Nordenskiöld, L'Apiculture Indienne. Journ. Soc. Amér. Paris. N.S. 21. 1929. p. 169–182, figs. 11–12, pl. 3.

N. E. H. Nordenskiöld, L'Apiculture en Amérique. L'Anthropologie. 39. 1929. p. 285/6.

H. Noyes, Man and the termite. London. 1937. p. 226 ff.

A. G. Olivier, Travels in the Ottoman Empire, Egypt and Persia. Engl. Transl. London. 2 vol. 1813. I, p. 139.

L. Oliphant, Narrative of Earl of Elgin's Mission to China and Japan in 1857/9. New York. 1860. p. 273.

A. C. V. D. d'Orbigny, Voyage dans l'Amérique Méridionale. Partie Historique. Paris et Strasbourg 1839/43. vol. 2. p. 600, 624 ff.

A. S. Packard, Edible Mexican Insects. The Amer. Nat. 19. 1885. p. 893.

H. A. Pagenstecher, Über Myrmecocystus mexicanus. Verh. Nat.-Med. Ver. Heidelberg. 2. 1862. p. 72–82.

W. G. Palgrave, Narrative of a year's journey through Central and Eastern Arabia (1862/3). London and Cambridge. 1865. 2 vol.

E. Palmer, Food products of North American Indians. Rept. Commissioner of Agric., Washington. 1870. p. 404–428.

K. L. Parker, The Enahleyi Tribe. A study of aboriginal life in Australia. London. 1905.

T. Parsons, Grasshoppers and the Palmerworm. Mass. Soc. Prom. Agric. 1807. p. 32–35.

H. Patenostre, Ann. Méd. Pharmac. Colon. 1927. p. 51.

R. Paulian, Les Coléopteres. Paris. 1943. p. 351–352.

A. Pavie, Exposé des traveaux de la Mission. Vol. I. Paris. 1901. p. 118.

F. Pax, Praktische Verwendung von Termiten. Natur u. Volk. 80. 1951. p. 264–271.

T. Peckolt, Über brasilianische Bienen. Natur (Halle). 42. 1893. p. 579–781, 43. 1894. p. 87, 223, 233.

D. L. Phares, (Note on Cicada). Republican, Woodville. Miss. 5. 5. 1858.

E. Peau, Pourquoi ne mangeons-nous pas d'insectes? Tour du Monde. 1912. p. 85.

H. M. Pendlebury, Journ. M. S. Mus. 1923. p. 11.

R. Percival, Voyage à l'Ile de Ceylon (1797–1800). 2 vol. Paris. 1803. II, p. 62.

R. Percival, An account of the Island of Ceylon, containing its natural history. London. 1803.

W. W. A. Phillips, The Food of the Ceylon Slender Loris (*Loris tardigradus*) in Captivity. Spolia Zeylanica. 16. 1931. p. 205–208.

W. D. Pierce, The uses of certain weevils and weevil products in food and medicine. Proc. Ent. Soc. Washington. 17. 1915. p. 151–154.

J. Pinkerton, General Collection of Voyages and Travels in all parts of the World. 17 vol. London. 1808/14.

S. Pinto, Comment j'ai traversé l'Afrique 1877/78. Tour du Monde. 1881. I. p. 235.

W. Piso, De Indiae utriusque re naturali et medica libri XIV. Amsterdam. 1658. Lib. I, p. 9, 11, V, p. 291.

P. Poey, Memorias sobre la Historia naturel de la Isla de Cuba. 2 vol. Habana. 1851, 1858–1861.

G. Portevin, Les Insectes Comestibles. La Terre et la Vie. 1933. p. 336–340.

S. Postmus and A. G. Van Veen, Dietary surveys in Java and East-Indonesia. Ono Public. No. 1. Bandoeng. 1949. Also in Chron. Nat. 1949. 105. p. 229–236, 261–268, 316–323.

E. B. Poulton, Feeding Experiments with Lepidoptera conducted by W. W. A. Phillips on a Ceylon Lemur. Proc. Ent. Soc. London 7. 1932. p. 32–35, 49–50.

H. Priest, The Call of the Bush. London. 1932.

T. Pringle, Narrative of a Resident in S. Africa. New ed. London. 1851. p.

Provancher, Natural. Canadien 12. 1882. p. 30–31.

S. Purchas, Hakluytus Posthumus; or, Purchas his Pilgrimes. London. 5. vol. 1625/26.

W. L. Puxley, Wanderings in the Queensland Bush. London. 1923.

R. A. M., De wilde Indische honingbij (Apis dorsata). Algemeen Landbouw-weekblad voor Nederlandsch-Indië. 8. 1924. p. 1034–1036.

H. M. Ransome, The Sacred Bee in ancient times and folklore. London. 1937.

Carl C. Raswan, Im Lande der schwarzen Zelte. Berlin. 1934.

W. Raum, Insekten und andere Tracheaten als menschliche Nahrungsmittel. Entom. Jahrb. 33/34. 1924/25. p. 82–92.

C. Raveret-Wattel, Rapport sur les Mélipones. Bull. Soc. Acclim. (3) 2. 1875. p. 732–759.

R. A. F. de Réaumur, Mémoirs pour servir à l'Histoire des Insectes. 6 vol. Paris. 1734/42 .Quoted from Amsterdam edition 1737/48. II, 2p. 113 ff. III, p. 416.

R. Redfield and A. R. Villa, Chan Kom, a Maya village. Carnegie Inst. Washington. Publ. no. 448. 1934. Many references to bees and honey.

C. L. Remington, Insects as food in Japan. Entom. New 57. 1946. p. 119–121.

G. Renard, Le Travail dans le Préhistoire. Nouv. ed. Paris. 1931. p. 47–49.

J. Riley, Authentic Narrative of the Loss of the American Brigatte Commerce, wrecked on the western coast of Africa, in 1815. Hartford. 1850. p. 237.

C. V. Riley, The edible quality of ants. Insect Life. 5. 1893. p. 268.

W. H. Rivers, The Todas. London. 1906. p. 191, 198.

A. Robbins, Journal of Adventures in Africa in 1815/17. Hartford. 1851. p. 172.

C. H. Robinson, Hausaland 3rd. ed. London. 1900. p. 149.

C. de Rochefort, Histoire naturelle et morale des iles Antilles de l'Amérique. Rotterdam. 1658.

W. Roepke, Beobachtungen an Indischen Honigbienen, insbesondere an *Apis dorsata* F. Meded. Landbouwhoogeschool Wageningen. 34, no. 6. 1930. p. 1–28.

M. P. Roeser, Sur les sauterelles comestibles. Journ. de Pharmacie et de Chimie. (6) 10. 1899. no. 5.

G. Roheim, Australian Totemism. London. 1925. p. 65.

J. Rohmer, Contribution à l'étude du *Schistocera gregaria*. Thèse de Pharmacie. Strasbourg. 1936.

E. Roquette-Pinto, Rondonio. 2nd ed. Rio de Janeiro. 1919. p. 13, 341, fig. 3.

J. Roscoe, The Bagesu and other Tribes of the Uganda Protectorate. London. 1924. p. 171, 186.

Le Roux, L'Art Entomologique. Poème didactic en six chants. Versailles. 1814.

R. L. Roys, The book of Chilam Balam of Chumayel. Carnegie Inst. Washington. Publ. no. 438. 1933. Many references to honey and bees.

R. L. Roys, The titles of Ebtun. Middle Amer. Papers, Middle Amer. Res. Ser, Tulane Univ. Publ. no. 505. 1939.

B. de Sahagun, General History of Things of New Spain. (ab. 1560). Paris. 1880.

H. Salt, Voyage en Abyssinie (1809/10). 2 vol. Paris. 1816. II, p. 291, 371. I, p. 222; Voyage to Abyssinia. London. 1814.

G. A. J. Van der Sande, Nova Guinea. Vol. III. Ethnography and Anthropology. Leyden. 1907. p. 4.

A. Sandel, (Note). Mitchell and Millers Medical Repository. 4. 1715. p. 71.

K. T. Sapper, Bienenhaltung und Bienenzucht in Mittelamerika und Mexico. Ibero-Amer. Archiv 9. 1935. p. 183–193.

K. Sapper, Geographie und Geschichte der Indianischen Landwirtschaft. Hamburg. 1936.

M. Sarre, Les Fondaments biologiques de la Géographie Humaine. Paris. 1943.

F. Sartorius, Las Hormigas meliferas. La Naturaleza. Mexico. 7. 1885/86. p. 229–233. Extracted from C. Sterne in Die Gartenlaube.

P. Schebesta, Die Bambuti-Pygmaeen von Ituri. Inst. Roy. Col. Belge. Sci. Mor. et Polit. Mém. Vol. I. 1938.

P. Schebesta, Les Pygmées. Paris. 1940. p. 31 f., 42, 51 f.

A. M. P. A. Scheltma, De voeding van de Inlandsche bevolking van Nederlandsch-Indië. Koloniale Studien. 14, II. p. 368–394.

G. Schneider, Über eine Urwald-biene (*Apis dorsata* F.). Zeitschr. Wiss. Insektenbiol. 4. 1908. p. 447–453.

E. J. R. Scholz, Bienen und Wespen, ihre Lebensgewohnheiten und Bauten. Leipzig. 1913. p. 158 f.

Sir R. H. Schomburgk, History of Barbados. London. 1847. p. 646.

M. R. Schomburgk, Reisen in Britisch Guiana. 3 vol. Leipzig. 1847/48.

H. Schomburgk, Wild und Wilde im Herzen Afrikas. Berlin. 1910

G. Schumacher, Der Adschlun. Zeitschr. D. Pal. Ver. 1924.

H. F. Schwarz, The stingless bees; a paradox among the producers of honey. Nat. Hist. New York. 53. 1944. p. 414 ff., 426, illustr.

H. F. Schwarz, Stingless bees (Meliponidae) of the Western Hemisphere. Bull. Amer. Mus. Nat. Hist. 90. 1948. p. 1–546.

H. Scott, The High Yemen. London. 1942. p. 58.

C. C. and B. Z. Seligmann, The Veddas. Cambridge. 1911, p. 42, 163, 252 ff., 327.

R. Semon, Im Australischen Busch, Leipzig. 1903.

G. Seyffert, Biene und Honig im Volksleben der Afrikaner mit besonderer Berücksichtigung der Bienenzucht, ihrer Entstehung und Verbreitung. Leipzig. 1930.

J. Shakespeare, Folklore. 23. 1912. p. 466.

T. Shaw, Travels, or observations relating to several parts of Barbary and the Levant. Oxford. 1738. p. 258.

P. L. Simmonds. The sources of Manna. Technologist 1.1861 p. 225–228.

P. L. Simmonds, Les Richesses de la Nature. Lè Regne Animal. Transl. from the 2nd English edition. Gand. 1877.

P. L. Simmonds, The animal resources of different nations. London. 1885.

A. Skinner, The use of insects and other invertebrates as food by the North-American Indians. Journ. New York Entom. Soc. 18. 1910. p. 264–267 p. 333, 379.

H. Sloane, Voyage to the Islands of Madeira, Barbados, Nieves, S. Christophers, and Jamaica; with the Natural History of Jamaica. 2 vol. London. 1707, 1725. II, p. 193, 204.

T. Smith, Wonders of Nature and Art. 12 vol. Philadelphia. 1806/7. XII, p. 197.

F. Smith, Descriptions of Brasilian Honey bees Melipona and Trigona. Trans. Ent. Soc. London. 6. 1862. p. 497-512, 1 pl.

R. Brough Smyth, The Aborigines of Victoria. 1878. p. 206 f.

R. Southey, History of Brazil. 3 vol. London. 1817/22. I, p. 110.

A. C. Sowerby, A Naturalist's Note-book in China. Shanghai. 1925.

A. Sparrmann, Voyage au Cap de Bonne-Espérance. 3 vol. Paris. 1778.

A. Sparrmann, Voyage to the Cape of Good Hope towards the Antarctic Circle, and Round the World from 1772/76. 2 vol. Perth. 1789. I, p. 263.

C. Spegazzini, Al traves de Misiones. Rev. Faculdad Agr. Veterinaria, Univ. Nac. La Plata (2) 5. 1909. a. 40, 62 ff.

W. Spence, Observations on the Honey bees in Brazil. Proc. Ent. Soc. London 9. 1847. p. 12–14.

B. Spencer and F. J. Gillen, The Native Tribes of Central Australia. London 1899. X. 771.

B. Spencer, Native tribes of the Northern Territory of Australia. London. 1914.

B. Spencer and F. J. Gillen, Across Australia. London. vol. 2. 1912. 1, p. 121 ff, 166 f, 466 ff..

B. Spencer, Guide to the Australian Ethnological Collection in the National Museum of Victoria. 3rd. ed. Melbourne. 1922, p. 29, 66, 68.

Sir Baldwin Spencer, Wanderings in Wild Australia. 2 vol. London. 1928.

M. Spinola, Observations sur les Apiaires Méloponides. Ann. Sc. Nat. (2) 13. 1840. p. 116–140, 1 pl.

M. Spinola, Note sur les Hyménopteres de la tribu des Méliponides. Rev. Zoöl. 5. 1842. p. 216–218, 267–268.

R. L. Spittel, Wild Ceylon. Describing in particular the lives of the present day Veddas. Ceylon. 1924.

F. Starr, Indians of southern Mexico. An ethnographic album. Chicago. 1899. p. 22, pl. 58.

G. L. Staunton, An authentic account of an embassy of China. 2 vol. London. 1797.

J. G. Stedman, Narrative of a five years' Expedition against revolted Negroes of Surinam, in Guiana in 1772/77. 2 vol. London. 1796. II, p. 23.

A. Steedman, Wanderings and Adventures in the Interior of S. Africa. 2 vol. London. 1835.

D. Steinhardt, Vom wehrhaften Riesen und seinem Reiche. 3rd ed. Hamburg. 1922. p. 113 ff., 207.

D. G. Stilbe, Encyclopaedie van Nederlandsch Indië. 2nd ed. Haag–Leiden. 1921. vol. 4. p. 599; Supplement II. vol. 6. 1932. p. 937.

E. C. Stirling, Anthropology. In: Report on the work of the Horn Scientific Expedition to Central Australia. IV. Anthropology. 1896. p. 51–54.

Storck, Bienen und Bienenzucht in Australien und auf den Fidschi Inseln. Mittheil. Bienenzucht (Bensheim) 1. 1863. p. 85–87.

J. C. B. Stratham, Through Angola. Edinburgh and London. 1922. p. 291 f.

C. Strickland, Edible and paralysific bugs. Indian J. Med. Res. 19. 1932. p. 873/6, Pl. 49.

C. J. Van der Swaan, Wilde Bijen Nesten. Het bosch (Java) 1934. p. 124–127.

E. W. Teale, The Boys' Book of Insects. New York. 1940. p. 168.

L. Failla Tedaldi, Insetti commestibili, sacri, medicinali, industriali e d'armamento. Il Naturalisto Siciliano. 1. 1881/2. p. 232–240.

E. Temple, Travels in various parts of Peru. 2 vol. London. 1830.

J. E. Tennent, Sketches of the Natural History of Ceylon. London. 1861. p. 407.

G. Tessmann, Die Pangwe. 2 vol. Berlin. 1913/14.

G. Tessmann, Die Indianer Nordost-Perus. Hamburg. 1930. Many quotations.

J. Théodoridès, Les Coléoptères comestibles. Natur. Belg. 30. 1949. p. 126–137.

C. Thomas, A study of the manuscript Troana. U.S. Geogr. and Geol. Survey of the Rocky Mountain Region. Washington. 1882. vol. 5 pt. 3, pp. 114 ff., 127, 150, 221.

N. W. Thomas, Natives of Australia. London. 1906 p. 111. f., 134.

J. Thomson, Voyage au Gabon. Arch. Entom. 2. 1858. p. 1–469.

J. K. R. Thorp, African beekeepers: Notes on methods and customs relating to the bee-culture of the Akamba tribe in Kenya colony. Journ. East Africa Nat. Hist. Soc. 17. 1943. pp. 257–269.

L. Tihon, A propos des termites au point de vue alimentaire. Bull. Agric. du
Congo Belge. 37. 1946. p. 865–868.

R. J. Tillyard, The Insects of Australia and New Zealand. Sydney. 1926.

J. Timon-David, Recherches sur les matières grasses des insectes. Ann. Fac. Sci.
Marseille (2) 4. 1930. p. 29–207.

N. B. Tindale, Revision of the Australian Ghost Moths (Hepialidae), I. Records
S. Austral. Mus. 4. 1932. p. 497–536.

N. B. Tindale, Ghost Moths of the Family Hepialidae. S. Austral. Natur. 19.
1938. p. 1–6.

H. P. Trewithick and R. R. Lewis, Fat from locusts. Oil and Soap 16. 1939.
p. 128.

B. P. Uvarov, Recent advances in Acridology. London. 1948.

E. B. Uvarov, The ash content of insects. Bull. Ent. Res. 22. 1931. p. 453–457.

F. Le Vaillant, Voyages dans l'Intérieur de l'Afrique (1781/5). 2 vol. Paris.
1931.

J. Vellard, Une civilisation du miel. Les Indiens Guayakis du Paraguay.
Paris. 1939.

A. Hyatt Verrill, Moeurs étranges des insectes. Paris. 1938. p. 159–164.

A. Villiers, Une Manne Africaine: Les Termites. La Nature. 1947. p. 239–240.

J. J. Virey, Histoire Naturelle du Genre Humain. on IX. Paris. 2 vol. I, p. 256.
2nd ed. Paris. 1824. II, p. 319, 329.

C. A. de Walkenaer, Histoire Naturelle des Insectes Aptères. Vol. I. Paris.
1837. p. 181 f.

A. R. Wallace, On the Insects used for food by the Indians of the Amazon.
Trans. Ent. Soc. London. N.S. 2. 1852/3. p. 241/4.

A. R. Wallace, The Malay Archipelago. London. 1902.

N. Wanley, Wonders of the Little World; or a General History of Man. 2 vol.
London. 1806. II, p. 273.

K. Ward, In Farthest Burma.

G. Warren, An impartial description of Surinam. London. 1667.

E. Wasmann, Die Honigameise des Göttergartens. Stimmen aus Maria Laach.
1884. p. 275–285.

C. N. De Wavrin, Les derniers Indiens primitifs du bassin du Paraguay. Paris.
1926. p. 19 f.

C. N. De Wavrin, Moeurs et coûtumes des Indiens Sauvages de l'Amérique
du Sud. Paris. 1937.

C. D. F. Weinland, Mexicanische Honigameise. Zool. Gart. Frankfurt. 2. 1861.
p. 135.

B. W. Westermann, Über die Lebensweise der Insecten in Ostindien und am
Cap. Germar's Magaz. Entom. 4. 1821. p. 411–427.

E. Westermarck, Folklore. 16. 1905. p. 28–37.

E. Westermarck, Ritual and Belief in Morocco. 2 vol. London. 1926. I, p. 104
II, p. 47 ff.

J. O. Westwood, Larva-nest of social Lepidopterous larvae from Africa. Proc.
Ent. Soc. London. 1863. p. 114.

C. M. WETHERILL, Chemical Investigations of the Mexican Honey Ant. Proc. Acad. Nat. Sci. Philadelphia. 6. 1852. p. 111–113.

W. M. WHEELER, Honey Ants. with a revision of the American Myrmecosysti. Bull. Amer. Mus. Nat. Hist. 24. 1908.

J. WHITE, Journal of a voyage to New South Wales. London. 1790.

G. WHITE, Natural History of Selborne. London. 1854. p. 293.

M. A. P. zu WIED-NEUWIED, Voyage au Brésil dans les années 1815/17. Paris. 1821/22. 3 vol. II, p. 49.

Sir G. H. WILKINS, Undiscovered Australia. London. 1928. p. 86, 246 ff.

C. B. WILLIAMS, The migration of butterflies. Edinburgh and London. 1930.

W. S. WILLIAMS, The Middle Kingdom; or, Survey of Chinese Empire. 3rd ed. 2 vol. New York. 1853. II, p. 50.

W. W. WITHERSPOON, Collection of honeydew by the Nevada Indians. Amer. Anthrop. Washington. 1889, II. p. 380.

WORTHINGTON, Science in Africa. Oxford. 1938. p. 572.

K. YAMAFUJI, Biologie der Seide. In: Tabulae Biologicae. 14. 1937. p. 36–50.

C. ZIMMERMANN, (Dietary survey of Siam). Siam Rural Econ. Survey 1930/31. Siamese Min. of Commerce and Communications. X. 1931.

C. J. VAN DER ZWAAN, Wilde Bijen-Nesten. Het Bosch. 2. 1934. p. 124–127.

Composition of some African foods and feeding stuffs mainly of vegetable origin. Imp. Bureau of Anim. Nutr. Techn. Commun. 6. 1936. Aberdeen.

The Koran. Transl. by G. SALE. London and New York. (N.D.).

Mirror of Literature, Amusement, and Instruction. 40 vol. London. 1823/42.

Locust Meal available in Argentina. World Trade Notes on Chemicals and Allied Products. Dept. of Commerce U.S.A. 9. 1935, no. 48. p. 42.

The Mishnah. Transl. by H. DANBY. London. 1933.

The New Testament in Hebrew and English. London. n.d.

Nutrition in the Colonial Empire. First Report. Part I and II. London. 1939.

The Holy Scriptures of the Old Testament. Hebrew and English. Berlin. 1903.

Report Commiss. Agriculture 1870. Washington. p. 426.

First Annual Report U.S. Entom. Commission 1877 relating to the Rocky Mountain Locust. Washington. 1878. p. 437–443.

This bibliography is an improvement over the bibliographies published so far. Yet quite a number of references were not available in the libraries of Paris, London and Amsterdam and could therefore neither be improved nor verified. In the text these are designed by vide.... Any improvement and completion of this bibliography will be gladly received by the writer.

CONTENTS

Contents